Transmembrane Signaling Protocols

METHODS IN MOLECULAR BIOLOGY™

John M. Walker, SERIES EDITOR

METHODS IN MOLECULAR BIOLOGY™

Transmembrane Signaling Protocols

Second Edition

Edited by

Hydar Ali

Department of Pathology
School of Dental Medicine
The University of Pennsylvania
Philadelphia, PA

Bodduluri Haribabu

Department of Microbiology and Immunology
James Graham Brown Cancer Center
University of Louisville
Louisville, KY

HUMANA PRESS ✳ TOTOWA, NEW JERSEY

© 2006 Humana Press Inc.
999 Riverview Drive, Suite 208
Totowa, New Jersey 07512

humanapress.com

This publication is printed on acid-free paper. ∞
ANSI Z39.48-1984 (American Standards Institute)

Permanence of Paper for Printed Library Materials.
Cover illustration: Fig. 3, Chapter 6, "Characterization of Constitutively Active Mutants of G Protein-Coupled Receptors," by Jean-Marc Navenot, Zi-xuan Wang, and Stephen C. Peiper (background image); and Fig. 1, Chapter 8, "Real-Time Analysis of G Protein-Coupled Receptor Signaling in Live Cells," by Venkatakrishna R. Jala and Bodduluri Haribabu (foreground image).

Production Editor: Amy Thau

Cover design by Patricia F. Cleary

For additional copies, pricing for bulk purchases, and/or information about other Humana titles, contact Humana at the above address or at any of the following numbers: Tel.: 973-256-1699; Fax: 973-256-8341; E-mail: orders@humanapr.com; or visit our Website: www.humanapress.com

Printed in the United States of America. 10 9 8 7 6 5 4 3 2 1

eISBN: 1-59745-048-0

ISSN: 1064-3745

Library of Congress Cataloging-in-Publication Data

Transmembrane signaling protocols / edited by Hydar Ali and Boddulur Haribabui.-- 2nd ed.
p. ; cm. -- (Methods in molecular biology ; 332)
Includes bibliographical references and index.
ISBN 1-58829-546-X (alk. paper)
1. Cellular signal transduction--Laboratory manuals. [DNLM: 1. Signal Transduction--Laboratory Manuals. 2. GTP-Binding Proteins--physiology--Laboratory Manuals. QU 375 T772
2006] I. Ali, Hydar. II. Haribabu, Bodduluri. III. Series: Methods in molecular biology (Clifton, N.J.) ; 332.
QP517.C45T736 2006
571.7'4--dc22

2005022678

Preface

The previous edition of *Transmembrane Signaling Protocols* was published in 1998. Since then the human genome has been completely sequenced and new methods have been developed for the use of microarrays and proteomics to analyze global changes in gene expression and protein profiles. These advances have increased our ability to understand transmembrane signaling processes in much greater detail. They have also simultaneously enhanced our ability to determine the role of a large number of newly identified molecules in signaling events. In addition, novel video microscopy methods have been developed to image transmembrane signaling events in live cells in real time.

In view of these major advances, it is time to update the previous edition. Because of the success of that volume, we have chosen to keep the essential character of the book intact. Introductory chapters from experts have been included to provide overall perspective and an overview of recent advances in signal transduction pathways. The individual chapters now include comprehensive detailed methods, studies in genetically tractable systems, fluorescence microscopy in live single cells, ex vivo analysis of primary cells from transgenic mice, as well as genomic and proteomic approaches to the analysis of transmembrane signaling events.

We would like to express our deep gratitude to the coauthors of this publication. We hope that *Transmembrane Signaling Protocols, Second Edition* will serve as a valuable resource for future progress in the study of signal transduction pathways.

Hydar Ali
Bodduluri Haribabu

Contents

Contributors

HYDAR ALI • *Department of Pathology, School of Dental Medicine, The University of Pennsylvania, Philadelphia, PA*

KATHLEEN BOESZE-BATTAGLIA • *Department of Biochemistry, School of Dental Medicine, The University of Pennsylvania, Philadelphia, PA*

BRADFORD C. BERK • *Center for Cardiovascular Research, Department of Medicine, University of Rochester, Rochester, NY*

GARY M. BOKOCH • *Departments of Immunology and Cell Biology, The Scripps Research Institute, La Jolla, CA*

JAMES R. BROACH • *Department of Molecular Biology, Princeton University, Princeton, NJ*

SHAMSHAD COCKCROFT • *Department of Physiology, Rockefeller Building, University College London, London, UK*

CELINE DERMARDIROSSIAN • *Department of Immunology and Cell Biology, The Scripps Research Institute, La Jolla, CA*

RITU GARG • *Ludwig Institute for Cancer Research, University College London, London, UK*

AMANDA FENSOME-GREEN • *Department of Physiology, Rockefeller Building, University College London, London, UK*

BODDULURI HARIBABU • *Department of Microbiology and Immunology, James Graham Brown Cancer Center, University of Louisville, Louisville, KY*

PAUL A. INSEL • *Department of Pharmacology, School of Medicine, University of California, San Diego, La Jolla, CA*

TARIK ISSAD • *Institut Cochin, INSERM, CNRS, Université Paris, Paris, France*

SHANKAR S. IYER • *Inflammation Program, Department of Internal Medicine, University of Iowa Carver College of Medicine, Iowa City, IA*

VENKATAKRISHNA R. JALA • *James Graham Brown Cancer Center, University of Louisville, Louisville, KY*

RALF JOCKERS • *Institut Cochin, INSERM, CNRS, Université Paris, Paris, France*

MARTHA S. JORDAN • *Department of Cancer Biology, Abramson Family Cancer Research Institute, University of Pennsylvania, Philadelphia, PA*

KELLY L. JORDAN-SCIUTTO • *Department of Pathology, School of Dental Medicine, The University of Pennsylvania, Philadelphia, PA*

EUNJOON KIM • *Department of Biological Sciences, Korea Advanced Institute of Science and Technology, Daejeon, Korea*

JON B. KLEIN • *Departments of Medicine, University of Louisville, and VA Medical Center, Louisville, KY*

JAEWON KO • *Department of Biological Sciences, Korea Advanced Institute of Science and Technology, Daejeon, Korea*

DAVID J. KUSNER • *Inflammation Program, Department of Internal Medicine, University of Iowa Carver College of Medicine, Iowa City, IA*

HYUN WOO LEE • *Department of Biological Sciences, Korea Advanced Institute of Science and Technology, Daejeon, Korea*

MASSIMO LOCATI • *Institute of General Pathology, University of Milan, Milan, Italy*

GEORGE LOMINADZE • *Biochemistry and Molecular Biology, University of Louisville, Louisville, KY*

LOUIS M. LUTTRELL • *Department of Medicine and Biochemistry and Molecular Biology, Medical University of South Carolina, Charleston, SC*

MARCIN MAJKA • *James Graham Brown Cancer Center, University of Louisville, Louisville, KY*

CARLOS MARTÍNEZ-A. • *Department of Immunology and Oncology, Centro Nacional de Biotecnología/CSIC, UAM Campus de Cantoblanco, Madrid, Spain*

FERNANDO O. MARTINEZ • *Institute of General Pathology, University of Milan, Milan, Italy*

BRUCE J. MAYER • *Department of Genetics and Developmental Biology, University of Connecticut Health Center, Farmington, CT*

KENNETH R. MCLEISH • *Department of Medicine, University of Louisville, and VA Medical Center, Louisville, KY*

MARIO MELLADO • *Department of Immunology and Oncology, Centro Nacional de Biotecnología/CSIC, UAM Campus de Cantoblanco, Madrid, Spain*

MARSHALL B. MONTGOMERY • *Department of Pathology, School of Dental Medicine, The University of Pennsylvania, Philadelphia, PA*

KANCHANA NATARAJAN • *Center for Cardiovascular Research, Department of Medicine, University of Rochester, Rochester, NY*

JEAN-MARC NAVENOT • *Department of Pathology, Medical College of Georgia, Augusta, GA*

RENNOLDS S. OSTROM • *Department of Pharmacology and the Vascular Biology Center of Excellence, University of Tennessee Health Science Center, Memphis, TN*

STEPHEN C. PEIPER • *Department of Pathology, Medical College of Georgia, Augusta, GA*

MARIUSZ Z. RATAJCZAK• *James Graham Brown Cancer Center, University of Louisville, Louisville, KY*

ANNE J. RIDLEY • *Ludwig Institute for Cancer Research, University College London, London, UK*

JOSÉ MIGUEL RODRÍGUEZ-FRADE • *Department of Immunology and Oncology, Centro Nacional de Biotecnología/CSIC, UAM Campus de Cantoblanco, Madrid, Spain*

ANTONIO SERRANO • *Department of Immunology and Oncology, Centro Nacional de Biotecnología/CSIC, UAM Campus de Cantoblanco, Madrid, Spain*

MARY STOFEGA • *Department of Immunology and Cell Biology, The Scripps Research Institute, La Jolla, CA*

ZI-XUAN WANG • *Department of Pathology, Medical College of Georgia, Augusta, GA*

RICHARD A. WARD • *Department of Medicine, University of Louisville, Louisville, KY*

I

OVERVIEWS

1

Transmembrane Signaling by G Protein-Coupled Receptors

Louis M. Luttrell

Summary

G protein-coupled receptors (GPCRs) make up the largest and most diverse family of membrane receptors in the human genome, relaying information about the presence of diverse extracellular stimuli to the cell interior. All known GPCRs share a common architecture of seven membrane-spanning helices connected by intra- and extracellular loops. Most GPCR-mediated cellular responses result from the receptor acting as a ligand-activated guanine nucleotide exchange factor for heterotrimeric guanine nucleotide-binding (G) proteins whose dissociated subunits activate effector enzymes or ion channels. GPCR signaling is subject to extensive negative regulation through receptor desensitization, sequestration, and down regulation, termination of G protein activation by GTPase-activation proteins, and enzymatic degradation of second messengers. Additional protein–protein interactions positively modulate GPCR signaling by influencing ligand-binding affinity and specificity, coupling between receptors, G proteins and effectors, or targeting to specific subcellular locations. These include the formation of GPCR homo- and heterodimers, the interaction of GPCRs with receptor activity-modifying proteins, and the binding of various scaffolding proteins to intracellular receptor domains. In some cases, these processes appear to generate signals in conjunction with, or even independent of, G protein activation.

Key Words: G protein-coupled receptor; heterotrimeric guanine nucleotide-binding protein; second messenger; signal transduction; G protein-coupled receptor kinase; arrestin.

1. Introduction

The G protein-coupled, or seven membrane-spanning, receptors (GPCRs) constitute the largest and most diverse superfamily of cell surface receptors in the mammalian genome. Approximately 800 distinct genes encoding functional GPCRs make up greater than 1% of the human genome *(1,2)*. With alternative splicing, it is estimated that 1000 to 2000 discrete receptor proteins may be

From: *Methods in Molecular Biology, vol. 332: Transmembrane Signaling Protocols, Second Edition*
Edited by: H. Ali and B. Haribabu © Humana Press Inc., Totowa, NJ

expressed. In nematodes, the situation is even more dramatic. In *Caenorhabditis elegans*, genes encoding more than 1000 GPCRs comprise 5% of the genome *(3)*. Such evolutionary diversity generates GPCRs that detect an extraordinary array of extracellular stimuli, from neurotransmitters and peptide hormones to odorants and photons of light. Humans literally see, smell, and taste the world through GPCRs. Internally, GPCRs function in neurotransmission, direct neuroendocrine control of physiological homeostasis and reproduction, regulate hemodynamics and intermediary metabolism, and influence the growth, proliferation, differentiation, and death of multiple cell types. Not surprisingly then, it is estimated that more than half of all drugs in current clinical use target GPCRs, acting either to mimic endogenous GPCR ligands, to block ligand access to the receptor, or to modulate ligand production *(4)*.

The basic model of GPCR-signaling derives from the ability of these receptors to act as ligand-activated guanine nucleotide exchange factors (GEFs) for heterotrimeric guanine nucleotide-binding (G) proteins that transmit signals intracellularly through the activation of effector enzymes or ion channels. This fundamental paradigm of receptor biology accounts for most of the GPCR-mediated cellular responses described to date. Recent work, however, has indicated that GPCRs participate in numerous other protein–protein interactions that generate intracellular signals in conjunction with, or even independent of, G protein activation. Indeed, whole genome analyses suggest that some seven membrane-spanning receptors that are not G protein-coupled, such as the frizzled receptors, are nonetheless evolutionary branches of the GPCR phylogenetic tree *(5)*. This chapter will review the fundamentals of GPCR signaling and many of the processes that positively or negatively regulate GPCR function. In addition, we examine recent data supporting a role for other GPCR-binding proteins in the transduction of putatively "G protein-independent" signals.

2. The Receptor–G Protein–Effector Model of GPCR Signaling

The seminal work of Gilman and Rodbell and their colleagues established the hypothesis that a regulatory element was interposed between hormone receptors that controlled adenylate cyclase activity and the enzyme itself *(6–8)*. The reconstitution of hormone-sensitive adenylate cyclase activity in the UNC variant of S49 lymphoma cells *(9)*, followed by the purification of the G/F protein *(10)*, and the discovery that the regulator of rod outer segment cyclic guanosine monophosphate (cGMP)-specific phosphodiesterase (Gt or transducin), G/F (Gs), and the islet-activating protein (pertussis toxin) substrate (Gi) were members of a family of structurally homologous guanine nucleotide-binding regulatory proteins *(11)* lead to the now classic tripartite paradigm of GPCR signaling.

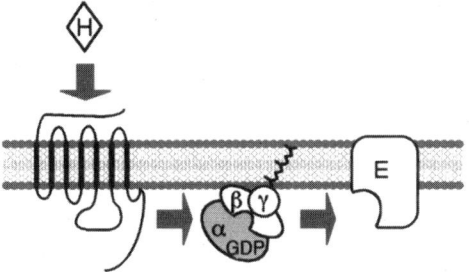

Heptahelical Receptor
- >800 known GPCR sequences.
- Detects the presence of hormone in the extracellular millieu.
- Catalyzes activation of heterotrimeric G proteins.

Heterotrimeric G protein
- 16 α subunits
  5 β subunits
  12 γ subunits
- Gα subunit has intrinsic GTPase activity.
- Upon activation dissociates into Gα-GTP and Gβγ subunits which regulate effector enzymes.

Effector
- Enzymatic production or degradation of small molecule second messengers.
- Ion channel.

Fig. 1. The basic G protein-coupled receptor (GPCR)–G protein–effector model of GPCR signaling. The three principle components of GPCR signaling are the heptahelical receptor, heterotrimeric G protein, and effector enzyme. The receptor detects the presence of a hormone (H) or "first messenger" in the extracellular millieu. The heterotrimeric G protein dissociates into a guanosine triphosphate (GTP)-bound Gα subunit and Gβγ heterodimer upon interaction with a ligand-bound receptor. The effector, which is typically an enzyme or ion channel, is activated by free Gα-GTP or Gβγ subunits, and produces small molecule "second messengers" that transmit signals intracellularly.

In the model, depicted schematically in **Fig. 1**, the binding of a "first-messenger" hormone to the extracellular or transmembrane domains of a GPCR triggers conformational changes that are transmitted through the intracellular receptor domains to promote coupling between the receptor and its cognate heterotrimeric G proteins. The receptor stimulates G protein activation by catalyzing the exchange of guanosine triphosphate (GTP) for guanosine diphosphate (GDP) on the Gα subunit and dissociation of the GTP-bound Gα subunit from the Gβγ subunit heterodimer. Once dissociated, free Gα-GTP and Gβγ subunits regulate the activity of enzymatic effectors, such as adenylate cyclases, phospholipase C (PLC) isoforms, and ion channels, to generate small molecule "second messengers." Second messengers, in turn, control the activity of protein kinases that regulate key enzymes involved in intermediary metabolism. Signaling continues until the intrinsic GTPase activity of the Gα subunit returns the G protein to the inactive heterotrimeric state.

2.1. The GPCR

In the early 1980s, the sequencing and subsequent cloning of the bovine retinal photoreceptor, rhodopsin, revealed a novel mammalian protein structure with similarity to bacteriorhodopsin, a light-sensitive proton pump found in halophilic bacteria. With the cloning of the β2 adrenergic receptor in 1986 and other receptor sequences that soon followed, it became clear that this basic architecture, consisting of an extracellular N-terminus, seven membrane-spanning α-helices connected by intracellular and extracellular loops, and an intracellular C-terminus, was representative of a large family of membrane receptors *(12)*.

2.1.1. Receptor Architecture

Although X-ray crystallographic data currently are available only for rhodopsin *(13)*, sequence similarities, hydropathy plots, and a large amount of biochemical and mutagenic data support the conclusion that all GPCRs exhibit a seven transmembrane architecture. As shown schematically in **Fig. 2A**, the GPCRs contain seven membrane-spanning α-helices (TMI-VII), linked by three alternating intracellular and extracellular loops (i1-3 and e1-3). The trans-

---

Fig. 2. *(opposite page)* Structure and phylogeny of G protein-coupled receptors (GPCRs). **(A)** Schematic diagram of the predicted heptahelical structure of the β2 adrenergic receptor. Seven membrane-spanning domains are connected by three extracellular (e1–e3) and three intracellular (i1–i3) loops. The approximate positions of posttranslational modification, including glycosylation of the extracellular N-terminus and palmitoylation of the intracellular C-terminus, are indicated. Residues involved in epinephrine binding, and predicted sites of cAMP-dependent protein kinase and G protein receptor kinase phosphorylation are shown. **(B)** Schematic representation of the phylogenetic relationships of the five main GPCR families according to the GRAFS system of classification (adapted from **ref. 5**). The glutamate receptor family contains the metabotropic glutamate, GABA$_B$, and calcium-sensing receptors, plus the Taste1 receptors. The Rhodopsin receptor family is the largest, with 241 non-olfactory and 701 total GPCRs divided into four groups. The α group is comprised five main branches: the prostaglandin, amine, opsin, melatonin, and melanocortin/*edg*/cannabinoid/adenosine receptor clusters. The β group has no main branches and contains receptors for peptide hormones. The γ group has three major branches: the somatostatin/opoiod/galanin, melanocortin-concentrating hormone, and chemokine receptor clusters. The δ group has four main branches: the Mas-related, glycoprotein, purin, and olfactory receptor clusters. The olfactory receptor cluster (arrow), containing approx 460 genes, is omitted for clarity. The adhesion receptor family consists of receptors with GPCR-like seven transmembrane-spanning domains fused to one or more functional domains with adhesion-like motifs in the N-terminus. The Frizzled/Taste2 receptor family contains the frizzled and Taste2 receptor clusters. The Secretin receptor family has no main branches and contains receptors for large peptides, such as vasoactive intestinal peptide, calcitonin, glucagon, parathyroid hormone, and secretin.

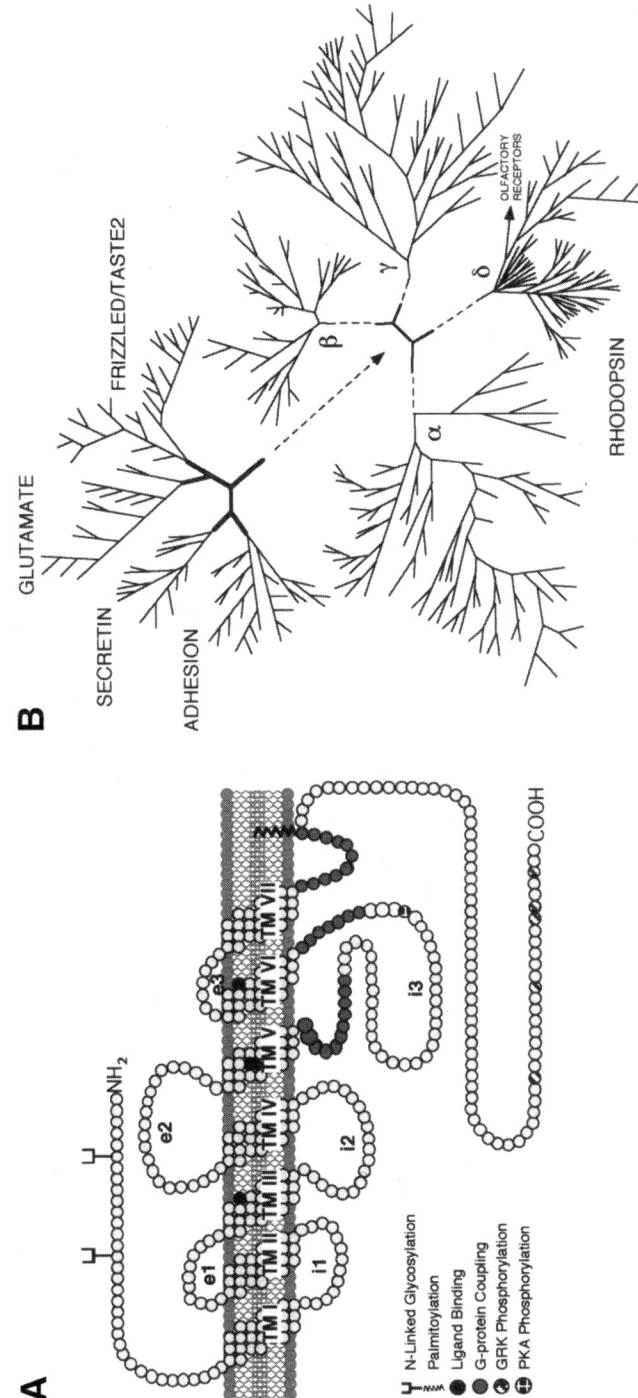

membrane domains share the highest degree of sequence conservation, whereas the intracellular and extracellular domains exhibit extensive variability in size and complexity. The extracellular and transmembrane regions of the receptor are involved in ligand binding, whereas the intracellular domains are important for signal transduction and for feedback modulation of receptor function. One or more sites for *N*-glycosylation are present within the N-terminus or, less often, the extracellular loops. Most GPCRs have in common two Cys residues that form a disulfide bridge between e1 and e2 that is critical for normal protein folding, and another Cys residue in the C-terminal domain that serves as a site for palmitoylation. This lipid modification leads to the formation of a putative fourth intracellular loop.

2.1.2. Receptor Taxonomy

Several classification systems have been devised that group GPCRs based on their ligands or sequence similarities. The widely used A through F classification system of Kolakowski *(14)*, for example, divides the GPCRs into six families, of which three (Families A, B, and C) contain the majority of known human receptors. In this system, Family A is made up of the rhodopsin-related receptors and is by far the largest group, containing the receptors for biogenic amines and other small nonpeptide ligands, chemokines, opioids and other small peptides, protease-activated receptors, and receptors for glycoprotein hormones. Family B GPCRs, the second largest group, contains receptors that bind to higher molecular-weight peptide hormones, such as glucagon, calcitonin, and parathyroid hormone. Family C, the smallest group, contains the metabotropic glutamate receptors, the γ-aminobutyric acid $(GABA)_B$ receptor, and the calcium-sensing receptor.

As genome-wide data from a number of species have become available, it has been possible to model the phylogeny of the GPCRs in some detail. Analysis of the chromosomal positions and sequence fingerprints of a large number of GPCRs has led Fredriksson et al. to propose the GRAFS classification system, in which the receptors are grouped into five families: glutamate, rhodopsin, adhesion, frizzled/taste2, and secretin *(5,15)*. This comprehensive system, which is diagrammed in simplified form in **Fig. 2B**, proposes that GPCRs in the GRAFS family arose from a common ancestor and evolved through gene duplication and exon shuffling. The GRAFS system contains some surprising relationships, such as the proposed link between Frizzled receptors, which generally are not thought to signal via heterotrimeric G proteins, and TAS2 group of taste receptors. Such phylogenetic linkages hint that the term "G protein-coupled receptor" may be a partial misnomer for a superfamily of seven transmembrane receptors that use diverse signaling mechanisms.

2.1.3. Receptor Structure–Function Relationships

GPCRs act as ligand-activated GEFs for heterotrimeric G proteins. As such, they have the potential to act catalytically in that one GPCR may activate multiple G proteins. In the rod outer segment, for example, a single light-activated activated rhodopsin catalyzes the activation of hundreds of transducin molecules *(16)*. Free G protein subunits then act on enzymatic effectors or ion channels, producing a pool of second messengers that control effector activity. The presence of multiple enzymatic steps in the GPCR signaling cascade creates the potential for tremendous signal amplification. As a result of this amplification process, a full biological response often can be obtained with as little as 5% receptor occupancy, a phenomenon that early pharmocologists referred to as "spare receptors."

The binding of an agonist to the transmembrane or extracellular domains of a GPCR produces conformational changes in the receptor that are transmitted to the intracellular domains in contact with the G protein *(17,18)*. Early study of the pharmacology of adrenergic receptors revealed the existence of two agonist affinity states of the receptor, the relative proportions of which are modulated by the presence of guanine nucleotides. The model developed to explain these phenomena predicts that in the presence of GDP, agonist promotes the formation of a high-affinity ternary complex among agonist, GPCR, and heterotrimeric G protein *(19)*. In the absence of the G protein, or when the presence of GTP allows for receptor-catalyzed G protein activation, the receptor resides in a low-affinity state. Subsequently, this model was refined to account for the phenomena of constitutively active GPCRs and the existence of full and partial agonists, neutral antagonists, and inverse agonists. In the extended ternary complex model, the GPCR is presumed to exist in a spontaneous equilibrium between two states, one inactive (R) and the other active (R* *[20]*). In the model, the efficacy of a ligand reflects its ability to alter the equilibrium between R and R*. A full agonist preferentially stabilizes the R* conformation, pulling the equilibrium toward the active state. In contrast, a neutral antagonist binds indiscriminately to both R and R* and exerts its effects only by excluding the binding of other ligands, whereas an inverse agonist binds preferentially to R and pulls the equilibrium toward the inactive state. This model explains the ability of some ligands to suppress the constitutive activity of mutationally activated GPCRs *(21)*.

Despite its power to predict the behavior of GPCRs binding to different classes of drug, the extended ternary complex model still presupposes that GPCRs exist in only two conformations. Several studies suggest that more complex models are necessary to account for all aspects of GPCR function *(22,23)*. In the case of the M2 muscarinic acetylcholine and *N*-formyl peptide receptors, the binding of receptor to arrestins, proteins that block receptor–G protein coupling, also induces a receptor conformation with increased affinity for agonists, but not antagonists *(24,25)*. Such findings have led to speculation that the agonist–receptor–arrestin complex represents an "alternative ternary complex." On the basis of experimental observations made by fluorescence lifetime spectroscopy, Swaminath et al. *(26)* recently have developed a model for β2-adrenergic receptor activation that predicts two distinct agonist-induced conformations, one of which supports G protein activation, and a second that is necessary for receptor internalization. Moreover, certain GPCR ligands appear to be capable of selectively inducing each of these states. The μ-opioid ligands DAMGO and morphine are equipotent with respect to G protein activation but differ in their ability to induce receptor desensitization *(27)*. Two endogenous ligands for the CCR7 chemokine receptor, CCL19 and CCL21, appear to exhibit similar properties, in that both stimulate G protein coupling, but only CCL19 promotes arrestin binding and desensitization *(28)*. Conversely, certain angiotensin II analogs that act as antagonists with respect to PLC signaling are nonetheless capable of inducing β-arrestin recruitment, receptor sequestration, and mitogen-activated protein (MAP) kinase activation *(29)*. The recent finding that angiotensin AT1a receptors can internalize and signal through a β-arrestin-dependent mechanism in the apparent absence of G protein coupling *(30)* suggests that the "alternative ternary complex" may in fact represent a distinct signaling state of the receptor.

2.2. The Heterotrimeric G Protein

Heterotrimeric G proteins provide the "missing link" between GPCRs in the plasma membrane and the enzymatic effectors that convey information about the presence of an external stimulus to the cell interior. Heterotrimeric G proteins are a group of GTPases that share a common multi-subunit structure. They are composed of a 39- to 52-kDa GTP-binding Gα subunit that possesses intrinsic GTPase activity and a tightly linked heterodimeric Gβγ subunit that is noncovalently bound to the Gα subunit in its inactive GDP-bound state.

2.2.1. G Protein Taxonomy

In contrast to the immense sequence diversity found in GPCRs, there are comparatively few genes encoding each of the heterotrimeric G protein subunits *(31)*. The 16 known mammalian Gα subunit genes are grouped by sequence homology into four families. With splice variants, approx 20 distinct Gα subunit proteins are expressed. The Gαs family contains the adenylate cyclase-stimulatory α subunit, Gαs, and the olfactory α subunit, Gαolf. The Gαi family includes the adenylate cyclase-inhibitory α subunits, Gαi1, Gαi2, Gαi3, and Gαo; two isoforms of the retinal α subunit, transducin or Gαt; the taste α subunit, gustducin or Gαgust; and Gαz, an α subunit whose function is not well understood. The Gαq family includes two α subunits that regulate PLC activity, Gαq and Gα11, as well as Gα14 and Gα15. Finally, the Gα12 family contains Gα12 and Gα13, two α subunits whose function also is poorly understood.

In addition to the Gα subunits, there are five known Gβ subunits and 12 Gγ subunits. Most, but not all, Gβ and Gγ subunit pairs can form stable heterodimers in vitro, suggesting that they may combine in vivo to generate a diverse array of Gβγ subunit heterodimers *(32)*. In combination with the Gα subunits, Gβγ subunit diversity creates the potential for upward of 1000 possible G protein heterotrimer combinations. Still, the role of subunit diversity in heterotrimer formation and its effect on signaling by G proteins are not well understood *(33)*. Tissue-specific variation in the complement of accessible G protein heterotrimers could expand the diversity of GPCR-mediated responses. It is clear that some combinations form the bulk of the heterotrimeric G protein in specific tissues, for example Gβ1γ1 is the predominant form associated with Gαt in the retina, and that specific Gβγ subunit heterodimers can differentially regulate certain effectors *(34)*.

2.2.2. G Protein Structure–Function Relationships

The Gα subunits share approx 40% sequence homology, corresponding principally to the regions of the protein that form the guanine nucleotide binding pocket. Most of the known Gα subunit–effector interactions involve this central core. More divergence is found in the N-terminus, which is required for Gβγ subunit binding, and in the C-terminus, which participates in the association with both receptors and effectors. Crystallographic studies of Gαt and Gαi1 reveal that the Gα subunit is composed of two major domains: a GTP–GDP binding domain that is similar to in structure to p21[ras], and an α-helical domain that in inserted into the Ras-like domain *(35,36)*. GTP hydrolysis is associated with order–disorder transitions in which the "switch II" and "switch III" regions of the Ras-like domain melt, whereas the N- and C-termini become ordered, consistent with the loss of an effector binding site and gain of a Gβγ binding site.

Gβ subunits are all approx 36 kDa in size, and share a common three-dimensional structure with an N-terminal coiled-coil structure that associates tightly with the Gγ subunit *(31)*. The remainder of the protein resembles a seven-bladed propeller, with each blade composed of interlocking β-sheets. Gγ subunits are all 7–8 kDa and possess a C-terminal Cys-Ala-Ala-x motif, which serves a site for prenylation, a lipid modification that is essential for membrane localization. In the heterotrimeric state, the Gβ subunit contacts the N-terminal and switch II regions of the Gα subunit and precludes access to the guanine nucleotide-binding pocket. The mostly α-helical Gγ subunit is tightly embedded in the surface of the Gβ subunit and does not make contact with Gα *(35,36)*. Although none of the Gα subunits possess membrane-spanning domains, all associate with the plasma membrane *(37)*. Gαi and Gαo are posttranslationally modified by the addition of the fatty acid myristate to an N-terminal glycine residue, and this lipid modification is required for membrane association. All Gα subunits, with the exception of Gαt, also undergo palmitoylation of one or more N-terminal cysteine residues, which is also important for membrane localization. Unlike myristoylation, however, this is a reversible modification. Upon binding of GTP and Gβγ subunit dissociation, palmitate is cleaved. Reassociation of the Gα and Gβγ subunits promotes repalmitoylation.

2.3. The G Protein-Regulated Effector

The third component of the GPCR signaling system is the G protein-regulated effector, which is typically an enzyme or ion channel. Although it was originally thought that only Gα subunits interacted with effectors, it is now clear that both GTP-bound Gα and Gβγ subunits regulate effector activity *(38)*.

The most widely studied G protein-regulated enzymes are the adenylate cyclases, which catalyze the conversion of adenosine triphosphate (ATP) to the intracellular second messenger cyclic-adenosine 3',5'-monophosphate or cyclic adenosine monophosphate (cAMP). cAMP regulates the activity of the cAMP-dependent protein kinase (PKA), as well as certain cAMP-regulated guanine nucleotide exchange factors, such as Epac1 and Epac2, that activate low-molecular-weight GTPases *(39)*. The 10 cloned adenylate cyclases all share a common 12 membrane-spanning domain architecture but vary significantly in tissue distribution and in regulation by G protein subunits *(40)*. All of the adenylate cyclases are stimulated by Gαs. Some are also stimulated by Gβγ subunits or by calcium–calmodulin, but only in the presence of active Gαs, allowing for "conditional stimulation," in which activation of more than one type of GPCR might be required for full cyclase activity. Gαi mediates inhibition of some, but not all, adenylate cyclase isoforms, as do Gαo and intracellular calcium. The activity of some adenylate cyclases also is modulated through phosphorylation by the second messenger-dependent protein kinases PKA and protein kinase C (PKC).

The four PLC-β isoforms are also regulated by G proteins *(41)*. PLC catalyzes the hydrolysis of membrane phosphatidylinositol to yield two intracellular second messengers, inositol 1,4,5-trisphosphate, which controls calcium efflux from the endoplasmic reticulum, and diacylglycerol, which along with calcium, controls the activity of several isoforms of PKC. PLCβ1–3 are activated independently by both Gα subunits of the Gq family and by Gβγ subunits. The PLCβ2 and PLCβ3 isoforms are more sensitive to Gβγ subunit regulation than PLCβ1, and usually account for the activation of PLC by GPCRs coupled to Gi family proteins. Gβγ subunits may also directly regulate phospholipase A2 activity in some settings. In the rod cells of the retina, activated transducin stimulates a cGMP phosphodiesterase, leading to closure of plasma membrane sodium channels, hyperpolarization of the plasma membrane, and signaling of second-order retinal neurons *(16,17)*. Other ion channels function as direct G protein-regulated effectors. Direct interaction between channel components and G protein subunits may account for the inhibition of high voltage "N-type" calcium channels by GPCRs acting via subtypes of Gαo, and the stimulation of "L-type" calcium channels by GPCRs acting via Gαs *(42,43)*. Gβγ subunits are the physiological activator of the inward-rectifying muscarinic-gated potassium channel, IKACh, which mediates the cardiac response to vagal stimulation *(42)*.

Finally, some kinases, notably the GPCR kinases (GRK)2 and GRK3, are directly regulated by Gβγ subunits *(44)*. These GRKs, which play a central role in receptor desensitization and also may phosphorylate nonreceptor substrates, possess C-terminal regulatory domains that bind free Gβγ subunits. Binding to the Gβγ subunits of activated G proteins allows the kinase to translocate from the cytosol to the plasma membrane where it gains access to substrate.

3. Negative Regulation of GPCR Signaling

Mechanisms to dampen GPCR signals exist at every level, from receptor to G protein to effector. Second messengers are degraded enzymatically by cAMP phosphodiesterases, phosphatidylinositol phosphatases, and diacylglycerol kinases. In some cases, the degradative enzymes are recruited into the proximity of the activated effector enzyme by binding to other GPCR regulatory proteins. For example, some isoforms of cAMP phosphodiesterase (PDE) bind directly to β-arrestins, proteins involved the in physical uncoupling of activated receptors from G proteins *(45,46)*. Stimulation of β2-adrenergic receptors causes β-arrestin-dependent recruitment of PDE4D3 and PDE4D5 to the receptor, promoting cAMP degradation and accelerated termination of membrane-associated PKA activity. Effector activity is modulated both by the availability of free G protein subunits and by feedback inhibition, as in the regulation of certain isoforms of adenylate cyclase by PKA or PKC phosphorylation. Gα subunit activity is modulated by protein–protein interactions that accelerate the normally slow intrinsic GTPase activity of the protein. In some cases this GTPase-activating protein (GAP) activity is an inherent property of the effector, such that interaction of the Gα–GTP complex with the effector leads to an acceleration of GTP hydrolysis and a return to the inactive state *(47,48)*. In other cases, GAP activity is conferred by members of a large family of proteins called regulators of G protein signaling, or RGS proteins, that interact with the

Gα subunit and stabilize the transition state for GTP hydrolysis *(48,49)*. Some 19 mammalian genes containing RGS core domains are known. Most act as GAPs toward Gi proteins, and some additionally act as GAPs for Gq proteins. Gβγ subunit function is also under control. In the retina, the Gβγ subunit-binding protein phosducin can sequester free Gβγ subunits, thereby modulating the Gβγ–Gαt interaction *(50)*.

The GPCR itself is the target for extensive negative regulation. The processes that regulate GPCR responsiveness at the receptor level typically are divided, based on mechanism, into receptor desensitization, sequestration, and downregulation. Together, they lead sequentially to the uncoupling of receptor from G protein, removal of receptors from the plasma membrane, receptor recycling or degradation, and reduced synthesis of new receptors. **Figure 3** depicts these processes schematically.

3.1. Heterologous vs Homologous Desensitization

Desensitization begins within seconds of agonist exposure and is initiated by phosphorylation of the receptor. Second messenger-dependent protein kinases, including PKA and PKC, phosphorylate serine and threonine residues within the cytoplasmic loops and C-terminal tail domains of many GPCRs. Phosphorylation of these sites directly impairs receptor–G protein coupling. For example, PKA phosphorylation of purified β2 adrenergic receptors in vitro is sufficient to impair receptor-stimulated Gs activation in the absence of other proteins *(51)*. Agonist occupancy of the target GPCR is not required for this process. Thus, receptors that have not bound agonist, including receptors for other ligands, can be desensitized by the activation of second messenger-dependent protein kinases. This lack of requirement for receptor occupancy has led to the use of the term heterologous desensitization to describe the process *(52)*. In some cases, such as the β2-adrenergic and murine prostacyclin receptors, PKA phosphorylation also alters the G protein-coupling selectivity

---

Fig. 3. *(opposite page)* Desensitization, sequestration, and recycling of G protein-coupled receptors (GPCRs). Within seconds of hormone (H) binding, GPCRs are phosphorylated by second messenger-dependent protein kinases (PKA/C) or by GPCR kinases (GRKs). PKC/A phosphorylation produces heterologous desensitization, whereas GRK phosphorylation promotes β-arrestin (β-arr) binding and homologous desensitization. β-Arrs engage clathrin and β2-adaptin (AP-2), causing agonist-occupied receptors to cluster in clathrin-coated pits, where they undergo dynamin (Dyn)-dependent endocytosis. Once internalized, GPCRs traffic from clathrin-coated vesicles into early endosomes, where they are sorted either into recycling endosomes for resensitization and return to the plasma membrane, or into late endosomes, from which they are either slowly recycled or targeted to lysosomes for degradation.

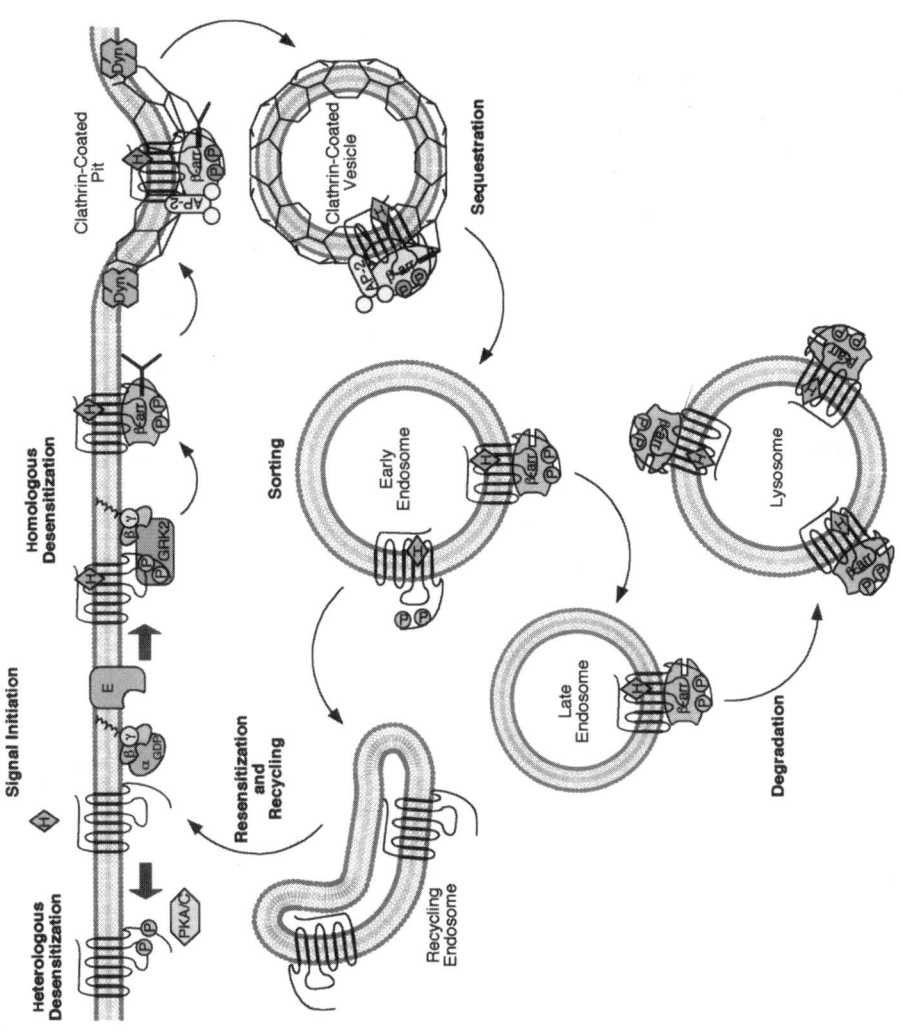

of the receptor to favor coupling to the adenylate cyclase inhibitory Gi protein, over the stimulatory Gs protein, causing the PKA-phosphorylated receptor to "reverse direction" with respect to cAMP production *(53–55)*. The phosphorylation-induced "switch" in G protein coupling also may couple the receptor to alternative signaling pathways, leading for example, to activation of MAP kinases *(56)*.

Like heterologous desensitization, homologous desensitization involves receptor phosphorylation, in this case by specialized GRKs. Unlike heterologous desensitization, homologous desensitization is a two-step process in which receptor phosphorylation is followed by the binding of an arrestin protein, whose role is to physically uncouple the receptor and the G protein.

There are seven known GRKs *(44)*. Rhodopsin kinase (GRK1) and GRK7 are retinal kinases involved in the regulation of rhodopsin photoreceptors, whereas GRK2–GRK6 are expressed more widely. Membrane targeting of all of the GRKs apparently is critical to their function and is conferred by a C-terminal tail domain. GRK1 and GRK7 each possess a C-terminal Cys-Ala-Ala-x motif. Light-induced translocation of GRK1 from the cytosol to the plasma membrane is facilitated by the posttranslational farnesylation of this site. The β-adrenergic receptor kinases (GRK2 and GRK3) have C-terminal Gβγ subunit-binding and pleckstrin-homology domains and translocate to the membrane as a result of interactions between these domains and free Gβγ-subunits and inositol phospholipids. Palmitoylation of GRK4 and GRK6 on C-terminal cysteine residues leads to constitutive membrane localization. Targeting of GRK5 to the membrane involves the electrostatic interaction of a highly basic 46 residue C-terminal domain with membrane phospholipids.

GRKs phosphorylate GPCRs on serine and threonine residues in the i3 loop and C-terminal tail. In contrast to PKA and PKC, GRKs preferentially phosphorylate receptors that are in the agonist-occupied conformation. Furthermore, GRK phosphorylation alone has little direct effect on receptor–G protein coupling. Rather, the principal function of the GRK in GPCR desensitization is to the increase receptor affinity for arrestins. In vitro, GRK2 phosphorylation of the β2 adrenergic receptor increases receptor affinity for β-arrestin 1 by 10- to 30-fold *(57)*. It is the binding of arrestin to receptor domains involved in G protein coupling rather than GRK phosphorylation *per se* that leads to homologous desensitization.

Four functional members of the vertebrate arrestin gene family have been cloned *(58,59)*. The two arrestins expressed in the retina, visual arrestin (S antigen or arrestin 1) and cone arrestin (X-arrestin or C-arrestin), exist primarily to regulate photoreceptor function. The nonvisual arrestins, β-arrestin 1 (arrestin 2) and β-arrestin 2 (arrestin 3), regulate the activity of most of the other GPCRs in the genome. All four arrestins bind specifically to activated,

GRK-phosphorylated GPCRs and block the receptor–G protein interaction. In addition to their role in desensitization, β-arrestins have functions that are not shared with the visual arrestins. The β-arrestin C-terminal tail contains binding motifs for clathrin *(60)* and the β2-adaptin subunit of the AP-2 complex *(61)* that allows β-arrestin to act as an adapter protein that targets GPCRs to clathrin-coated pits for endocytosis. It is these additional interactions that distinguish the two arrestin subfamilies and make β-arrestin binding integral to the processes of GPCR endocytosis, intracellular trafficking, resensitization, and downregulation.

In some cases, GRKs may mediate GPCR desensitization through a phosphorylation-independent mechanism. In addition to its C-terminal Gβγ binding domain, GRK2 possesses an N-terminal domain with homology to RGS proteins *(62)*. Although this RGS homology domain has little or no GAP activity, it can bind directly to free Gαq/11 subunits. The recently solved crystal structure of the GRK2-Gβγ subunit complex demonstrates that the kinase domain, RGS homology domain, and Gβγ binding domain of GRK2 reside at the three vertices of the protein, suggesting that GRK2 may be able to simultaneously interact with the receptor, free Gαq/11, and Gβγ subunits *(63)*. In the case of the mGluR1a metabotropic glutamate receptor, receptor desensitization does not require GRK2 kinase activity; desensitization occurs because receptor-bound GRK2 sequesters Gβγ and prevents its reassociation with GDP-bound Gαq/11 once GTP hydrolysis has occurred *(64)*.

3.2. Sequestration

Internalization of GPCRs, also termed *receptor sequestration* or *endocytosis*, occurs more slowly than desensitization, involving a period of several minutes after agonist exposure. Most, but not all, GPCRs undergo sequestration, and for many the process can be blocked by inhibiting the function of dynamin, a large GTPase necessary for the fission of clathrin-coated vesicles from the plasma membrane. It is now clear that GRK-mediated GPCR phosphorylation and β-arrestin binding provide the mechanism by which many GPCRs are targeted for clathrin-dependent endocytosis *(58,59)*. Once β-arrestins bind GRK-phosphorylated GPCRs, LIEF/L and RxR motifs near the β-arrestin C-terminus engage clathrin and β2 adaptin, respectively, leading to the clustering of receptors in clathrin-coated pits and removal from the plasma membrane.

GPCRs can be grouped on the basis of their pattern of interaction with the two β-arrestin isoforms *(65–67)*. "Class A" receptors, which include the β2 and α1B adrenergic, μ-opioid, endothelin A and dopamine D1A receptors, bind to β-arrestin 2 with higher affinity than β-arrestin 1. In addition, their interaction with β-arrestin is transient. β-Arrestin is recruited to the receptor at the plasma membrane and translocates with it to clathrin-coated pits. Upon internalization of the receptor, the receptor-β-arrestin complex dissociates. β-Arrestin then recycles to the plasma membrane while the internalized receptor proceeds into an endosomal pool. "Class B" receptors, represented by the angiotensin AT1a, neurotensin 1, vasopressin V2, thyrotropin-releasing hormone and neurokinin NK-1 receptors, bind β-arrestin 1 and β-arrestin 2 with equal affinity. These receptors form stable complexes with β-arrestin, such that the receptor-β-arrestin complex internalizes as a unit that is targeted to endosomes. Studies done by reintroducing β-arrestin 1 or 2 singly into a β-arrestin 1/2 null background have shown that β2 adrenergic receptor endocytosis involves primarily β-arrestin 2, whereas either β-arrestin can support AT1a receptor sequestration, consistent with the hypothesis that β-arrestin 1 and β-arrestin 2 differentially regulate GPCR sequestration *(68)*.

β-Arrestin function is further regulated by posttranslational modification, notably phosphorylation and ubiquitination. Cytoplasmic β-arrestin 1 is stoichiometrically phosphorylated on S412, which lies within the C-terminal regulatory domain *(69)*. Dephosphorylation of S412 occurs on receptor binding, and phosphomimetic mutations at these sites impair clathrin binding and GPCR endocytosis without affecting β-arrestin 2-mediated receptor desensitization, suggesting that phosphorylation of S412 regulates the ability of β-arrestin 1 to engage the endocytic machinery. A similar story applies to β-arrestin 2, except that the phosphorylated residues, S361 and T383, are further removed from the clathrin and AP-2 binding motifs *(70)*.

Regulated ubiquitination of β-arrestin 2 also plays an important role in GPCR endocytosis and trafficking. Both β-arrestin 2 and β2 adrenergic receptors are rapidly and transiently ubiquitinated in response to agonist *(71)*. β-Arrestin 2 ubiquitination is catalyzed by the E3 ubiquitin ligase, Mdm2, which binds directly to β-arrestin. Ubiquitination of the receptor is catalyzed by an as-yet-unidentified ubiquitin ligase but still requires the presence of β-arrestin. β-Arrestin 2 ubiquitination is apparently required for β2 receptor internalization, whereas ubiquitination of the receptor is involved in receptor degradation, but not internalization. The V2 vasopressin receptor also undergoes β-arrestin-dependent ubiquitination. Unlike the β2 adrenergic receptor, the V2 receptor-associated β-arrestin 2 remains stably ubiquitinated as it traffics with the receptor into the endosomal pool. Like the β2 adrenergic receptor, ubiquitination of the V2 receptor itself is not required for endocytosis, but does accelerate receptor degradation *(72)*. Expression of a β-arrestin 2-ubiquitin chimera causes the β-arrestin to remain associated with both β2 and V2 receptors as they internalize, suggesting that de-ubiquitination of β-arrestin 2 is a trigger for dissociation of the receptor-β-arrestin complex *(73)*.

Despite the broad applicability of the preceding model, it is clear that β-arrestin-dependent desensitization and sequestration are not inextricably linked. In the case of the PAR1 receptor, phosphorylation and β-arrestin 1 binding are required for receptor desensitization but not for clathrin-dependent endocytosis, which proceeds normally when the receptor is expressed in a β-arrestin 1/2 null background. Interestingly, a C-terminal phosphorylation site mutant of PAR-1 fails to internalize in either a β-arrestin-replete or β-arrestin-null background, suggesting that PAR1 receptors use a phosphorylation-dependent, β-arrestin-independent mechanism for endocytosis *(72)*. Similar results have been obtained using the *N*-formyl peptide receptor and the somatostatin receptor type 2A *(75,76)*. Other GPCRs that undergo β-arrestin-mediated desensitization appear to internalize via clathrin-independent mechanisms. Internalization of angiotensin AT1a and m2 muscarinic acetylcholine receptors is insensitive to expression of dominant inhibitory mutants of either β-arrestin or dynamin *(77,78)*. β1-Adrenergic receptor mutants lacking GRK phosphorylation sites exhibit impaired β-arrestin recruitment and clathrin-mediated endocytosis but still internalize through an alternative pathway that involves PKA phosphorylation and can be blocked by pharmacological disruption of caveloae *(79)*. Collectively, these data indicate that, at least under certain circumstances, β-arrestin-dependent receptor desensitization and sequestration are dissociable processes.

3.3. Resensitization, Recycling, and Downregulation

After endocytosis, GPCRs enter either a rapid recycling pathway leading to return to the plasma membrane or a "slow-recycling" pathway from which most receptors are targeted for degradation. The stability of the GPCR–β-arrestin interaction is a major determinant of path taken by the internalized receptor. Receptors like the β2 adrenergic receptor that dissociate from β-arrestin on endocytosis transit to an acidified endosomal vesicle fraction that is enriched in GPCR-specific protein phosphatase PP2A activity *(80)*. After receptor dephosphorylation and ligand dissociation in this acidified environment, β2 receptors return to the plasma membrane via recycling endosomes. In contrast, receptors like the V2 vasopressin receptor, which remain β-arrestin-bound as they internalize, recycle slowly and tend to be degraded. Switching the C-terminal tails of the β2 and V2 receptors, which converts the β2 receptor into a Class B receptor and the V2 receptor into a Class A receptor, reverses the pattern of dephosphorylation and recycling *(81)*.

The Rab family of small Ras-like GTPases regulates the budding, transport, docking, and fusion of intracellular vesicles. Several of the Rab GTPases, including Rab4, Rab5, Rab7, and Rab11, are involved in regulating GPCR endocytosis and sorting between early, late, and recycling endosomes and lysosomes *(82)*. Many GPCRs rapidly accumulate in Rab5 positive early endosomes, and Rab5 is required for GPCR endocytosis, as well as vesicle fusion and GPCR sorting. Interestingly, Rab5 binds directly to the C-terminal tail of the angiotensin AT1a receptor, and angiotensin binding leads to Rab5 GDP/GTP exchange, suggesting that some GPCRs directly regulate their own intracellular trafficking *(83)*. Rab4 also localizes to early and recycling endosomes and appears to be involved in rapid plasma membrane recycling but not in the initial endocytosis of GPCRs. Rab11 appears to regulate the slow recycling of GPCRs, whereas Rab7 controls GPCR trafficking between late endosomes and lysosomes.

Other protein–protein interactions involving GPCRs modulate endocytic sorting and influence the fate of the internalized receptor. Binding between a DSLL motif at the extreme C-terminus of the β2 adrenergic receptor and the PDZ-domain containing protein NHERF-1/EBP50, has been implicated in directing the β2 adrenergic receptor into the recycling pathway *(84)*. Conversely, a candidate GPCR-associated sorting protein (GASP) has been identified that binds to the C-terminus of the δ-opioid receptor and preferentially targets the receptor to lysosomes for proteolytic degradation *(85)*. Transferring the PDZ domain-binding motif of the β2 receptor onto the δ-opioid receptor, causes it to reroute from the degradative pathway into the rapid recycling pathway *(86)*.

Other components of the GPCR desensitization machinery also recruit proteins involved in GPCR internalization and trafficking. GRK2 has been shown to associate with actin and with a novel ARF6 GTPase-activating protein called GIT-1 *(87)*. β-Arrestins complex with ARNO, an ARF guanine nucleotide exchange factor, which along with GIT-1, regulates ARF function *(88)*. ARFs are small GTPases involved in vesicle trafficking and sorting, and both GIT-1 and ARF6 are involved in GPCR endocytosis.

Downregulation of GPCRs, the persistent loss of cell surface receptors that occurs during a period of hours to days, is the least understood of the processes controlling GPCR responsiveness. Control of cell surface receptor density occurs partially at the transcriptional level, but the removal of agonist-occupied receptors from the cell surface and their sorting for either degradation or recycling to the membrane is also important, at least in the early stages of downregulation.

4. Positive Regulation of GPCR Signaling

Interactions between GPCRs and numerous other proteins modulate the specificity, selectivity, and time course of signaling by the basic GPCR–G protein–effector module. Recent data indicate that GPCRs frequently are organized into multi-receptor or multiprotein "signalsomes" that influence the response to receptor activation. Protein–protein interactions that positively modulate GPCR signaling by influencing ligand binding, coupling to G proteins and effectors, or targeting to specific subcellular locations, include GPCR

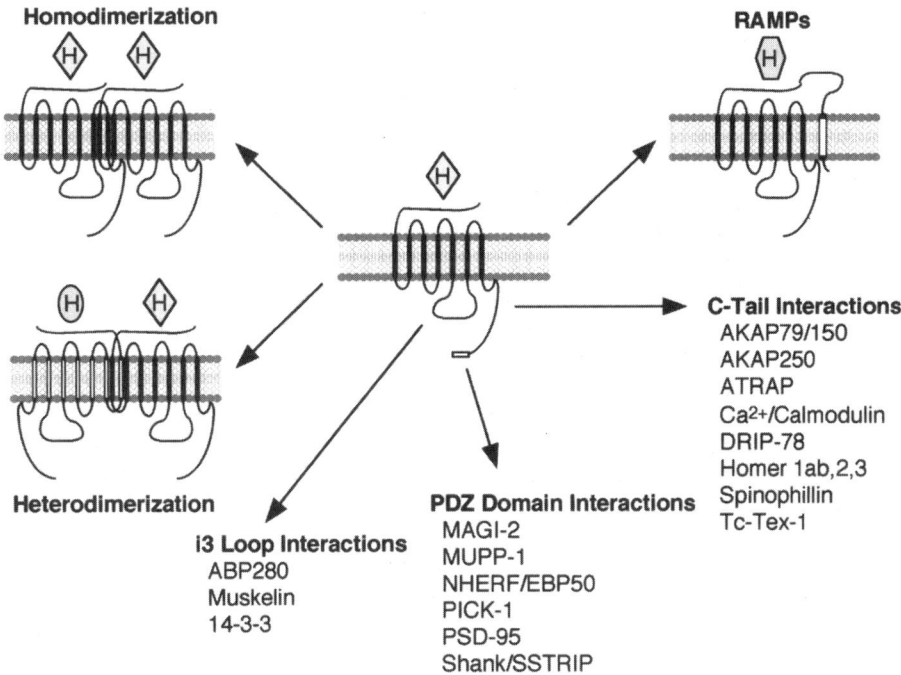

Fig. 4. Protein–protein interactions that regulate G protein-coupled receptor (GPCR) localization, trafficking, and signaling specificity. Many GPCRs form either homodimers or heterodimers with other GPCRs. Dimerization affects trafficking of nascent GPCRs to the plasma membrane, receptor pharmacology, and signaling specificity. The interaction of calcitonin receptor-like receptor and calcitonin receptors with single transmembrane receptor activity-modifying proteins is necessary for receptor trafficking and confers unique ligand-binding properties on the receptor. Interactions between the intracellular domains of GPCRs and numerous other proteins target receptors to specific intracellular loci, or modulate the specificity, selectivity and time course of signaling by the basic GPCR–G protein–effector module.

dimerization, the interaction of GPCRs with receptor activity-modifying proteins (RAMPs), and the binding of various scaffolding proteins to the GPCR i3 loop and C-terminus. Many of these regulatory interactions are depicted in **Fig. 4**.

4.1. Dimerization

Coprecipitation approaches, complementation studies using mutated or chimeric receptors, and fluorescence energy transfer measurements all support the conclusion that many, if not most, GPCRs can exist as homodimers, heterodimers, or higher order multimers *(89–91)*. For example, the finding that the internal tethered ligand of one protease-activated receptor can "transactivate" other family members *(92,93)* fits the notion that the PAR receptors are very closely associated. Fluorescence or bioluminescence resonance energy transfer methods demonstrate that many homodimeric or heterodimeric GPCR combinations are allowed *(89–91)*. Recently, direct measurements performed on reconstituted functional leukotriene B4 receptor:Gαi2β1γ2 complexes indicate an 2R:Gαβγ stoichiometry, consistent with a dimeric receptor interacting with a single heterotrimeric G protein *(94)*. Comparison of the crystallographic structures of rhodopsin and transducin indicates that a monomeric receptor cannot account for all of the known contact points between receptor and G protein *(95)*, whereas atomic force microscopy has shown that native rhodopsin in the murine rod outer segment exists primarily as dimers *(96)*.

In a number of cases, GPCR dimers appears to form shortly after synthesis, and in some cases, dimerization is the prerequisite for trafficking nascent receptors to the plasma membrane. The clearest example of such an obligatory role for receptor dimerization is the GABA$_B$ receptor *(97–99)*. The GABA$_B$R1, which contains the structural determinants necessary for ligand binding but not for G protein coupling *(99)*, is retained in the endoplasmic reticulum as an immature glycoprotein unless it is coexpressed with a second GABA$_B$ receptor transcript, the GABA$_B$R2. The GABA$_B$R2 alone can reach the cell surface and is capable of G protein coupling but cannot bind ligand *(100)*. Dimerization of the two receptors, which is mediated by their C-terminal tails, masks an endoplasmic reticulum retention sequence located in the tail of the GABA$_B$R1, and allows for correct processing and membrane transport of GABA$_B$R1 *(101)*. Thus, the GABA$_B$R2 functions as a chaperone for GABA$_B$R1 through the formation of an obligatory heterodimer. Similarly, coexpression of truncated D2 dopamine or V2 vasopressin receptors with their intact counterparts inhibits the trafficking and function of the intact receptor, suggesting that dimer formation prior to membrane delivery may be a general process *(102,103)*. In other cases, dimerization appears to be regulated on the plasma membrane, positively or negatively, by agonist binding. Ligand binding to the CCR2b, CCR5, and CXCR4 chemokine receptors, for example, promotes the formation of receptor dimers from essentially undetectable basal levels *(104–106)*.

The potential for receptor oligomerization to increase the diversity of GPCR signaling is enormous. Receptor pharmacology, G protein-coupling efficiency, downstream signaling, and the endocytosis of activated receptors are all influenced by the formation of receptor multimers. For example, in cells expressing both the δ-opioid and κ-opioid receptors, high-affinity binding of subtype selective ligands is enhanced when ligands for both receptors are available, suggesting that the receptors exhibit cooperative binding when heterodimerized *(107)*. Similarly, interaction between μ-opioid and δ-opioid receptors leads to altered pharmacology suggestive of cooperative ligand binding *(108)*. Agonist-induced trafficking of opioid receptors is also influenced by heterodimerization. The nonselective opioid agonist etorphine, which causes internalization of δ-opioid, but not κ-opioid receptors, does not cause δ-opioid receptor internalization when it is coexpressed with κ-opioid receptor *(107)*.

Heterodimerization between more distantly related receptors also modulates GPCR function. For example, the AT1a angiotensin receptor heterodimerizes with several other GPCRs, including the B2 bradykinin and β adrenergic receptors. Interaction between AT1a and B2 receptors increases the efficiency of Gi and Gq activation, and increases potency and efficiency of angiotensin II, a vasopressor, while decreasing that of bradykinin, a vasodepressor *(109,110)*. In cardiomyocytes, blockade of β receptors inhibits the signaling and trafficking of the AT1 receptor by inducing functional uncoupling of the AT1 receptor from its cognate Gq protein without affecting angiotensin binding. In a reciprocal manner, selective blockade of the AT1 receptor uncouples the β receptor from Gs and inhibits downstream signaling. Inhibition of the two receptors by a single antagonist apparently results from direct interactions occurring at the receptor level *(111)*.

As the preceeding examples illustrate, the potential for complex crossregulation generated by GPCR heterodimerization is vast and as yet poorly understood. However the pharmacotherapeutic implications are clear *(112–114)*. In vivo studies of preeclampsia in humans *(110)*, of morphine analgesia in rats *(115)*, and of cardiac function in mice *(111)* suggest that formation of GPCR heterodimers leads to physiologically relevant changes in drug pharmacology that present opportunities for selective enhancement or dampening of receptor function.

4.2. RAMPs

The pharmacology of at least two GPCRs is determined not strictly by the intrinsic structure of receptor but by their interaction with members of a family of novel transmembrane proteins, called RAMPs. The three known RAMP proteins are 148- to 174-amino acid single transmembrane domain glycoproteins with large extracellular domains and short cytoplasmic domains *(116,117)*. RAMPs form complexes with the calcitonin receptor-like receptor (CRLR) and calcitonin receptor and control receptor trafficking and function. RAMP binding to the CRLR is required for transport of nascent receptors to the plasma membrane. Furthermore, the specific CRLR–RAMP complex determines the ligand specificity of the receptor. The CRLR–RAMP1 complex acts as a receptor for the calcitonin gene-related peptides, a pleiotropic family of neuropeptides with homology to calcitonin, amylin and adrenomedullin. When CRLR is complexed with RAMP2 or RAMP3, it functions as an adrenomedullin receptor. Similarly, complexes between a naturally occurring splice variant of the calcitonin receptor and RAMP1 or RAMP3 yields a functional amylin receptor. RAMP expression is modified under various forms of physiological stress and in response to glucocorticoids, suggesting that cellular responsiveness to certain hormones may be regulated through control of accessory proteins.

4.3. PDZ Domain-Containing Proteins

Postsynaptic density of 95-kDa (PSD95)-disc large zona occludens (PDZ) domains are protein–protein recognition domains that specifically bind to short peptide motifs usually located at the C-terminus of proteins. PDZ domain-containing proteins often possess other protein interaction domains, and commonly serve as scaffolds for protein localization, trafficking, or complex assembly. The C-terminal tails of several GPCRs, including the β2 adrenergic receptor, 5HT2a-c serotonin receptors, SSTR2 somatostatin receptor, the mGluR1a and mGluR5 metabotropic glutamate receptors, and the parathyroid hormone PTH/PTHrP receptor, end in a PDZ domain-binding motif. Not surprisingly, a number of PDZ domain-containing proteins have been shown to interact with these GPCRs and to play important roles in modifying receptor function *(118,119)*.

The Na$^+$/H$^+$ exchanger regulatory factor/ezrin binding protein 50 (NHERF/EBP50) is a 55-kDa protein with two tandem PDZ domains that originally was identified as a required protein for PKA-mediated inhibition of the NHE3 Na$^+$/H$^+$ exchanger in renal and gastrointestinal epithelial cells. The first PDZ domain of NHERF/EBP50 binds to the β2 adrenergic receptor via the last four amino acids of the receptor C-terminus *(120)*. This interaction results in agonist-dependent clustering of NHERF and contributes to β2 adrenergic receptor-mediated stimulation of NHE3 via a potentially cAMP-independent mechanism. In addition, the NHERF1/EBP-50 protein, through its ezrin/radixin/moesin (ERM) motif, links β2 adrenergic receptors to the actin cytoskeleton. As previously mentioned, disruption of this interaction leads to incorrect postendocytic sorting of the receptor and accelerated receptor degradation *(84)*. PLC-β1 and the PTH1 parathyroid hormone receptor interact with the first and second PDZ domains of NHERF1/EBP-50, respectively. In cells that express NHERF1/EBP-50, like the brush border of the proximal renal tubular epithelium, the PTH1 receptor stimulates phosphatidylinositol hydrolysis and inhibits adenylate cyclase by coupling to Gi/o family G proteins. In cells that lack NHERF1/EBP-50, adenylate cyclase activation through Gs coupling predominates *(121)*. The pattern of G protein and effector coupling can be reversed by overexpression of NHERF1/EBP-50. Disrupting the interaction between NHERF1/EBP-50 and actin blocks apical localization of PTH1 receptors and inhibits PLCβ-induced calcium influx, indicating that NHERF1/EBP-50 both controls receptor localization and specifies G protein and effector coupling *(122)*.

Additional GPCR–PDZ domain-containing protein interactions influence trafficking and localization of GPCRs. PSD-95 binds to the β1 adrenergic receptor C-terminus and decreases receptor endocytosis *(123)*, whereas another

PDZ domain-containing protein that binds the same C-terminal epitope, membrane-associated guanylate kinase inverted-2 (MAGI-2), a multidomain scaffolding protein containing six PDZ domains, markedly enhances β1 receptor internalization *(124)*. The PDZ domain of the *src* homology (SH)3 multiple ankrin domain-containing protein (Shank)/somatostatin receptor interacting protein (SSTRIP) associates with the C-terminus of the somatostatin SST2 receptor. Shank/SSTRIP contains multiple protein interacting domains, including six N-terminal ankrin repeats, single SH3 and PDZ domains, seven proline rich domains, and a C-terminal sterile α motif. Both SST2 receptors and Shank/SSTRIP colocalize to neuronal postsynaptic densities, suggesting that Shank/SSTRIP plays a role in SST2 receptor targeting. In addition, the sterile α motif domain may mediate SST2 receptor dimerization *(125)*.

Protein interacting with C kinase 1 (PICK-1) is a single PDZ domain-containing protein that binds to PKCα. Binding of the PICK-1 PDZ domain to the C-terminus of the mGluR7a receptor is required for clustering the receptor at presynaptic terminals, and for coupling the receptor to inhibition of P/Q type calcium channels via a Go-PLCβ-PKC dependent pathway *(126,127)*. The multi-PDZ domain protein 1 (MUPP1), which contains 13 PDZ domains, interacts with the C-termini of the serotonin 5HT2a, 5HT2b, and 5HT2c subtypes, via PDZ domain 10. Coexpression of MUPP1 with 5HT2c receptors leads to receptor clustering. Although no functional role of the interaction has been identified, the pattern of expression of the two proteins coincides in the rat brain, suggesting that MUPP1 may function in GPCR signalsome assembly *(128)*.

Spinophillin is a PDZ domain-containing protein that localizes to dendritic spines and binds to protein phosphatase 1 (PP1) and F-actin. Spinophillin interacts with both D2 dopamine and α2a-c adrenergic receptors via a non-PDZ-domain-mediated mechanism involving the i3 loop of the receptor and a region of the protein located between the F-actin binding and PDZ domains *(129,130)*. Spinophillin can bind to both the D2 receptor and PP1 simultaneously, suggesting that it plays a role in scaffolding GPCRs and their regulators to specific cell surface microdomains.

4.4. Other GPCR-Interacting Proteins

A number of other protein–protein interactions involving GPCRs recently have been identified. Some appear to control receptor trafficking or subcellular localization, whereas others scaffold protein complexes that dictate the specificity of the GPCR–G protein–effector interaction *(119)*.

The Homer proteins share a 120 amino acid N-terminal Ena/VASP homology 1/WASP homology 1 (EVH) domain that has been implicated in the control of actin filament dynamics. This domain interacts with polyproline sequences in several signaling and scaffolding proteins located at glutaminergic postsynaptic sites, including mGluR1α, mGluR5, Shank/SSTRIP, IP3 receptors, ryanodine receptors, and P/Q type calcium channels *(131)*. Homer 1a expression in the hippocampus is upregulated by excitatory synaptic activity and triggers the targeting of mGluR5 to axon and dendrites. Homer 1b/c, Homer 2, and Homer 3, but not Homer 1a, have C-terminal coiled-coil motifs that confer homo- and hetero-oligomerization properties. Both the mGluR1α and mGluR5 receptors terminate with PDZ domain-binding motifs that, along with a polyproline sequence in the mGluR C-terminal tail, mediates mGluR binding to Homer. In neurons, nascent mGluRs accumulate in the endoplasmic reticulum in association with Homer 1b. Excitation-induced expression of Homer 1a, which cannot oligomerize with the other Homers proteins and therefore acts as a "dominant-negative" for Homer-induced receptor clustering, promotes mGluR trafficking to the plasma membrane. At the membrane, the receptors appear to dissociate from Homer 1a and engage in interactions with Homer 1c and other mGluR-interacting proteins *(132)*.

The dopamine receptor-interacting protein (DRIP78) of 78 kDa binds to a highly conserved C-terminal hydrophobic motif in GPCRs that functions as an export sequence from the endoplasmic reticulum. DRIP78 binding to D1 dopamine receptors leads to receptor retention in the endoplasmic reticulum and delayed posttranslational processing of the receptor *(133)*. Conversely, interaction of the t-complex testis-expressed 1 (Tctex-1) protein with the C-terminus of rhodopsin is required for receptor transport to the rod outer seqment *(134)*. Mutations in the rhodopsin sequence that binds Tctex-1 are associated with retention of rhodopsin in the cell body and the development of retinitis pigmentosa *(135)*.

Other GPCR-binding proteins affect signaling by targeting effector enzymes or regulatory proteins to the receptor. A kinase anchoring proteins (AKAPs) localize PKA to its substrates by interacting with the RII regulatory subunit of the kinase. The β2 adrenergic receptor C-terminus interacts with a complex containing Gravin, an AKAP of 250 kDa, PKA, PP2A, and PP2B *(136)*. Agonist stimulation of the receptor brings additional proteins into the complex, including GRK2, β-arrestin, and clathrin. Suppression of Gravin expression does not affect receptor desensitization, but markedly reduces the sequestration and resensitization of β2 receptors. Similarly, AKAP150, the rat homolog of human AKAP 79, interacts with the β2 adrenergic receptor C-terminus and coprecipitates with the receptor from rat brain in complex with PKA, PKC, and PP2B *(137)*. The AKAP 79/150 interaction appears to facilitate receptor desensitization, both by targeting PKA to the receptor and by increasing GRK2 activity *(138)*. The angiotensin receptor-associated protein (ATRAP) selectively binds to the C-terminus of the angiotensin AT1a receptor *(139)*. ATRAP localizes to endoplasmic reticulum, Golgi and endocytic vesicles, and appears to constitutively shuttle toward the plasma membrane. Binding of AT1a receptors to ATRAP inhibits coupling to PLC and markedly reduces AngII stimulated transcriptional activity and cell proliferation. Calcium–calmodulin and Gβγ subunits compete for a common binding site on the C-terminus of the mGluR7, and are involved in the inhibition of Ca^{2+} channels by this receptor *(140)*.

Actin-binding protein 280 (ABP-280 or filamin A) is a cytoplasmic protein that contains an actin-binding domain at its N-terminus. ABP-280 interacts with the i3 loop of the D2 and D3 dopamine receptors. Its expression fosters D2 receptor clustering at the plasma membrane and enhances the ability of D2 receptors to inhibit adenylate cyclase. The interaction with ABP-280 may be inhibited by PKC phosphorylation of the receptor i3 loop, allowing for PKC activation to modulate D2 receptor coupling *(141)*. Similarly, Muskelin binds to the α splice variant of the prostaglandin E2 receptor and affects coupling to Gi and receptor internalization *(142)*.

The 14-3-3 proteins are a family of at least seven acidic brain proteins that have been shown to modulate the function of a variety of signaling molecules by binding to phosphorylated serine/threonine motifs. 14-3-3 proteins exist as dimers, which may allow coordination of signal transduction. The 14-3-3 ζ isoform interacts with the i3 loop of α2 adrenergic receptors *(143)*, whereas the $GABA_BR1$ receptor interacts with the 14-3-3 η and ζ isoforms *(144)*. These interactions may be involved in regulating GPCR dimerization, activation of the Ras/Raf cascade, and the localization of RGS proteins. Additional direct or indirect interactions between the C-terminus of GPCRs and cytoskeletal or cytoskeleton-associated proteins that may regulate GPCR localization or function have been reported, including binding to myosin heavy chain IIa, 4.1N, actin, spectrin, and CapZ *(119)*.

5. Novel Mechanisms of GPCR Signaling

As evident from the preceding discussion, many GPCRs engage in protein–protein interactions that control spatial organization of GPCRs and modify the function of the basic GPCR–G protein–effector module. Additional lines of evidence, however, suggest that the full diversity of GPCR-mediated signals cannot be explained solely on the basis of traditional models of GPCR signaling. Most prominent are the effects of GPCR stimulation of cellular growth, proliferation, and differentiation, which in many systems are demonstrably independent of the activation of second messenger-dependent protein kinases. Study of the mechanisms underlying these responses has led to the discovery that GPCRs employ a number of novel signal transduction mechanisms, some of which appear not to require G protein activation. Several of these interactions are shown schematically in **Fig. 5**.

5.1. Crosstalk Between GPCRs and Other Membrane Receptors

Stimulation of most GPCRs leads to the rapid activation of MAP kinases. Mammalian cells contain three major classes of MAP kinase, the extracellular

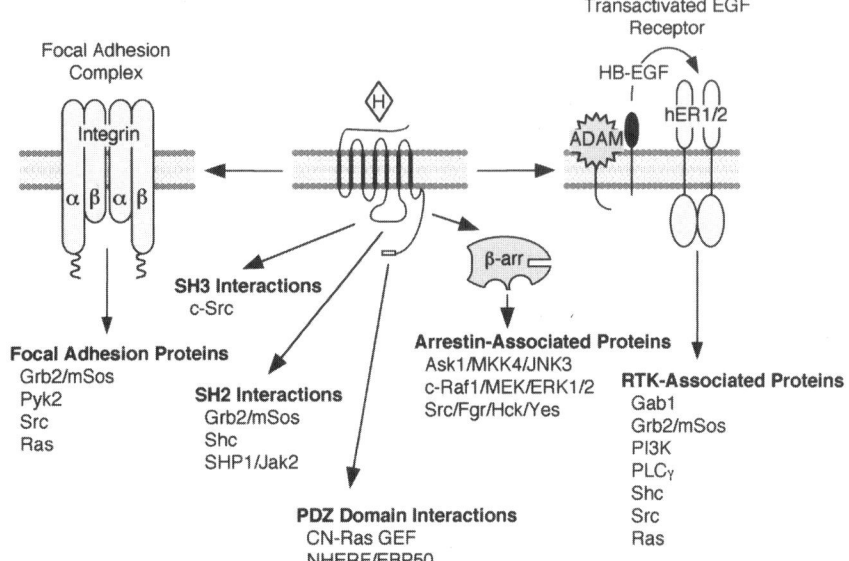

Fig. 5. Novel mechanisms of G protein-coupled receptor (GPCR) signaling. Recent data indicate that GPCRs can use multiple intermediates to signal intracellularly. Crosstalk between GPCRs and focal adhesions or "transactivated" receptor tyrosine kinases activates tyrosine kinase signaling cascades and initiates Ras-dependent signaling. The intracellular domains of some GPCRs may interact directly with src homology (SH)2, SH3, or PDZ domain-containing adapters or enzymes. Arrestins also function as signaling adapters by recruiting Src family kinases or components of mitogen-activated protein kinase pathways to GPCRs. Because arrestin binding uncouples GPCRs from G proteins, these "arrestin-dependent" signals may represent a form of second-wave signaling by desensitized GPCRs.

signal-regulated kinases (ERKs), c-Jun N-terminal kinase/Stress-activated protein kinases (JNK/SAPK), and p38/HOG1 MAP kinases. The ERK pathway is important for control of the G_0-G_1 cell cycle transition and the passage of cells through mitosis or meiosis, whereas the JNK/SAPK and p38/HOG1 MAP kinases are involved in regulation of growth arrest, apoptosis, and activation of immune and reticuloendothelial cells in response to a variety of environmental and hormonal stresses *(145,146)*. MAP kinases are regulated through a series of modular kinase cascades, each of which is composed of three kinases that successively phosphorylate and activate the downstream component. It is now clear that a complex set of GPCR-derived signals converge to determine the activity of the MAP kinase modules, and that the extent to which different inputs contribute to MAP kinase activation varies widely between different receptors and cell types. Moreover, the exact mechanism employed to activate the MAP kinase appears to determine where, and for how long, the MAP kinase is active, which in turn affects the functional consequences of MAP kinase activation.

Early studies of the mechanisms of GPCR-stimulated ERK activation found that these signals often involved the activation of Src family nonreceptor tyrosine kinases and the low-molecular-weight G protein, Ras, intermediates in signaling by classical receptor tyrosine kinase (RTK) growth factor receptors *(147,148)*. One basis for this convergence lies in crosstalk between GPCRs and either RTKs or focal adhesion complexes, either of which can scaffold the assembly of a Ras activation complex.

At least two RTKs, those for platelet-derived growth factor and epidermal growth factor (EGF), can be "transactivated" by GPCRs *(149,150)*. The best-understood mechanism of RTK transactivation is through the autocrine/paracrine generation of EGF receptor ligands via a process termed ectodomain shedding *(151)*. Each of the known ligands for the EGF receptor, including EGF, transforming growth factor (TGF)-α, heparin-binding (HB)-EGF, amphiregulin, betacellulin, and epiregulin, is synthesized as a transmembrane precursor that is proteolyzed to produce a soluble growth factor. Of these, amphiregulin, HB-EGF and TGF-α have been shown to undergo GPCR-stimulated release in various cell types *(152)*. In fibroblasts, both Gi/o-coupled and Gq/11-coupled receptors activate EGF receptors by generating HB-EGF. For Gi/o-coupled GPCRs, HB-EGF shedding and EGF receptor transactivation are mediated by G$\beta\gamma$ subunits, whereas HB-EGF shedding in response to stimulation of Gq/11-coupled receptors is mediated by Gq/11α subunits. GPCR-mediated transactivation does not typically involve activation of either PLCβ isoforms or ion channels. The G protein effectors that regulate ectodomain shedding are poorly defined, although phosphatidylinositol-3' kinases *(153)* and Src family kinases have been proposed as early intermediates in the pathway *(154,155)*. Proteolysis of the HB-EGF precursor is mediated by members of the ADAM family of matrix metalloproteases (MMPs), one of which, ADAM 12, has been implicated GPCR-mediated HB-EGF shedding in the heart *(156)*. Several of the ADAMs, notably ADAM-9, -10, -12, -15, -17, and -19, possess consensus SH3 domain binding motifs within their short intracellular domains that might mediate interaction with Src family kinases. Although HB-EGF released in response to GPCR stimulation can cause paracrine activation of EGF receptors in adjacent cells *(151,155)*, β2 adrenergic receptors coprecipitate with transactivated EGF receptors, suggesting that transactivation normally occurs over very short distances in the context of GPCR–RTK complexes *(157)*.

Because EGF receptors can be transactivated by many extracellular stimuli in addition to those transmitted by GPCRs, their principal function may be to integrate mitogenic signals from multiple sources *(158)*. In some cases, the mitogenic response to GPCR stimulation has been attributed to Ras-dependent signals arising from transactivated EGF receptors. In cardiac fibroblasts, for example, angiotensin II-stimulated ERK activation and DNA synthesis are both EGF receptor-dependent *(159)*. Similarly, NK1 receptor-mediated ERK activation and DNA synthesis in U-373 MG cells is blocked by either pharmacological inhibition of the EGF receptor or expression of a dominant-negative EGF receptor mutant *(160)*.

A second form of GPCR crosstalk involves the focal adhesion complex. Focal adhesions form when integrin heterodimers, which serve as extracellular matrix receptors, cluster at points of contact between the cell surface and specific matrix proteins, such as fibronectin. Gαq/11-dependent activation of a calcium- and cell adhesion-dependent focal adhesion kinase family member, Pyk2, leads to Ras-dependent ERK activation *(161)*. In this system, intracellular calcium, released as a result of PLCβ-mediated inositol trisphosphate production, triggers Pyk2 autophosphorylation, recruitment of c-Src, tyrosine phosphorylation of Shc, and Ras-dependent ERK activation *(162)*.

The involvement of a second cell surface receptor in this form of GPCR signaling imparts a conditional nature to the response and allows for crossregulation of receptor classes. Because Pyk2 activation requires both the presence of focal adhesions and a GPCR-derived calcium signal, the pathway is functional only in adherent cells that express Pyk2 *(163)*. Prior stimulation of 5HT receptors in primary renal mesangial cells renders cells insensitive to subsequent EGF exposure owing to the rapid transactivation-dependent internalization of nearly the entire complement of EGF receptors on the cell surface *(164)*.

5.2. Alternative Mechanisms of GPCR Signaling

The intracellular domains of several GPCRs can bind, either directly or through adapter proteins, to enzymatic effectors, including GEFs for small G proteins, nonreceptor tyrosine kinases, and components of several of the MAP kinase pathways, raising the prospect that some GPCR signals may arise from "G protein-independent" activation of effectors.

5.2.1. GPCRs as Signaling Scaffolds

A number of direct associations between GPCRs and putative effector enzymes or adapters have been reported, including PDZ domain-mediated, as well as SH2 and SH3 domain-mediated, protein–protein interactions. As previously noted, binding of the PDZ domain of NHERF/EBP50 to the C-terminus of β2 adrenergic and PTH/PTHrP receptors may confer the ability to regulate NHE3 activity *(120)*. A novel mechanism of GPCR-mediated activation of Ras family GTPases involves direct binding of the C-terminus of the β1 adrenergic receptor to the PDZ domain of cAMP-regulated Ras GEF (CN-Ras GEF *[165]*).

Some GPCRs are substrates for tyrosine phosphorylation, which can lead to their association with SH2 domain-containing proteins. For example, the β2-adrenergic receptor can be phosphorylated on tyrosine residues by the insulin receptor tyrosine kinase *(166)*. This reportedly leads to direct association of the receptor with the adapter proteins Grb2 and Shc, central elements in the control Ras activity. Stimulation of the JAK-STAT pathway of transcriptional regulation by angiotensin AT1a receptors involves tyrosine phosphorylation of Tyr319 in the AT1a receptor tail by an Src family kinase, followed by association of JAK2 with the receptor *(167)*. In this case, the binding of JAK2, which does not have an SH2 domain, appears to be indirect, and may be mediated by members of the SHP family of SH2 domain-containing tyrosine phosphatases *(168)*. JAK2 recruitment to the receptor appears to be necessary, but not sufficient, for activation of the JAK–STAT pathway by angiotensin, which still requires Gq/11-dependent PKC activation *(169)*.

Several GPCRs contain relatively large proline-rich inserts in their intracellular loops that could mediate binding to proteins containing SH3 domains. The β3 adrenergic receptor, for example, which does not recruit β-arrestin, binds to the c-Src SH3 domain in an agonist-dependent manner via proline-containing motifs in its i3 loop and C-terminus *(170)*. This interaction may contribute to the ability of β3 adrenergic receptors to activate the ERK pathway, a response that still requires activation of pertussis toxin-sensitive G proteins.

5.2.2. GPCR-Binding Proteins as Signaling Scaffolds

Many GPCR-associated proteins, such as Shank/SSTRIP, PICK-1, MUPP1, Spinophillin, Homer proteins, AKAP 79/150 and 14-3-3, bind other signaling molecules, suggesting the hypothesis that GPCRs might signal by "coupling" to non-G protein-regulated effectors. However, the most compelling data indicating that GPCRs can act through G protein-independent signaling mechanisms concern the role of β-arrestins in forming an alternative ternary complex for GPCR signaling *(171,172)*.

5.2.2.1. β-Arrestins as Scaffolds for MAP Kinase Activation

In COS-7 cells, overexpression of β-arrestin 2 paradoxically enhances angiotensin AT1a receptor-mediated ERK1/2 activation while predictably attenuating G protein-mediated phosphatidylinositol hydrolysis *(173)*. Conversely, depletion of β-arrestins in HEK-293 cells using RNA interference (RNAi) inhibits AT1a receptor mediated ERK1/2 activation and receptor sequestration, while markedly enhancing second messenger production. Stimulation of a G protein-uncoupled mutant AT1a receptor (DRY/AAY) with angiotensin II fails to induce detectable G protein loading, but still promotes β-arrestin 2 recruitment, receptor sequestration, and ERK1/2 activation that is abolished when β-arrestin 2 is selectively depleted by RNA interference

(30,174). In an analogous fashion, exposure of the wild-type AT1a receptor to the synthetic peptide angiotensin antagonist [Sarcosine1,Ile4,Ile8] AngII induces β-arrestin 2 recruitment and ERK1/2 activation in the absence of detectable G protein activation. This signal too is abolished by depletion of β-arrestin 2 by RNAi. Analogous results have been reported for the β2-adrenergic receptor, where ligands such as propranolol and ICI118551, which function as inverse agonists for Gs-stimulated adenylate cyclase activation, nonetheless act as partial agonists for ERK1/2 activation *(175)*. ERK1/2 activation by β2 receptor inverse agonists is absent in β-arrestin 1/2 null MEFs, but can be restored by expression of β-arrestin 2, suggesting that signaling in the absence of G protein activation is mediated through β-arrestins. Collectively, these data suggest that β-arrestin binding to a GPCR, in the absence of G protein activation, is both necessary and sufficient for ERK1/2 activation.

The β-arrestin-dependent activation of ERK1/2 results from scaffolding of the MAP kinase pathway by GPCR-bound β-arrestin. In KNRK cells, stimulation of the PAR2 protease-activated receptor induces assembly of a complex containing the internalized receptor, β-arrestin 1, Raf-1 and activated ERK1/2 *(176)*. Similar results have been obtained for the angiotensin AT1a receptor *(177)*. Angiotensin II stimulation triggers the formation of complexes containing AT1a receptor, β-arrestin 2, and the component kinases of the ERK cascade, cRaf-1, MEK1, and ERK2. The NK1 neurokinin receptor provides a third example. Activation of NK1 receptors causes the formation of complexes comprising internalized receptor, barrestin, c-Src, and ERK1/2 *(178)*. When associated with Class B GPCRs, such as the PAR2, AT1a, and NK-1 receptors, β-arrestin–ERK complexes localize to endosomal vesicles along with the sequestered receptor. The C-terminus of β-arrestin 2 contains a classical leucine-rich nuclear export sequence, which probably contributes to the nuclear exclusion of β-arrestin-bound MAP kinases *(179)*. Because ERK1/2 activated in this manner is spatially constrained, it does not induce transcription of Elk-1 driven reporters, and unlike Ras-dependent ERK1/2 activation by transactivated EGF receptors, does not enhance mitogenesis.

Presumably, the existence of distinct mechanisms of ERK1/2 activation allows GPCRs to exert control over the substrate specificity and function of these multifunctional kinases. In addition to phosphorylating nuclear transcription factors, ERK1/2 phosphorylates numerous plasma membrane, cytoplasmic and cytoskeletal substrates, including several proteins involved in GPCR signaling, such as β-arrestin 1 *(180)* , GRK2 *(181)*, and GAIP *(182)*. Phosphorylation of GRK2 by ERK1/2 enhances its rate of degradation, and the process is accelerated by overexpression of β-arrestin 1, suggesting that β-arrestins may target ERK1/2 to GRK2 *(183)*. In NIH-3T3 cells, PAR2 receptors stimulate prolonged activation of a discrete pool of ERK1/2 that is retained in receptor-β-arrestin-ERK1/2 complexes. In a chemotactic gradient, these complexes are enriched in pseudopodia. PAR-2 receptor-mediated cytoskeletal reorganization, polarized pseudopod extension and chemotaxis are ERK1/2-dependent and inhibited by expression of a dominant-negative mutant of β-arrestin 1, suggesting that the formation of β-arrestin-ERK1/2 signaling complexes at the leading edge of a cell may direct localized actin assembly and drive chemotaxis *(184)*. Consistent with this hypothesis, T- and B-cells from β-arrestin 2 knockout mice are strikingly impaired in their ability to chemotax in response to CXCL12 in transwell and in transendothelial migration assays *(185)*. Thus, the different mechanisms of ERK1/2 activation may permit selective regulation of cell proliferation, GPCR desensitization, and cell migration.

β-Arrestins also scaffold other MAP kinase modules. β-Arrestin 2 binds to Ask 1 and the neuronal JNK/SAPK isoform, JNK3 *(186)*. Ask1 binds the β-arrestin 2 *N*-terminus, whereas JNK3 binding is conferred by an RRSLHL motif in the C-terminal half of β-arrestin 2 *(187)*. β-Arrestin 2 forms complexes with Ask1, MKK4, and JNK3 but not JNK1 or JNK2 and dramatically increases Ask1-dependent phosphorylation of JNK3. As with ERK1/2, the active JNK3 pool is retained in the cytosol by the nuclear export sequence contained within β-arrestin2 *(179)*. In HeLa and HEK293 cells, overexpression of β-arrestin 2 enhances ERK1/2 and p38 MAP kinase activation, as well as the chemotactic response to activation of CXCR4 and CXCR5 chemokine receptors *(188)*. Conversely, suppression of endogenous β-arrestin 2 expression by antisense or RNAi attenuates CXCR4-mediated cell migration. In this system, inhibition of p38 MAP kinase, but not ERK1/2, blocked the effect of β-arrestin 2 on chemotaxis, suggesting that β-arrestin may act as a positive regulator of chemokine receptor-mediated chemotaxis by enhancing activation of the ASK1/p38 MAP kinase pathway.

5.2.2.2. β-Arrestins as Activators of Src Family Tyrosine Kinases

Another signaling role of β-arrestins involves the activation of Src family tyrosine kinases. β-Arrestin 1 and β-arrestin 2 bind directly to Src kinases and recruit them to agonist-occupied GPCRs. In HEK-293 cells, stimulation of β2 adrenergic receptors leads to the assembly of a complex containing activated c-Src, β-arrestin and the receptor, and colocalization of the receptor with both proteins in clathrin-coated pits *(189)*. Similar results have been obtained in KNRK cells, where β-arrestins are involved in recruiting c-Src to the NK-1 receptor *(178)*; in neutrophils, where β-arrestins recruit the Src family kinases Hck and Fgr to the CXCR-1 receptor *(190)*; and in the rod outer segment, where bleached rhodopsin, visual arrestin, and Src assemble to form a multimeric complex *(191)*. Src binding to arrestins is mediated in part by an interaction between the SH3 domain of the kinase and proline-rich PXXP motifs in the globular β-arrestin N-domain. β-Arrestin 1 has three such motifs, spanning residues 88–91, 121–124, and 175–178, whereas visual arrestin has only a single such motif. All three of the β-arrestin 1 PXXP motifs reside on the solvent exposed surface of the molecule, where they would be available for SH3 domain binding, and a P91G/P121E mutant of β-arrestin 1 is impaired in c-Src binding *(189,192)*, suggesting that at least two of the motifs contribute to Src binding. A second major site of interaction appears to involve the N-terminal portion of the catalytic (SH1) domain of Src and additional epitopes located within the N-domain of β-arrestin 1 *(193)*.

GPCR stimulation results in the Src-mediated phosphorylation of several proteins directly involved in the modulation of GPCR signaling, including dynamin, GRK2, and Gαq/11 subunits. Activation of β2 adrenergic receptors results in rapid Src-dependent tyrosine phosphorylation of dynamin on Y597, which stimulates dynamin self-assembly, increases its GTPase activity, and enhances GPCR endocytosis *(194)*. These effects are blocked by expression of a β-arrestin dominant-negative. GRK2 is another potential substrate for β-arrestin-bound Src *(195)*. β2 Adrenergic or CXCR4 receptor stimulation in HEK-293, Jurkat, or C6 cells stimulates Src-mediated GRK2 phosphorylation, which is followed by rapid ubiquitination and proteosomal degradation of the GRK. This may represent a feedback mechanism for regulating GRK levels and GPCR signaling.

β-Arrestin-dependent Src activation also affects endocytic and exocytic vesicle trafficking. β-Arrestin 1 recruits Src to ET_A endothelin receptors in 3T3-L1 cells, leading to Src activation, Src-dependent tyrosine phosphorylation of Gαq/11 subunits, translocation of glucose transporter 4 (GLUT4)-containing vesicles to the plasma membrane, and an increase in GLUT4-mediated glucose uptake. Treatment with Src inhibitors, or microinjection of antibodies against either the Src family kinase c-Yes or β-arrestin1, blocks ET_A-stimulated glucose uptake *(196)*. In granulocytic neutrophils, activation of the chemokine receptor CXCR1 by interleukin-8 stimulates the rapid formation of complexes containing endogenous β-arrestin and the Src family kinases, Hck or Fgr *(190)*. The formation of β-arrestin-Hck complexes leads to Hck activation, trafficking of the complexes to granule-rich regions, and enhanced chemoattractant-stimulated granule release.

Ras-dependent activation of the ERK1/2 MAP kinase cascade by many GPCRs requires Src kinase activity *(197)*. In some cases, the interaction between β-arrestin and Src appears to play a role in the process. In HEK-293 cells, overexpression of β-arrestin 1 mutants that exhibit either impaired Src binding or that are unable to target receptors to clathrin-coated pits, blocks β2 adrenergic receptor-mediated activation of ERK1/2 *(189)*. In KNRK cells, activation of NK1 receptors by Substance P leads to assembly of a scaffolding complex containing the internalized receptor, β-arrestin, Src and ERK1/2. Expression of either a dominant-negative β-arrestin 1 mutant or a truncated NK1 receptor that fails to bind β-arrestin blocks complex formation and inhibits both substance P-stimulated endocytosis of the receptor and activation of ERK1/2 *(178)*.

5.2.2.3. OTHER SIGNALING ROLES OF β-ARRESTINS

Another downstream signaling event that may involve β-arrestin-dependent Src activation is activation of nuclear factor-κB. Dopamine D2 receptors expressed in HeLa cells activate nuclear factor-κB through a pathway involving pertussis toxin-sensitive G proteins, Gβγ subunits, and Src family tyrosine kinases. The signal is independent of PLC and phosphatidyinositol 3-kinase activity, but is enhanced by overexpression of β-arrestin 1, suggesting that β-arrestin recruitment, rather than activation of traditional heterotrimeric G protein effectors, may be the initiating event in the process *(198)*.

The frizzled family of seven membrane-spanning receptors is distantly related to the GPCRs *(5)*. Frizzleds bind ligands called Wnts, and are important regulators of development in many organisms. Unlike the classical GPCRs, frizzled receptors do not appear to couple to heterotrimeric G proteins. Rather, they recruit the cytoplasmic proteins Dishevelled 1 and Dishevelled 2 (Dvl 1 and Dvl 2), which link the receptor to several signalling cascades, including inhibition of glycogen synthase kinase-3β, stabilization of β-catenin, and activation of lymphoid enhancer factor. Interestingly, β-arrestin 1 has been identified as a binding partner of both Dvl 1 and Dvl 2 *(199)*. Phosphorylation of Dvl 1 enhances its binding to β-arrestin 1, and overexpression of β-arrestin 1 along with Dvl1 synergistically activates lymphoid enhancer factor transcription. In addition, endocytosis of Frizzled 4 in response to the Wnt5A protein is mediated by β-arrestin 2 that is recruited to the receptor by binding to phosphorylated Dvl2 *(200)*. These data further suggest that the arrestin proteins may function as independent signal transducers, as well as mediators of receptor endocytosis.

6. Conclusions

As the most diverse type of cell surface receptor, the importance GPCR signaling to clinical medicine cannot be overestimated. Visual, olfactory, and gustatory sensation, intermediary metabolism, and cell growth and differentiation are all influenced by GPCR signals. The basic receptor–G protein-effector mechanism of GPCR signaling is tuned by a complex interplay of positive and negative regulatory events that amplify the effect of a hormone binding the receptor or that dampen cellular responsiveness. The association of heptahelical receptors with a variety of intracellular partners other than G proteins has to the discovery of potential mechanisms of GPCR signaling that extend beyond the classical paradigms. Although the physiologic relevance of many of these novel mechanisms of GPCR signaling remains to be established, their existence suggests that the mechanisms of GPCR signaling are even more diverse than previously imagined.

References

1. Lander, E. S., Linton, L. M., Birren, B., Nusbaum, C., Zody, M. C., Baldwin, J., et al. (2001) Initial sequencing and analysis of the human genome. *Nature* **409,** 860–921.
2. Venter, J. C., Adams, M. D., Myers, E. W., Li, P. W., Mural, R. J., Sutton, G. G., *et al.* (2001) The sequence of the human genome. *Science* **291,** 1304–1351.
3. Bargmann, C. (1998) Neurobiology of the *Caenorhabditis elegans* genome. *Science.* **282,** 2028–2033.
4. Flower, D. R. (1999) Modelling G-protein-coupled receptors for drug design. *Biochim. Biophys. Acta.* **1422,** 207–234.
5. Fredriksson, R., Lagerstrom, M. C., Lundin, L. G., and Schioth, H. B. (2003) The G-protein-coupled receptors in the human genome form five main families. Phylogenetic analysis, paralogon groups, and fingerprints. *Mol. Pharmacol.* **63,** 1256–1272.
6. Birnbaumer, L., Pohl, S. L., Michiel, H., Krans, M. J., and Rodbell, M. (1970) The actions of hormones on the adenyl cyclase system. *Adv. Biochem. Psychopharmacol.* **3,** 185–208.
7. Insel, P. A., Maguire, M. E., Gilman, A. G., Bourne, H. R., Coffino, P., and Melmon, K. L. (1976) Beta adrenergic receptors and adenylate cyclase: products of separate genes? *Mol. Pharmacol.* **12,** 1062–1069.
8. Gilman, A. G. (1987) G proteins: transducers of receptor-generated signals. *Ann. Rev. Biochem.* **56,** 615–649.
9. Sternweis, P. C. and Gilman, A. G. (1979) Reconstitution of catecholamine-sensitive adenylate cyclase. Reconstitution of the uncoupled variant of the S49 lymphoma cell. *J. Biol. Chem.* **254,** 3333–3340.
10. Northup, J. K., Sternweis, P. C., Smigel, M. D., Schleifer, L. S., Ross, E. M., and Gilman, A. G. (1980) Purification of the regulatory component of adenylate cyclase. *Proc. Natl. Acad. Sci. USA* **77,** 6516–6520.
11. Manning, D. R. and Gilman, A. G. (1983) The regulatory components of adenylate cyclase and transducin. A family of structurally homologous guanine nucleotide-binding proteins. *J. Biol. Chem.* **258,** 7059–7063.
12. Lefkowitz, R. J. (2000) The superfamily of heptahelical receptors. *Nat. Cell Biol.* **2,** E133–E136.
13. Palczewski, K., Kumasaka, T., Hori, T., Behnke, C. A., Motoshima, H., Fox, B. A., et al. (2000) Crystal structure of rhodopsin: A G protein-coupled receptor. *Science* **289,** 739–745.
14. Kolakowski, L. F., Jr. (1994) GCRDb: A G-protein coupled receptor database. *Recept. Channels* **2,** 1–7.
15. Perez, D. M. (2003) The evolutionarily triumphant G protein-coupled receptor. *Mol. Pharmacol.* **63,** 1202–1205.
16. Arshavsky, V. Y., Lamb, T. D., and Pugh, E. N., Jr. (2002) G proteins and phototransduction. *Ann. Rev. Physiol.* **64,** 153–187.

17. Ridge, K.D., Abdulaev, N. G., Sousa, M., and Palczewski, K. (2003) Phototransduction: Crystal clear. *Trends Biochem. Sci.* **28,** 479–487

18. Gether, U. and Kobilka, B. K. (1998) G protein-coupled receptors. II. Mechanism of agonist activation. *J. Biol. Chem.* **273,** 17,979–17,982.

19. De Lean, A., Stadel, J. M., and Lefkowitz, R. J. (1980) A ternary complex model explains the agonist-specific binding properties of the adenylate cyclase-coupled beta-adrenergic receptor. *J. Biol. Chem.* **255,** 7108–7117.

20. Samama, P., Cotecchia, S., Costa, T., and Lefkowitz, R. J. (1993) A mutation-induced activated state of the beta 2-adrenergic receptor. Extending the ternary complex model. *J. Biol. Chem.* **268,** 4625–4536.

21. Lefkowitz, R. J., Cotecchia, S., Samama, P., and Costa, T. (1993) Constitutive activity of receptors coupled to guanine nucleotide regulatory proteins. *Trends Pharmacol. Sci.* **14,** 303–307.

22. Kenakin, T. (2002) Drug efficacy at G protein-coupled receptors. *Ann. Rev. Pharmacol. Toxicol.* **42,** 349–379.

23. Kenakin, T. (2003) Ligand-selective receptor conformations revisited: the promise and the problem. *Trends Pharmacol. Sci.* **24,** 346–354.

24. Gurevich, V. V., Pals-Rylaarsdam, R., Benovic, J. L., Hosey, M. M., and Onorato, J. J. (1997) Agonist-receptor-arrestin, an alternative ternary complex with high agonist affinity. *J. Biol. Chem.* **272,** 28,849–28,852.

25. Key, T. A., Bennett, T. A., Foutz, T. D., Gurevich, V. V., Sklar, L. A., and Prossnitz, E. R. (2001) Regulation of formyl peptide receptor agonist affinity by reconstitution with arrestins and heterotrimeric G proteins. *J. Biol. Chem.* **276,** 49,204–49,212.

26. Swaminath, G., Xiang, Y., Lee, T. W., Steenhuis, J., Parnot, C., and Kobilka, B. K. (2004) Sequential binding of agonists to the beta2 adrenoceptor. Kinetic evidence for intermediate conformational states. *J. Biol. Chem.* **279,** 686–691.

27. Whistler, J. L. and von Zastrow, M. (1998) Morphine-activated opioid receptors elude desensitization by beta-arrestin. *Proc. Natl. Acad. Sci. USA* **95,** 9914–9919.

28. Kohout, T. A., Nicholas, S. L., Perry, S. J., Reinhart, G., Junger, S., and Struthers, R. S. (2004) Differential desensitization, receptor phosphorylation, beta-arrestin recruitment, and ERK1/2 activation by the two endogenous ligands for the CC chemokine receptor 7. *J. Biol. Chem.* **279,** 23,214–23,222.

29. Holloway, A. C., Qian, H., Pipolo, L., Ziogas, J., Miura, S., Karnik, S., et al. (2002) Side-chain substitutions within angiotensin II reveal different requirements for signaling, internalization, and phosphorylation of type 1a angiotensin receptors. *Mol. Pharmacol.* **61,** 768–777.

30. Wei, H., Ahn, S., Shenoy, S. K., Karnik, S. S., Hunyady, L., Luttrell, L. M., and Lefkowitz, R. J. (2003) Independent beta-arrestin 2 and G protein-mediated pathways for angiotensin II activation of extracellular signal-regulated kinases 1 and 2. *Proc. Natl. Acad. Sci. USA* **100,** 10,782–10,787.

31. Downes, G. B. and Gautam, N. (1999) The G protein subunit gene families. *Genomics* **62,** 544–552.

32. Schmidt, C. J., Thomas, T. C., Levine, M. A., and Neer, N. J. (1992) Specificity of G protein beta and gamma subunit interactions. *J. Biol. Chem.* **267,** 13,807–13,810.

33. Hildebrandt, J. D. (1997) Role of subunit diversity in signaling by heterotrimeric G proteins. *Biochem. Pharmacol.* **54,** 325–339.

34. Ford, C. E., Skiba, N. P., Bae, H., Daaka, Y., Reuveny, E., Shekter, L. R., et al. (1998) Molecular basis for interactions of G protein betagamma subunits with effectors. *Science* **280,** 1271–1274.

35. Sprang, S. R. (1997) G protein mechanisms: Insights from structural analysis. *Ann. Rev. Pharmacol, Toxicol.* **36,** 461–480.

36. Coleman, D. E. and Sprang, S. R. (1996) How G proteins work: A continuing story. *Trends. Biochem. Sci.* **21,** 41–44.

37. Casey, P. J. (1994) Lipid modifications of G proteins. *Curr. Opin. Cell Biol.* **6,** 219–225.

38. Clapham, D. E. and Neer, E. J. (1993) New roles for G-protein beta gamma-dimers in transmembrane signalling. *Nature* **365,** 403–406.

39. Zwartkruis, F. J. and Bos, J. L. (1999) Ras and Rap1: Two highly related small GTPases with distinct function. *Exp. Cell Res.* **253,** 157–165.

40. Sunahara, R. K., Dessauer, C. W., and Gilman, A. G. (1996) Complexity and diversity of mammalian adenylyl cyclases. *Ann. Rev. Pharmacol. Toxicol.* **36,** 461–480.

41. Morris, A. J. and Scarlata, S. (1997) Regulation of effectors by G-protein alpha- and beta gamma-subunits. Recent insights from studies of the phospholipase c-beta isoenzymes. *Biochem. Pharmacol.* **54,** 429–435.

42. Wickman, K. D. and Clapham, D. E. (1995) G-protein regulation of ion channels. *Curr. Opin. Neurobiol.* **5,** 278–285.

43. Albert, P. R. and Robillard, L. (2002) G protein specificity: Traffic direction required. *Cell. Signal.* **14,** 407–418.

44. Stoffel, R. H. 3rd, Pitcher, J. A., and Lefkowitz, R. J. (1997) Targeting G protein-coupled receptor kinases to their receptor substrates. *J. Membr. Biol.* **157,** 1–8.

45. Perry, S. J., Baillie, G. S., Kohout, T. A., McPhee, I., Magiera, M. M., Ang, K. L., et al. (2002) Targeting of cyclic AMP degradation to beta 2-adrenergic receptors by beta-arrestins. *Science* **298,** 834–836

46. Baillie, G. S., Sood, A., McPhee, I., et al. (2003) Beta-Arrestin-mediated PDE4 cAMP phosphodiesterase recruitment regulates beta-adrenoceptor switching from Gs to Gi. *Proc. Natl. Acad. Sci. USA* **100,** 940–945.

47. Ross, E. M. (1995) G protein GTPase-activating proteins: Regulation of speed, amplitude, and signaling selectivity. *Recent Prog. Horm. Res.* **50,** 207–221.

48. Ross, E. M. and Wilkie, T. M. (2000) GTPase-activating proteins for heterotrimeric G proteins: regulators of G protein signaling (RGS) and RGS-like proteins. *Annu. Rev. Biochem.* **69,** 795–827.

49. Berman, D. M. and Gilman, A. G. (1998) Mammalian RGS proteins: Barbarians at the gate. *J. Biol. Chem.* **273,** 1269–1272.

50. Schulz, R. (2001) The pharmacology of phosducin. *Pharmacol. Res.* **43,** 1–10.

51. Pitcher, J., Lohse, M. J., Codina, J., Caron, M. G., and Lefkowitz, R. J. (1992) Desensitization of the isolated beta 2-adrenergic receptor by beta-adrenergic receptor kinase, cAMP-dependent protein kinase, and protein kinase C occurs via distinct molecular mechanisms. *Biochemistry* **31,** 3193–3197.

52. Freedman, N. J. and Lefkowitz, R. J. (1996) Desensitization of G protein-coupled receptors. *Recent Prog. Horm. Res.* **51,** 319–351

53. Daaka, Y., Luttrell, L. M., and Lefkowitz, R. J. (1997) Switching of the coupling of the beta2-adrenergic receptor to different G proteins by protein kinase A. *Nature* **390,** 88–91.

54. Zamah, A. M., Delahunty, M., Luttrell, L. M., and Lefkowitz, R. J. (2002) Protein kinase A-mediated phosphorylation of the beta2-adrenergic receptor regulates its coupling to Gs and Gi. Demonstration in a reconstituted system. *J. Biol. Chem.* **277,** 31,249–31,256.

55. Lawler, O. A., Miggin, S. M., and Kinsella, B. T. (2001) Protein kinase A-mediated phosphorylation of serine 357 of the mouse prostacyclin receptor regulates its coupling to Gs-, to Gi- and to Gq-coupled effector signaling. *J. Biol. Chem.* **276,** 33,596–33,607.

56. Lefkowitz, R. J., Pierce, K. L., and Luttrell, L. M. (2002) Dancing with different partners: Protein kinase A phosphorylation of seven membrane-spanning receptors regulates their G protein-coupling specificity. *Mol. Pharmacol.* **62,** 971–974.

57. Lohse, M. J., Andexinger, S., Pitcher, J., et al. (1993) Receptor specific desensitization with purified proteins. Kinase dependence and receptor specificity of β–arrestin and arrestin in the β2-adrenergic receptor and rhodopsin systems. *J. Biol. Chem.* **267,** 8558–8564.

58. Ferguson, S. S. (2001) Evolving concepts in G protein-coupled receptor endocytosis: the role in receptor desensitization and signaling. *Pharm. Rev.* **53,** 1–24.

59. Luttrell, L. M., and Lefkowitz, R. J. (2002) The role of beta-arrestins in the termination and transduction of G-protein-coupled receptor signals. *J. Cell. Sci.* **115,** 455–465.

60. Goodman, O. B., Jr., Krupnick, J. G., Santini, F., et al. (1996) Beta-arrestin acts as a clathrin adaptor in endocytosis of the beta2-adrenergic receptor. *Nature* **383,** 447–450.

61. Laporte, S. A., Oakley, R. H., Zhang, J., et al. (1999) The beta2-adrenergic receptor/beta-arrestin complex recruits the clathrin adaptor AP-2 during endocytosis. *Proc. Natl. Acad. Sci. USA* **96,** 3712–3717.

62. Carman, C. V., Parent, J. L., Day, P. W., et al. (1999) Selective regulation of Galpha(q/11) by an RGS domain in the G protein-coupled receptor kinase, GRK2. *J. Biol. Chem.* **274,** 34,483–34,492.

63. Lodowski, D. T., Pitcher, J. A., Capel, W. D., Lefkowitz, R. J., and Tesmer, J. J. (2003) Keeping G proteins at bay: a complex between G protein-coupled receptor kinase 2 and G beta gamma. *Science* **300,** 1256–1262.

64. Dhami, G. K., Dale, L. B., Anborgh, P. H., O'Connor-Halligan, K. E., Sterne-Marr, R., and Ferguson, S. S. (2004) G Protein-coupled receptor kinase 2 RGS homology domain binds to both metabotropic glutamate receptor 1a and G alpha q to attenuate signaling. *J. Biol. Chem.* **279,** 16,614–16,620.

65. Barak, L. S., Ferguson, S. S., Zhang, J., and Caron, M. G. (1997) A beta-arrestin/green fluorescent protein biosensor for detecting G protein-coupled receptor activation. *J. Biol. Chem.* **272,** 27,497–27,500.

66. Oakley, R. H., Laporte, S. A., Holt, J. A., Barak, L. S., and Caron, M. G. (2001) Molecular determinants underlying the formation of stable intracellular G protein-coupled receptor-beta-arrestin complexes after receptor endocytosis. *J. Biol. Chem.* **276,** 19,452–19,460.

67. Oakley, R. H., Laporte, S. A., Holt, J. A., Caron, M. G., and Barak, L. S. (2000) Differential affinities of visual arrestin, beta-arrestin1, and beta-arrestin2 for G protein-coupled receptors delineate two major classes of receptors. *J. Biol. Chem.* **275,** 17,201–17,210.

68. Kohout, T. A., Lin, F-T., Perry, S. J., Conner, D. A., and Lefkowitz, R. J. (2001) Beta-Arrestin 1 and 2 differentially regulate heptahelical receptor signaling and trafficking. *Proc. Natl. Acad. Sci. USA* **98,** 1601–1606.

69. Lin, F-T., Krueger, K. M., Kendall, H. E., et al. (1997) Clathrin-mediated endocytosis of the beta-adrenergic receptor is regulated by phosphorylation/dephosphorylation of beta-arrestin1. *J. Biol. Chem.* **272,** 31,051–31,057.

70. Lin, F-T., Chen, W., Shenoy, S., Cong, M., Exum, S. T., and Lefkowitz, R. J. (2002) Phosphorylation of beta-arrestin2 regulates it function in internalization of beta(2)-adrenergic receptors. *Biochemistry* **41,** 10,692–10,699.

71. Shenoy, S. K., McDonald, P. H., Kohout, T. A., and Lefkowitz, R. J. (2001) Regulation of receptor fate by ubiquitination of activated β2-adrenergic receptor and β-arrestin. *Science* **294,** 1307–1313.

72. Martin, N. P., Lefkowitz, R. J., and Shenoy, S. K. (2003) Regulation of V2 vasopressin receptor degradation by agonist-promoted ubiquitination. *J. Biol. Chem.* **278,** 45,954–45,959.

73. Shenoy, S. K. and Lefkowitz, R. J. (2003) Trafficking pattern of beta-arrestin and G protein-coupled receptors determined by the kinetics of beta-arrestin deubiquitination. *J. Biol. Chem.* **278,** 14,498–14,506.

74. Paing, M. M., Stutts, A. B., Kohout, T. A., Lefkowitz, R. J., and Trejo, J. (2002) Beta-arrestins regulate protease-activated receptor-1 desensitization but not internalization or down-regulation. *J. Biol. Chem.* **277,** 1292–1300.

75. Vines, C. M., Revankar, C. M., Maestas, D. C., et al. (2003) N-formyl peptide receptors internalize but do not recycle in the absence of arrestins. *J. Biol. Chem.* **278,** 41,581–41,584.

76. Brasselet, S., Guillen, S., Vincent, J. P., and Mazella, J. (2002) Beta-arrestin is involved in the desensitization but not in the internalization of the somatostatin receptor 2A expressed in CHO cells. *FEBS Lett.* **10,** 124–128.

77. Zhang, J., Ferguson, S. S., Barak, L. S., Menard, L., and Caron, M. G. (1996) Dynamin and beta-arrestin reveal distinct mechanisms for G protein-coupled receptor internalization. *J. Biol. Chem.* **271,** 18,302–18,305.

78. Vogler, O., Nolte, B., Voss, M., Schmidt, M., Jakobs, K. H., and van Koppen, C. J. (1999) Regulation of muscarinic acetylcholine receptor sequestration and function by beta-arrestin. *J. Biol. Chem.* **274,** 12,333–12,338.

79. Rapacciuolo, A., Suvarna, S., Barki-Harrington, L., et al. (2003) Phosphorylation sites of the beta-1 adrenergic receptor determine the internalization pathway. *J. Biol. Chem.* **278,** 35,403–35,411.

80. Pitcher, J. A., Payne, E. S., Csortos, C., DePaoli-Roach, A. A., and Lefkowitz, R. J. (1995) The G-protein-coupled receptor phosphatase: a protein phosphatase type 2A with a distinct subcellular distribution and substrate specificity. *Proc. Natl. Acad. Sci. USA* **92,** 8343–8347.
81. Oakley, R. H., Laporte, S. A., Holt, J. A., Barak, L. S., and Caron, M. G. (1999) Association of beta-arrestin with G protein-coupled receptors during clathrin-mediated endocytosis dictates the profile of receptor resensitization. *J. Biol. Chem.* **274,** 32,248–32,257.
82. Dale, L. B., Seachrist, J. L., Babwah, A. V., and Ferguson, S. S. (2004) Regulation of angiotensin II type 1A receptor intracellular retention, degradation, and recycling by Rab5, Rab7, and Rab11 GTPases. *J. Biol. Chem.* **279,** 13,110–13,118.
83. Seachrist, J. L. and Ferguson, S. S. (2003) Regulation of G protein-coupled receptor endocytosis and trafficking by Rab GTPases. *Life Sci.* **74,** 225–235.
84. Cao, T. T., Deacon, H. W., Reczek, D., Bretscher, A., and von Zastrow M. (1999) A kinase-regulated PDZ-domain interaction controls endocytic sorting of the beta 2-adrenergic receptor. *Nature* **401,** 286–290.
85. Whistler, J. L., Enquist, J., Marley, A., et al. (2002) Modulation of postendocytic sorting of G protein-coupled receptors. *Science* **297,** 529–531.
86. Gage, R. M., Kim, K. A., Cao, T. T., and von Zastrow, M. (2001) A transplantable sorting signal that is sufficient to mediate rapid recycling of G protein-coupled receptors. *J. Biol. Chem.* **276,** 44,712–44,720.
87. Premont, R. T., Claing, A., Vitale, N., et al. (1998) Beta2-Adrenergic receptor regulation by GIT1, a G protein-coupled receptor kinase-associated ADP ribosylation factor GTPase-activating protein. *Proc. Natl. Acad. Sci. USA* **95,** 14,082–14,087.
88. Claing, A., Chen, W., Miller, W. E., et al. (2001) Beta-Arrestin-mediated ADP-ribosylation factor 6 activation and beta 2-adrenergic receptor endocytosis. *J. Biol. Chem.* **276,** 42,509–42,513.
89. Devi, L. (2001) Heterodimerization of G-protein-coupled receptors: pharmacology, signaling and trafficking. *Trends Pharmacol. Sci.* **22,** 532–537.
90. Milligan, G. (2001) Oligomerisation of G-protein-coupled receptors. *J. Cell Sci.* **114,** 1265–1271.
91. Angers, S., Salahpour, A., and Bouvier, M. (2002) Dimerization: An emerging concept for G protein-coupled receptor ontogeny and function. *Ann. Rev. Pharmacol. Toxicol.* **42,** 409–435.
92. Nakanishi-Matsui, M., Zheng, Y. W., Sulciner, D. J., et al. (2000) PAR3 is a cofactor for PAR4 activation by thrombin. *Nature* **404,** 609–613.
93. O'Brien, P. J., Prevost, N., Molino, M., et al. (2000) Thrombin responses in human endothelial cells. Contributions from receptors other than PAR1 include the transactivation of PAR2 by thrombin-cleaved PAR1. *J. Biol. Chem.* **275,** 13,502–13,509.
94. Baneres, J. L. and Parello, J. (2003) Structure-based analysis of GPCR function: Evidence for a novel pentameric assembly between the dimeric leukotriene B4 receptor BLT1 and the G-protein. *J. Mol. Biol.* **329,** 815–829.

95. Marshall, G. R. (2001) Peptide interactions with G-protein coupled receptors. *Biopolymers* **60,** 246–277.

96. Fotiadis, D., Liang, Y., Filipek, S., Saperstein, D. A., Engel, A., and Palczewski, K. (2003) Atomic-force microscopy: Rhodopsin dimers in native disc membranes. *Nature* **421,** 127–128.

97. Jones, K. A., Borowsky, B., Tamm, J. A., et al. (1998) GABA(B) receptors function as a heteromeric assembly of the subunits GABA(B)R1 and GABA(B)R2. *Nature* **396,** 674–679.

98. Kaupmann, K., Malitschek, B., Schuler, V., et al. (1998) GABA(B)-receptor subtypes assemble into functional heteromeric complexes. *Nature* **396,** 683–687.

99. Kniazeff, J., Galvez, T., Labesse, G., and Pin, J. P. (2002) No ligand binding in the GB2 subunit of the GABA(B) receptor is required for activation and allosteric interaction between the subunits. *J. Neurosci.* **22,** 7352–7361.

100. Robbins, M. J., Calver, A. R., Filippov, A. K., et al. (2001) GABA(B2) is essential for G-protein coupling of the GABA(B) receptor heterodimer. *J. Neurosci.* **21,** 8043–8052.

101. Margeta-Mitrovic, M., Jan, Y. N., and Jan, L. Y. (2000) A trafficking checkpoint controls GABA(B) receptor heterodimerization. *Neuron* **27,** 97–106.

102. Ng, G. Y., O'Dowd, B. F., Lee, S. P., et al. (1996) Dopamine D2 receptor dimers and receptor-blocking peptides. *Biochem. Biophys. Res. Commun.* **227,** 200–204.

103. Schulz, A., Grosse, R., Schultz, G., Gudermann, T., and Schoneberg, T. (2000) Structural implication for receptor oligomerization from functional reconstitution studies of mutant V2 vasopressin receptors. *J. Biol. Chem.* **275,** 2381–2389.

104. Vila-Coro, A. J., Rodriguez-Frade, J. M., Martin de Ana, A., Moreno-Ortiz, M. C., Martinez, A. C., and Mellado, M. (1999) The chemokine SDF-1alpha triggers CXCR4 receptor dimerization and activates the JAK/STAT pathway. *FASEB J.* **13,** 1699–1710.

105. Rodriguez-Frade, J. M., Vila-Coro, A. J., Martin de Ana, A. M., Albar, J. P., Martinez, A. C., and Mellado, M. (1999) The chemokine monocyte chemoattractant protein-1 induces functional responses through dimerization of its receptor CCR2. *Proc. Natl. Acad. Sci. USA* **96,** 3628–3633.

106. Vila-Coro, A. J., Mellado, M., Martin de Ana, A., et al. (2000) HIV-1 infection through the CCR5 receptor is blocked by receptor dimerization. *Proc. Natl. Acad. Sci. USA* **97,** 3388–3393.

107. Jordan, B. A. and Devi, L. A. (1999) G-protein-coupled receptor heterodimerization modulates receptor function. *Nature* **399,** 697–700.

108. George, S. R., Fan, T., Xie, Z., et al. (2000) Oligomerization of mu- and delta-opioid receptors. Generation of novel functional properties. *J. Biol. Chem.* **275,** 26,128–26,135.

109. AbdAlla, S., Lother, H., and Quitterer, U. (2000) AT1-receptor heterodimers show enhanced G-protein activation and altered receptor sequestration. *Nature* **407,** 94–98.

110. AbdAlla, S., Lother, H., el Massiery, A., and Quitterer, U. (2001) Increased AT(1) receptor heterodimers in preeclampsia mediate enhanced angiotensin II responsiveness. *Nat. Med.* **7,** 1003–1009.

111. Barki-Harrington, L., Luttrell, L. M., and Rockman, H. A. (2003) Dual inhibition of beta-adrenergic and angiotensin II receptors by a single antagonist: a functional role for receptor-receptor interaction in vivo. *Circulation* **108,** 1611–1618.

112. Kroeger, K. M., Pfleger, K. D., and Eidne, K. A. (2003) G-protein coupled receptor oligomerization in neuroendocrine pathways. *Front. Neuroendocrinol.* **24,** 254–278.

113. Breitwieser, G. E. (2004) G protein-coupled receptor oligomerization: Implications for G protein activation and cell signaling. *Circ. Res.* **94,** 17–27.

114. Terrillon, S. and Bouvier, M. (2004) Roles of G-protein-coupled receptor dimerization. *EMBO Rep.* **5,** 30–34.

115. Whistler, J. L., Chuang, H. H., Chu, P., Jan, L. Y., and von Zastrow, M. (1999) Functional dissociation of mu opioid receptor signaling and endocytosis: implications for the biology of opiate tolerance and addiction. *Neuron* **23,** 737–746.

116. Sexton, P. M., Albiston, A., Morfis, M., and Tilakaratne, N. (2001) Receptor activity modifying proteins. *Cell. Signal.* **13,** 73–83.

117. Foord S. M. and Marshall, F. H. (1999) RAMPs: accessory proteins for seven transmembrane domain receptors. *Trends Pharmacol. Sci.* **20,** 184–187.

118. Brady, A. E. and Limbird, L. E. (2002) G protein-coupled receptor interacting proteins: Emerging roles in localization and signal transduction. *Cell. Signal.* **14,** 297–309.

119. Bockaert, J., Marin, P., Dumuis, A., and Fagni, L. (2003) The "magic tail" of G protein-coupled receptors: an anchorage for functional protein networks. *FEBS Lett.* **546,** 65–72.

120. Hall, R. A., Premont, R. T., Chow, C. W., et al. (1998) The beta2-adrenergic receptor interacts with the Na+/H+-exchanger regulatory factor to control Na+/H+ exchange. *Nature* **392,** 626–630.

121. Mahon, M. J., Donowitz, M., Yun, C. C., and Segre, G. V. (2002) Na(+)/H(+) exchanger regulatory factor 2 directs parathyroid hormone 1 receptor signalling. *Nature* **417,** 858–861.

122. Mahon, M. J. and Segre, G. V. (2004) Stimulation by parathyroid hormone of a NHERF-1 assembled complex consisting of the parathyroid hormone I receptor, PLC-beta and actin increases intracellular calcium in OK cells. *J. Biol. Chem.* **279,** 23,550–23,558.

123. Hu, L. A., Tang, Y., Miller, W. E., et al. (2000) Beta 1-Adrenergic receptor association with PSD-95. Inhibition of receptor internalization and facilitation of beta 1-adrenergic receptor interaction with N-methyl-D-aspartate receptors. *J. Biol. Chem.* **275,** 38,659–38,666.

124. Xu, J., Paquet, M., Lau, A. G., Wood, J. D., Ross, C. A., and Hall, R. A. (2001) Beta 1-Adrenergic receptor association with the synaptic scaffolding protein membrane-associated guanylate kinase inverted-2 (MAGI-2). Differential regulation of receptor internalization by MAGI-2 and PSD-95. *J. Biol. Chem.* **276,** 41,310–41,317.

125. Zitzer, H., Honck, H. H., Bachner, D., Richter, D., and Kreienkamp, H. J. (1999) Somatostatin receptor interacting protein defines a novel family of multidomain proteins present in human and rodent brain. *J. Biol. Chem.* **274,** 32,997–33,001.

126. Boudin, H., Doan, A., Xia, J., et al. (2000) Presynaptic clustering of mGluR7a requires the PICK1 PDZ domain binding site. *Neuron* **28**, 485–497.

127. Perroy, J., Prezeau, L., De Waard, M., Shigemoto, R., Bockaert, J., and Fagni, L. (2000) Selective blockade of P/Q-type calcium channels by the metabotropic glutamate receptor type 7 involves a phospholipase C pathway in neurons. *J. Neurosci.* **20**, 7896–7904.

128. Becamel, C., Figge, A., Poliak, S., et al. (2001) Interaction of serotonin 5-hydroxytryptamine type 2C receptors with PDZ10 of the multi-PDZ domain protein MUPP1. *J. Biol. Chem.* **276**, 12,974–12,982.

129. Smith, F. D., Oxford, G. S., and Milgram, S. L. (1999) Association of the D2 dopamine receptor third cytoplasmic loop with spinophilin, a protein phosphatase-1-interacting protein. *J. Biol. Chem.* **274**, 19,894–19,900.

130. Richman, J. G., Brady, A. E., Wang, Q., Hensel, J. L., Colbran, R. J., and Limbird, L. E. (2001) Agonist-regulated Interaction between alpha2-adrenergic receptors and spinophilin. *J. Biol. Chem.* **276**, 15,003–15,008.

131. Fagni, L., Worley, P. F., and Ango, F. (2002) Homer as both a scaffold and transduction molecule. *Sci. STKE.* **2002(137)**, RE8.

132. Ciruela, F., Soloviev, M. M., and McIlhinney, R. A. (1999) Co-expression of metabotropic glutamate receptor type 1alpha with homer-1a/Vesl-1S increases the cell surface expression of the receptor. *Biochem. J.* **341**, 795–803.

133. Bermak, J. C., Li, M., Bullock, C., and Zhou, Q. Y. (2001) Regulation of transport of the dopamine D1 receptor by a new membrane-associated ER protein. *Nat. Cell Biol.* **3**, 492–498.

134. Tai, A. W., Chuang, J. Z., Bode, C., Wolfrum, U., and Sung, C. H. (1999) Rhodopsin's carboxy-terminal cytoplasmic tail acts as a membrane receptor for cytoplasmic dynein by binding to the dynein light chain Tctex-1. *Cell* **97**, 877–887.

135. Sung, C. H., Makino, C., Baylor, D., and Nathans, J. (1994) A rhodopsin gene mutation responsible for autosomal dominant retinitis pigmentosa results in a protein that is defective in localization to the photoreceptor outer segment. *J. Neurosci.* **14**, 5818–5833.

136. Shih, M., Lin, F., Scott, J. D., Wang, H. Y., and Malbon, C. C. (1999) Dynamic complexes of beta2-adrenergic receptors with protein kinases and phosphatases and the role of gravin. *J. Biol. Chem.* **274**, 1588–1595

137. Fraser, I. D., Cong, M., Kim, J., et al. (2000) Assembly of an A kinase-anchoring protein-beta(2)-adrenergic receptor complex facilitates receptor phosphorylation and signaling. *Curr. Biol.* **10**, 409–412.

138. Cong, M., Perry, S. J., Lin, F. T., et al. (2001) Regulation of membrane targeting of the G protein-coupled receptor kinase 2 by protein kinase A and its anchoring protein AKAP79. *J. Biol. Chem.* **276**, 15,192–15,199.

139. Lopez-Ilasaca, M., Liu, X., Tamura, K., and Dzau, V. J. (2003) The angiotensin II type I receptor-associated protein, ATRAP, is a transmembrane protein and a modulator of angiotensin II signaling. *Mol. Biol. Cell.* **14**, 5038–5050.

140. O'Connor, V., El Far, O., Bofill-Cardona, E., et al. (1999) Calmodulin dependence of presynaptic metabotropic glutamate receptor signaling. *Science* **286**, 1180–1184.

141. Li, M., Bermak, J. C., Wang, Z. W., and Zhou, Q. Y. (2000) Modulation of dopamine D(2) receptor signaling by actin-binding protein (ABP-280). *Mol. Pharmacol.* **57,** 446–452.

142. Hasegawa, H., Katoh, H., Fujita, H., Mori, K., and Negishi, M. (2000) Receptor isoform-specific interaction of prostaglandin EP3 receptor with muskelin. *Biochem. Biophys. Res. Commun.* **276,** 350–354.

143. Prezeau, L., Richman, J. G., Edwards, S. W., and Limbird, L. E. (1999) The zeta isoform of 14–3-3 proteins interacts with the third intracellular loop of different alpha2-adrenergic receptor subtypes. *J. Biol. Chem.* **274,** 13,462–13,469.

144. Couve, A., Kittler, J. T., Uren, J. M., et al. (2001) Association of GABA(B) receptors and members of the 14–3-3 family of signaling proteins. *Mol. Cell. Neurosci.* **17,** 317–328.

145. Kryiakis, J. M., and Avruch, J. (1996) Sounding the alarm: Protein kinase cascades activated by stress and inflammation. *J. Biol. Chem.* **271,** 24,313–24,316.

146. Pearson, G., Robinson, F., Beers Gibson, T., et al. (2001) Mitogen-activated protein (MAP) kinase pathways: Regulation and physiologic functions. *Endocr. Rev.* **22,** 153–183.

147. van Biesen, T., Hawes, B. E., Luttrell, D. K., et al. (1995) Receptor-tyrosine-kinase- and Gβγ-mediated MAP kinase activation by a common signalling pathway. *Nature* **376,** 781–784.

148. Luttrell, L. M., Hawes, B. E., van Biesen, T., Luttrell, D. K., Lansing, T. J., and Lefkowitz, R. J. (1996) Role of c-Src in G protein-coupled receptor- and Gβγ subunit-mediated activation of mitogen activated protein kinases. *J. Biol. Chem.* **271,** 19,443–19,450.

149. Hackel, P. O., Zwick, E., Prenzel, N., and Ullrich, A. (1999) Epidermal growth factor receptors: critical mediators of multiple receptor pathways. *Curr. Opin. Cell Biol.* **11,** 184–189.

150. Shah, B. H., and Catt, K. J. (2004) GPCR-mediated transactivation of RTKs in the CNS: Mechanisms and consequences. *Trends Neurosci.* **27,** 48–53.

151. Prenzel, N., Zwick, E., Daub, H., et al. (1999) EGF receptor transactivation by G-protein-coupled receptors requires metalloproteinase cleavage of proHB-EGF. *Nature* **402,** 884–888.

152. Schafer, B., Gschwind, A., and Ullrich, A. (2004) Multiple G-protein-coupled receptor signals converge on the epidermal growth factor receptor to promote migration and invasion. *Oncogene* **23,** 991–999.

153. Yart, A., Roche, S., Wetzker, R., et al. (2002) A function for phosphoinositide 3-kinase beta lipid products in coupling beta gamma to Ras activation in response to lysophosphatidic acid. *J. Biol. Chem.* **277,** 21,167–21,178.

154. Luttrell, L. M., Della Rocca, G. J., van Biesen, T., Luttrell, D. K., and Lefkowitz, R. J. (1997) Gβγ subunits mediate Src-dependent phosphorylation of the epidermal growth factor receptor. *J. Biol. Chem.* **272,** 4637–4644.

155. Pierce, K. L., Tohgo, A., Ahn, S., Field, M. E., Luttrell, L. M., and Lefkowitz, R. J. (2001) Epidermal growth factor receptor dependent ERK activation by G protein-coupled receptors: A co-culture system for identifying intermediates upstream and downstream of HB-EGF shedding. *J. Biol. Chem.* **276,** 23,155–23,165.

156. Asakura, M., Kitakaze, M., Takashima, S., et al. (2002) Cardiac hypertrophy is inhibited by antagonism of ADAM12 processing of HB-EGF: metalloproteinase inhibitors as a new therapy. *Nat. Med.* **8,** 35–40.

157. Maudsley, S., Pierce, K. L., Zamah, A. M., et al. (2000) The β2-adrenergic receptor mediates MAP kinase activation via assembly of a multireceptor complex including the EGF receptor. *J. Biol. Chem.* **275,** 9572–9580.

158. Gschwind, A., Zwick, E., Prenzel, N., Leserer, M., and Ullrich, A. (2001) Cell communication networks: epidermal growth factor receptor transactivation as the paradigm for interreceptor signal transmission. *Oncogene* **20,** 1594–1600.

159. Murasawa, S., Mori, Y., Nozawa, Y., et al. (1998) Angiotensin II type 1 receptor-induced extracellular signal-regulated protein kinase activation is mediated by Ca2+/calmodulin-dependent transactivation of epidermal growth factor receptor. *Circ. Res.* **82,** 1338–1348.

160. Castagliuolo, I., Valenick, L., Liu, J., and Pothoulakis, C. (2000) Epidermal growth factor receptor transactivation mediates substance P-induced mitogenic responses in U-373 MG cells. *J. Biol. Chem.* **275,** 26,545–26,550.

161. Lev, S., Moreno, H., Martinez, R., et al. (1995) Protein tyrosine kinase PYK2 involved in Ca(2+)-induced regulation of ion channel and MAP kinase functions. *Nature* **376,** 737–745.

162. Dikic, I., Tokiwa, G., Lev, S., Courtneidge, S. A., and Schlessinger, J. (1996) A role for PYK2 and Src in linking G-protein-coupled receptors with MAP kinase activation. *Nature* **383,** 547–550.

163. Della Rocca, G. J., Maudsley, S., Daaka, Y., Lefkowitz, R. J., and Luttrell, L. M. (1999) Pleiotropic coupling of G-protein-coupled receptors to the MAP kinase cascade: Role of focal adhesions and receptor tyrosine kinases. *J. Biol. Chem.* **274,** 13,978–13,984.

164. Grewal, J. S., Luttrell, L. M., and Raymond, J. R. (2001) G protein-coupled receptors desensitize and downregulate EGF receptors in renal mesangial cells. *J. Biol. Chem.* **276,** 27,335–27,344.

165. Pak, Y., Pham, N., and Rotin, D. (2002) Direct binding of the beta1 adrenergic receptor to the cyclic AMP-dependent guanine nucleotide exchange factor CNrasGEF leads to Ras activation. *Mol. Cell. Biol.* **22,** 7942–7952.

166. Karoor, V. and Malbon, C. C. (1998) G-protein-linked receptors as substrates for tyrosine kinases: cross-talk in signaling. *Adv. Pharmacol.* **42,** 425–428

167. Ali, M. S., Sayeski, P. P., Dirksen, L. B., Hayzer, D. J., Marrero, M. B., and Bernstein, K. E. (1997) Dependence on the motif YIPP for the physical association of Jak2 kinase with the intracellular carboxyl tail of the angiotensin II AT1 receptor. *J. Biol. Chem.* **272,** 23,382–23,388.

168. Marrero, M. B., Venema, V. J., Ju, H., Eaton, D. C., and Venema, R. C. (1998) Regulation of angiotensin II-induced JAK2 tyrosine phosphorylation: roles of SHP-1 and SHP-2. *Am. J. Physiol.* **275,** C1216–C1223.

169. Hunt, R. A., Bhat, G. J., and Baker, K. M. (1999) Angiotensin II-stimulated induction of sis-inducing factor is mediated by pertussis toxin-insensitive G(q) proteins in cardiac myocytes. *Hypertension* **34,** 603–608.

170. Cao, W., Luttrell, L. M., Medvedev, A. V., et al. (2000) Direct binding of activated c-Src to the beta 3-adrenergic receptor is required for MAP kinase activation. *J. Biol. Chem.* **275**, 38,131–38,134.

171. Miller, W. E. and Lefkowitz, R. J. (2001) Expanding roles for beta-arrestins as scaffolds and adapters in GPCR signaling and trafficking. *Curr. Opin. Cell Biol.* **13**, 139–145.

172. Perry, S. J. and Lefkowitz, R. J. (2002) Arresting developments in heptahelical receptor signaling and regulation. *Trends Cell Biol.* **12**, 130–138.

173. Tohgo, A., Pierce, K. L., Choy, E. W., Lefkowitz, R. J., and Luttrell, L. M. (2002) Beta-Arrestin scaffolding of the ERK cascade enhances cytosolic ERK activity but inhibits ERK-mediated transcription following angiotensin AT1a receptor stimulation. *J. Biol. Chem.* **277**, 9429–9436.

174. Ahn, S., Wei, H., Garrison, T. R., and Lefkowitz, R. J. (2004) Reciprocal regulation of angiotensin receptor-activated extracellular signal-regulated kinases by beta-arrestins 1 and 2. *J. Biol. Chem.* **279**, 7807–7811

175. Azzi, M., Charest, P. G., Angers, S., et al. (2003) Beta-arrestin-mediated activation of MAPK by inverse agonists reveals distinct active conformations for G protein-coupled receptors. *Proc. Natl. Acad. Sci. USA* **100**, 11,406–11,411.

176. DeFea, K. A., Zalevsky, J., Thoma, M. S., Dery, O., Mullins, R. D., and Bunnett, N. W. (2000) β-Arrestin-dependent endocytosis of proteinase-activated receptor 2 is required for intracellular targeting of activated ERK1/2. *J. Cell Biol.* **148**, 1267–1281.

177. Luttrell, L. M., Roudabush, F. L., Choy, E. W., et al. (2001) Activation and targeting of extracellular signal-regulated kinases by β-arrestin scaffolds. *Proc. Natl. Acad. Sci. USA* **98**, 2449–2454.

178. DeFea, K. A., Vaughn, Z. D., O'Bryan, E. M., Nishijima, D., Dery, O., and Bunnett, N. W. (2000) The proliferative and antiapoptotic effects of substance P are facilitated by formation of a β-arrestin-dependent scaffolding complex. *Proc. Natl. Acad. Sci. USA* **97**, 11,086–11,091.

179. Scott, M. G., Le Rouzic, E., Perianin, A., et al. (2002) Differential nucleocytoplasmic shuttling of beta-arrestins. Characterization of a leucine-rich nuclear export sequence in beta-arrestin2. *J. Biol. Chem.* **277**, 37,693–37,701.

180. Lin, F.-T., Miller, W. E., Luttrell, L. M., and Lefkowitz, R. J. (1999) Feedback regulation of beta-arrestin1 function by extracellular signal-regulated kinases. *J. Biol. Chem.* **274**, 15,971–15,974.

181. Pitcher, J. A., Tesmer, J. J., Freeman, J. L., Capel, W. D., Stone, W. C., and Lefkowitz, R. J. (1999) Feedback inhibition of G protein-coupled receptor kinase 2 (GRK2) activity by extracellular signal-regulated kinases. *J. Biol. Chem.* **274**, 34,531–34,534.

182. Ogier-Denis, E., Pattingre, S., El Benna, J., and Codogno, P. (2000) Erk1/2-dependent phosphorylation of Galpha-interacting protein stimulates its GTPase accelerating activity and autophagy in human colon cancer cells. *J. Biol. Chem.* **275**, 39,090–39,095.

183. Elorza, A., Penela, P., Sarnago, S., and Mayor, F., Jr. (2003) MAPK-dependent degradation of G protein-coupled receptor kinase 2. *J. Biol. Chem.* **278,** 29,164–29,173.

184. Ge, L., Ly, Y., Hollenberg, M., and DeFea, K. (2003) A beta-arrestin-dependent scaffold is associated with prolonged MAPK activation in pseudopodia during protease-activated receptor-2-induced chemotaxis. *J. Biol. Chem.* **278,** 34,418–34,426.

185. Fong, A. M., Premont, R. T., Richardson, R. M., Yu, Y. R., Lefkowitz, R. J., and Patel, D. D. (2002) Defective lymphocyte chemotaxis in beta-arrestin2- and GRK6-deficient mice. *Proc. Natl. Acad. Sci. USA* **99,** 7478–7483.

186. McDonald, P. H., Chow, C-W., Miller, W. E., et al. (2000) β-Arrestin 2: a receptor-regulated MAPK scaffold for the activation of JNK3. *Science* **290,** 1574–1577.

187. Miller, W. E., McDonald, P. H., Cai, S. F., Field, M. F., Davis, R. J., and Lefkowitz, R. J. (2001) Identification of a motif in the carboxy terminus of β–arrestin2 responsible for activation of JNK3. *J. Biol. Chem.* **276,** 27,770–27,777.

188. Sun, Y., Cheng, Z., Ma, L., and Pei, G. (2002) Beta-arrestin 2 is critically involved in CXCR4-mediated chemotaxis, and this is mediated by its enhancement of p38 MAPK activation. *J. Biol. Chem.* **277,** 49,212–49,219.

189. Luttrell, L. M., Ferguson, S. S. G., Daaka, Y., et al. (1999) β-Arrestin-dependent formation of β2 adrenergic receptor/Src protein kinase complexes. *Science* **283,** 655–661.

190. Barlic, J., Andrews, J. D., Kelvin, A. A., et al. (2000) Regulation of tyrosine kinase activation and granule release through β-arrestin by CXCRI. *Nat. Immunol.* **1,** 227–233.

191. Ghalayini, A. J., Desai, N., Smith, K. R., Holbrook, R. M., Elliott, M. H., and Kawakatsu, H. (2002) Light-dependent association of Src with photoreceptor rod outer segment membrane proteins in vivo. *J. Biol. Chem.* **277,** 1469–1476.

192. Milano, S. K., Pace, H. C., Kim, Y. M., Brenner, C., and Benovic, J. L. (2002) Scaffolding functions of arrestin-2 revealed by crystal structure and mutagenesis. *Biochemistry* **41,** 3321–3328.

193. Miller, W. E., Maudsley, S., Ahn, S., Kahn, K. D., Luttrell, L. M., and Lefkowitz, R. J. (2000) β-Arrestin1 interacts with the catalytic domain of the tyrosine kinase c-SRC. *J. Biol. Chem.* **275,** 11,312–11,319.

194. Ahn, S., Kim, J., Lucaveche, C. L., et al. (2002) Src-dependent tyrosine phosphorylation regulates dynamin self-assembly and ligand-induced endocytosis of the epidermal growth factor receptor. *J. Biol. Chem.* **277,** 26,642–26,651.

195. Penela, P., Elorza, A., Sarnage, S., and Mayor, F., Jr. (2001) Beta-arrestin and c-Src-dependent degradation of G-protein-coupled receptor kinase 2. *EMBO J.* **20,** 5129–5138.

196. Imamura, T., Huang, J., Dalle, S., et al. (2001) Beta-Arrestin-mediated recruitment of the Src family kinase Yes mediates endothelin-1-stimulated glucose transport. *J. Biol. Chem.* **276,** 43,663–43,667.

197. Luttrell, L. M. (2003) Location, Location, Location. Spatial and temporal regulation of MAP kinases by G protein-coupled receptors. *J. Mol. Endocrinol.* **30,** 117–126.

198. Yang, M., Zhang, H., Voyno-Yasenetskaya, T., and Ye, R. D. (2003) Requirement of G beta-gamma and c-Src in D2 dopamine receptor-mediated nuclear factor-kappa B activation. *Mol. Pharmacol.* **64,** 447–455.

199. Chen, W., Hu, L. A., Semenov, M. V., et al. (2001) Beta-Arrestin1 modulates lymphoid enhancer factor transcriptional activity through interaction with phosphorylated dishevelled proteins. *Proc. Natl. Acad. Sci. USA* **98,** 14,889–14,894.

200. Chen, W., ten Berge, D., Brown, J., et al. (2003) Dishevelled 2 recruits beta-arrestin 2 to mediate Wnt5A-stimulated endocytosis of Frizzled 4. *Science* **301,** 1391–1394.

2

Crosstalk Coregulation Mechanisms of G Protein-Coupled Receptors and Receptor Tyrosine Kinases

Kanchana Natarajan and Bradford C. Berk

Summary

G protein-coupled receptors (GPCRs) and receptor tyrosine kinases (RTKs) are transmembrane receptors that initiate intracellular signaling cascades in response to a diverse array of ligands. Recent studies have shown that signal transduction initiated by GPCRs and RTKs is not organized in distinct signaling cassettes where receptor activation leads to cell division and gene transcription in a linear manner. In fact, signal integration and diversification arises from a complex network involving crosscommunication between separate signaling units. Several different styles of crosstalk between GPCR- and RTK-initiated pathways exist, with GPCRs or components of GPCR-induced pathways being either upstream or downstream of RTKs. Activation of GPCRs sometimes results in a phenomenon known as "transactivation" of RTKs, which leads to the recruitment of scaffold proteins, such as Shc, Grb2, and Sos in addition to mitogen-activated protein kinase activation. In other cases, RTKs use different components of GPCR-mediated signaling, such as β-arrestin, G protein-receptor kinases, and regulator of G protein signaling to integrate signaling pathways. This chapter outlines some of the more common mechanisms used by both GPCRs and RTKs to initiate intracellular crosstalk, thereby creating a complex signaling network that is important to normal development.

Key Words: G protein-coupled receptor; growth factor receptor; crosstalk; transactivation; MAPK.

1. Introduction

Cells use a wide array of biochemical mechanisms to respond to extracellular signals, such as hormones, neurotransmitters, chemokines, odorants, and light. Three major classes of receptors on the surface of the cell detect these signals. The first class of receptor proteins is peripheral membrane proteins, which adhere only loosely to the biological membrane with which they are

From: *Methods in Molecular Biology, vol. 332: Transmembrane Signaling Protocols, Second Edition*
Edited by: H. Ali and B. Haribabu © Humana Press Inc., Totowa, NJ

associated. These molecules do not span the lipid bilayer core of the membrane but attach indirectly, typically by binding to integral membrane proteins, or by interactions with the lipid polar head. Another major class of receptors is represented by intracellular receptors, such as those for steroid hormones. A third major class of receptors includes transmembrane proteins, which reside and operate typically within a cell's plasma membrane but also are found in the membranes of some subcellular compartments and organelles. Binding of a signaling molecule to the receptor on the extracellular domain helps transduce the signal through the transmembrane domain to the intracellular space of the cell. There are several types of transmembrane receptors including integrins, G proteins, and protein tyrosine kinases.

All G protein-coupled receptors (GPCRs) identified to date share a typical structural motif of seven membrane-spanning helices and are coupled with heterotrimeric G proteins. Agonist-stimulated GPCRs function as guanosine diphosphate (GDP)/guanosine triphosphate (GTP) exchange factors and promote the release of GDP and binding of GTP to the α-subunits. This process activates the G protein by dissociating GTP-bound Gα from the heterodimeric G$\beta\gamma$ subunit. Both GTP-Gα and G$\beta\gamma$ subunits interact with a variety of effector systems, such as adenylyl cyclase, phospholipase (PL) C isoforms, and ion channels, thereby modulating cellular signaling pathways through second messengers cyclic adenosine monophosphate (cAMP), protein kinase (PK) C, and Ca^{2+} and other intermediate molecules, such as phosphatidylinositol 3-kinase (PI3K), reactive oxygen species (ROS), Pyk2, and Src *(1)*.

Receptor tyrosine kinases (RTKs) comprise another class of transmembrane proteins that span the membrane just once. Classically, RTKs are activated by ligands, such as growth factors and insulin. Upon ligand binding and receptor dimerization, the activated receptor acts as a tyrosine kinase, autophosphorylates itself on cytoplasmic tyrosine residues, and subsequently acts as a scaffold to assemble signaling partners. Classically these include Shc, Grb2, and Sos, which lead to Ras activation followed by an increase in mitogen-activated protein kinase (MAPK) activity *(2,3)*.

Initially, it was thought that GPCRs and RTKs, along with their respective downstream effectors, represented distinct and linear signaling units that converged on downstream targets, such as the MAPKs. Recently, it has become clear that GPCR- and RTK-mediated signaling pathways are not mutually exclusive of one another and often function as partners, with G protein participation being either upstream or downstream of the RTKs, stimulating interactions at multiple levels between various molecules downstream of the receptors *(4,5)*. For example, both pathways involve tyrosine phosphorylation of Shc and Ras activation upstream of MAPK activation *(6–8)*. The involvement of common molecules initiates an integration of diverse stimuli through complex

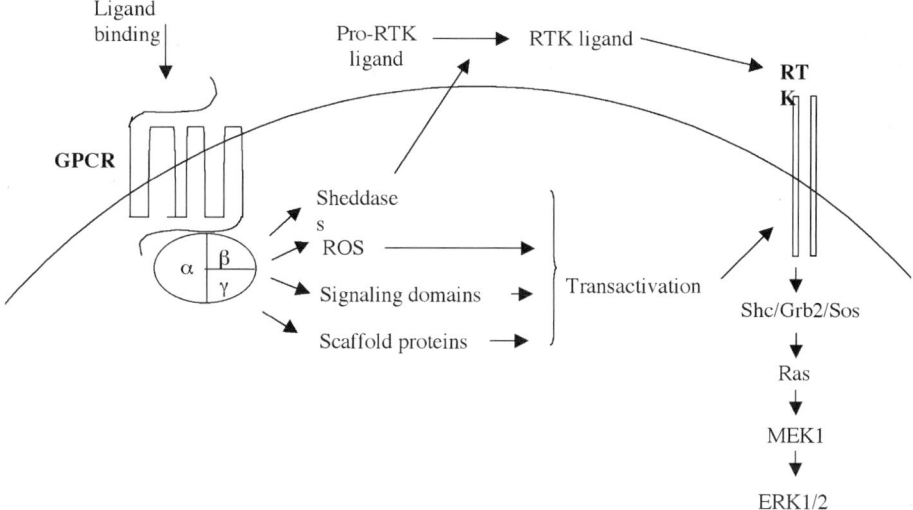

Fig. 1. Schematic showing G protein-coupled receptor–ligand-induced transactivation of receptor tyrosine kinase.

cross-communication and provides intricate control over regulatory mechanisms that affect cell proliferation, differentiation, growth, and survival. This chapter reviews the signaling pathways associated with crosstalk between GPCRs and RTKs that could be initiated by either GPCR or RTK ligands.

GPCRs initiate crosstalk in several different ways. In some cases, GPCRs can form homodimers and heterodimers in order to increase functional activity. Several such examples have been discovered, such as the heterodimerization of the γ-aminobutyric acid receptors, the homodimerization of the β2-adrenergic receptors, and the heterodimerization of the dopamine D2 and somatostatin SSTR5 receptor *(9–11)*. In addition, treatment of cells with ligands for GPCRs results in tyrosine phosphorylation and subsequent activation of RTKs, by a phenomenon known as "transactivation" *(12,13)*. In each case, increased dimerization of the RTKs leads to the recruitment of scaffold proteins, such as Shc, Grb2, and Sos, via their Src homology (SH)2 domains. Several GPCR agonists, such as angiotensin II (AngII), lysophosphatidic acid (LPA), bradykinin, and endothelin, transactivate RTKs such as the epidermal growth factor receptor (EGFR) and platelet-derived growth factor receptor (PDGFR).

In recent years, different concepts have emerged to explain mechanisms of transactivation as shown in **Fig 1**. Molecules such as PKC, Src, and ROS mediate RTK transactivation. In general, both calcium-dependent and -independent pathways leading to RTK transactivation have been suggested. One of the new concepts in transactivation mechanisms is that of GPCR ligands activating "sheddases," proteases that cleave an RTK ligand molecule to its RTK-binding form. This active ligand in turn activates the RTK. Another mechanism of transactivation involves the creation of signaling domains by GPCR–ligand interaction, where there is a movement of RTKs to a specific subcellular location, leading to RTK–GPCR association and downstream signaling. Several adaptor/ scaffold proteins such as Gab1, IRS-1, and GIT1, which serve as docking sites for multiprotein complexes at the RTK, also have been implicated as mediators of GPCR-ligand induced RTK transactivation, Activation of protein tyrosine phosphatases that "transinactivate" RTKs in response to GPCR activation also have been recently suggested as a mechanism of GPCR–RTK crosstalk.

In some cases, the RTK activation of downstream effector responses is sensitive to pertussis toxin, suggesting that G protein involvement is proximal to, and downstream of the RTKs. In this model, the RTKs use several different components of GPCR-mediated signaling, such as β-arrestin, regulator of G proteins (RGS), and G protein receptor kinases (GRKs). Studies by various groups have demonstrated two major models for G protein signaling downstream of RTKs. In the first scenario, activated RTKs have been shown to induce the activation of G proteins by dissociating the $G\alpha$ subunit from the $G\beta\gamma$ subunit leading to downstream signaling (**Fig. 2A**) Alternatively, stimulation of an RTK by a ligand leads to a direct association between GPCRs and RTKs through scaffold proteins, such as RGS, leading to the use of G protein-associated molecules such as β-arrestin and Grk2, as shown in **Fig. 2B**. These data indicate the involvement of GPCRs both upstream and downstream of the RTK signal transduction. Outlined in **Headings 2** and **3** are a few common examples of crosstalk between GPCRs and RTKs. The novel crosstalk that may occur between two different RTKs also will be discussed.

2. GPCR/G Protein Ligand-Initiated Receptor Crosstalk

2.1. Angiotensin II

AngII, a multifunctional octapeptide of the renin–angiotensin system, influences the function of cardiovascular cells via intracellular signaling that is initiated at the AngII type 1 and type 2 receptors (AT$_1$R and AT$_2$R), which are GPCRs that have opposing effects on cell growth and other physiological functions **(14,15)**. Crosstalk exists between AT$_1$R and AT$_2$R, and studies performed by Cui et al. demonstrate a role for SHP-1 tyrosine phosphatase in this cross talk that regulates survival of fetal vascular smooth muscle cells (VSMCs) **(16)**. Activation of $G_{q/11}$ by AngII stimulates PLC to generate inositol (1,4,5)-triphosphate and diacyglycerol, thereby increasing intracellular Ca^{2+} levels and activation of PKC. Downstream effectors of AngII signaling include the following:

1. Extracellular signal-regulated kinase (ERK) 1/2, p38 MAPK, and JNK.
2. Tyrosine kinases, such as Src and Pyk2.

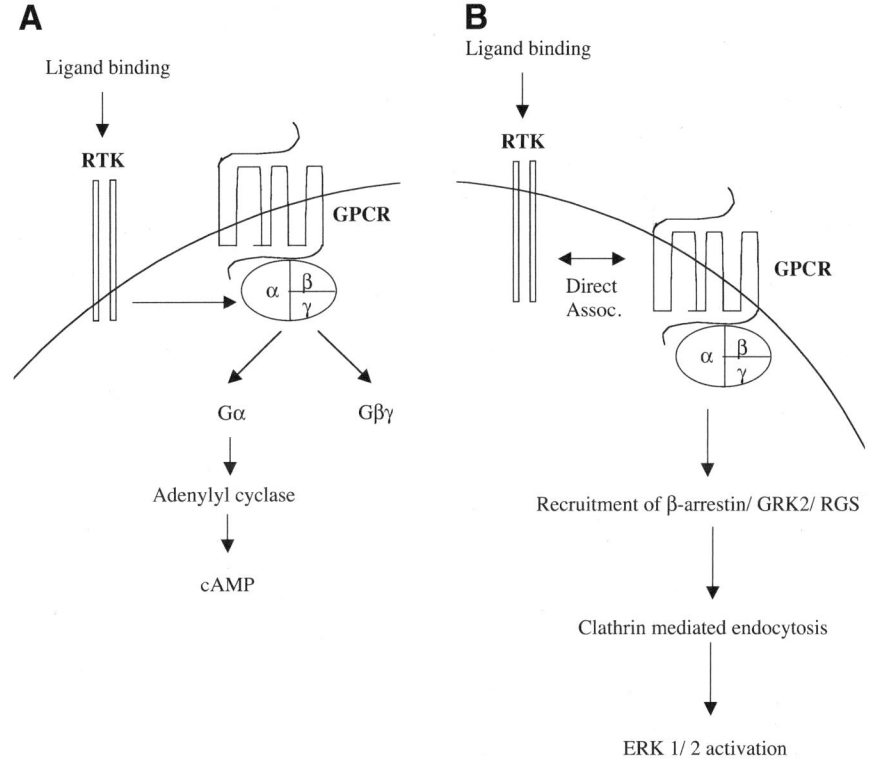

Fig. 2. Schematic showing receptor tyrosine kinase (RTK)–ligand-induced crosstalk with G protein-coupled receptors (GPCRs). (**A**) RTK–ligand-induced effect on G protein activation. (**B**) RTK–ligand-induced utilization of GPCR/G protein-regulating signaling components.

3. PI3K and PKB/Akt.
4. Janus-activating kinase (JAK) and signal transducers and activators of transcription (STATs).
5. RTKs, such as the EGFR and PDGFR *(17–23)*.

2.1.1. EGFR Transactivation

AngII induces transactivation of the EGFR and, in turn, the EGFR serves as a scaffold for assembling signaling molecules, such as MAPKs and Akt that are important for downstream signaling, as well as the expression of the AT_1R signaling repertoire in VSMCs *(20,24)*. Downstream, AngII-induced EGFR transactivation plays a role in inducing eukaryotic translation initiation factor 4E and 4E binding protein 1 phosphorylation, thereby playing a role in translational control and protein synthesis and this process upregulates proteins like the plasminogen activator inhibitor type 1 *(25,26)*. AngII induces EGFR transactivation by both Ca^{2+}-dependent and Ca^{2+}-independent processes *(21,23,24,27,28)*. Three major mechanisms are involved in AngII-induced EGFR transactivation—an upstream tyrosine kinase, ROS, or through the use of metalloproteases that generate EGF-like ligands (sheddases in **Fig. 1**) In addition, recent studies from our laboratory indicate a novel mechanism by which glucose-dependent EGFR *N*-glycosylation and, hence, transactivation, modulates AngII signal transduction *(29)*.

2.1.1.1. NON-RTKs

Two major non-RTKs have been shown to be involved in EGFR transactivation by AngII. Several studies done in VSMCs, cardiac myocytes, and rat anterior pituatory cells have shown that c-Src is necessary for the transactivation of the EGFR, and this in turn induces Ras/ERK activation downstream *(12,24,30–32)*. In rat liver epithelial cells, Li et al. proposed an AngII-stimulated EGFR-dependent signaling pathway to Ras only when PKC activity was inhibited *(33)*. Interestingly, in VSMCs, AngII-induced p70[rsk] activation is mediated via both the ERK and PI3K/Akt cascades that bifurcate at the point of EGFR-dependent Ras activation *(34)*.

Another non-RTK, the proline-rich kinase 2 (PYK2)/cell adhesion kinase β also is induced by several GPCR agonists. Its role in the transmission of mito-genic signals via EGFR transactivation is somewhat controversial as shown in AngII-stimulated VSMCs, cardiac fibroblasts, and PC12 cells *(27,35–37)*. Tyrosine phosphorylated Src is often found in association with the transactivated EGFR or with PYK2 on G_q-coupled receptor stimulation, suggesting activated Src to be the primary mediator of EGFR transactivation *(35,36,38)*.

In addition to activating Src and PYK2, AngII induces the JAK/STAT sig-naling pathway, which has been implicated in ERK activation and subsequent cell growth in VSMCs, cardiac fibroblasts, and cardiomyocytes *(39–41)*. Because JAK is involved in growth hormone-induced EGFR transactivation, the possibility of JAK-dependent EGFR transactivation by AngII also exists *(42)*.

2.1.1.2. REACTIVE OXYGEN SPECIES

The generation of ROS, such as superoxide and hydrogen peroxide (H_2O_2) that act as intercellular and intracellular second messengers, is regulated by cytokines and growth factors, including AngII, in several cell types *(43,44)*. AngII-induced transactivation of the EGFR is mediated, in part, through ROS derived from nicotinamide-adenine dinucleotide phosphate (NADPH) oxidase, and this transactivation is strongly inhibited by antioxidants, such as, tiron, and *N*-acetylcysteine *(28,45–47)*. Once produced, ROS activate several receptor- and non-RTKs, such as the JAK and Src families, PYK2, as well as the EGFR, stimulating the formation of the Shc–Grb2–Sos complex at the EGFR. This complex subsequently activates Ras followed by the p38 MAPK and Akt/PKB pathways downstream of the EGFR *(48,49)*. In addition, AngII promotes the movement of AT_1R to caveolae and lipid rafts leading to AT_1R–EGFR association in VSMCs through the tyrosine phosphatase SHP-2 *(20,50)*. Depletion of membrane cholesterol by β-cyclodextrin disrupts caveolae structure and inhibits tyrosine phosphorylation of the EGFR and subsequent activation of PKB induced by AngII.

2.1.1.3. Metalloproteinase Cleavage of Heparin-Binding EGF

Prenzel et al. first showed that a chimeric RTK in rat fibroblasts, consisting of the EGFR ectodomain and the PDGFR transmembrane and intracellular domain, was transactivated with GPCR ligands, whereas the endogenous PDGFR was not, by the cleavage of proheparin-binding (pro-HB)-EGF to its active form HB-EGF by matrix metalloproteinases (MMPs; *[51]*). Free HB-EGF subsequently binds to the EGFR, leading to EGFR transactivation. The role of MMPs in AngII-induced transactivation of the EGFR remains controversial; studies done in our laboratory on VSMCs did not show inhibition of EGFR transactivation with MMP inhibitors, whereas other studies have shown an inhibition by pharmacologically inhibiting the MMPs *(22,23,52)*. Eguchi et al. suggest that MMP-dependent EGFR transactivation by AngII activates the ERK and p38 MAPK pathways, whereas JNK activation is regulated independent of EGFR transactivation *(23)*.

Recent data suggests that different proteases (sheddases) may cleave pro-HB-EGF through PKC-dependent and PKC-independent mechanisms in response to different stimuli. Some data suggest that PKC mediates AngII-induced EGFR transactivation via activation of MMPs in response to GPCR agonists coupled to G_q *(26,51,53–56)*. However, other studies, such as those done by Frank et al., showed that ROS transactivate EGF receptors through the release of HB-EGF by metalloproteases in VSMCs and that this transactivation is independent of PKC *(57)*. In addition to the EGFR, the primary cognate HB-EGF receptor Erb1 has also shown to be transactivated by AngII in human prostate stromal cells, thereby promoting cell growth *(58)*.

2.1.2. PDGFR Transactivation

Although PDGFR has two distinct receptor subtypes, rapid tyrosine phosphorylation of only the PDGFβ receptor by AngII has been reported *(59–61)*. This transactivation induces association of the activated receptor with p66Shc, Grb2, and c-Src. In addition, PDGFR transactivation by AngII was not sensitive to BAPTA-AM, suggesting that this transactivation pathway was Ca^{2+}-independent *(59)*. Like AngII-induced EGFR transactivation, PDGFR transactivation is redox-sensitive and is abrogated by *N*-acetylcysteine and

Tiron. Recently, the potential downstream signaling of the PDGFR to ERK 1/2 via AngII-mediated transactivation was proposed in mesangial cells *(62)*. Additional studies by Conway et al. have shown that the activation of the MAPK pathway is dependent on both Src and complex formation of Grb2 with PI3K *(63)*. New studies indicate that, like the EGF-family of ligands, a new ligand for the PDGFRα, PDGF-C, could be another growth factor that is released from the cell surface after limited proteolysis leading to transactivation of the PDGFR *(64)*.

2.1.3. Insulin-Like Growth Factor 1 Receptor Activation

Another growth factor receptor that is transactivated by AngII is the insulin-like growth factor 1 receptor (IGF-1R) in VSMCs. IGF-1R becomes phosphorylated on its β-subunit and this in turn phosphorylates the adapter insulin receptor substrate-1 (IRS-1 *[65]*). Transactivation of the IGF-1R has been shown to play a critical role in PI3K activation by AngII, but does not seem to be required for stimulation of the MAPK cascade *(66)*. Touyz et al. demonstrated that AngII stimulates production of NADPH-inducible ROS partially through IGF-1R transactivation which leads to phosphorylation of p38 MAPK and ERK5, but not ERK 1/2 *(49)*. Also, the role of insulin receptor substrate (IRS)-1-mediated signaling in response to AngII in VSMCs remains controversial as inhibition of insulin and IGF-1 signaling by AngII at the levels of IRS-1 and PI3K have been reported *(67,68)*.

2.2. Lysophosphatidic Acid

LPA is an important component of serum that affects cell proliferation, survival, adhesion, and migration by transducing signaling through the Edg family of receptors that are coupled to Gi, Gq/11, and G12/13 proteins. LPA induces ERK 1/2 activation by mediator protein tyrosine kinases, such as Src, PYK2, and transactivated EGFR *(13,69–73)*. LPA-induced EGFR tyrosine phosphorylation is weak but functionally significant in several cell lines tested *(74)*. Inhibition of LPA induced EGFR transactivation suppressed tyrosine phosphorylation of adapter proteins Shc and Gab1, which in turn inhibited Shc-Grb2 and Gab1-SHP2 association that was necessary for ERK 1/2 activation. This indicates that LPA-induced transactivation is upstream of ERK 1/2 activation, *c-fos* induction and DNA synthesis *(13,74,75)*.

Several studies have shown that LPA-mediated EGFR is dependent on calcium and ROS *(76–80)*. In addition, LPA has been identified as a major serum factor for stimulating pro-HB-EGF ectodomain shedding via a Ras-Raf-MAPK/ERK pathway to transactivate the EGF receptor *(81,82)*. Recently, LPA also has been shown to transactivate the HB-EGF receptors ErbB1 and ErbB4 via a Ca^{2+}-dependent pathway *(83)*.

LPA receptors also interact with and transactivate the nerve growth factor receptor TrkA, stimulating translocation of the TrkA receptor to the nucleus and this regulates the ERK 1/2 pathway *(84)*. LPA also mediates phosphorylation of the PDGFR-β in human bronchial epithelial cells via phospholipase D *(85)*. In addition to transactivating these growth factor receptors, LPA induces phosphorylation of α_{1B}-adrenoreceptor phosphorylation through dissociated Gβγ subunits, EGFR transactivation, PI3K and PKC *(86)*.

2.3. Endothelin

Endothelin (ET) isopeptides (ET-1, ET-2, and ET-3) are potent vasoconstrictors that bind specific ET (ET_A and ET_B) receptors coupled to G_q proteins. Similar to the angiotensin receptors, crosstalk between the two ET receptors has also been reported in rat mesenteric arteries *(87)*. Activation of GPCRs by ET-1 phosphorylates the EGFR in a Ca^{2+}- and MMP-dependent manner, followed by an increased association of the phosphorylated EGFR with Shc and Grb2, subsequently leading to MAPK phosphorylation, p70[S6K] activation, *c-fos* induction, and cell proliferation *(13,51,88,89)*. In addition, Hua et al. have shown that ET-1 activates ERK 1/2 in mesangial cells predominantly through a pathway involving EGFR transactivation and its attachment to caveolin, leading to compartmentalization of these signaling molecules *(90)*. In a rat cardiac allograft model, Sihvola et al. demonstrated an increase in VSMC proliferation and migration via ET-1 induced PDGFR upregulation *(91)*. ET-1 also signals through other GPCRs. ET-1 and norepinephrine signaling crosstalk through differential pathways regulating myocardial contractility, and this is mediated by Ca^{2+} transients, PKA, PKC, PKG, and phosphatases *(92)*. PKC also plays a major role in ET-induced phosphorylation of the α_{1B}-adrenergic receptor *(93,94)*.

2.4. Bradykinin

Bradykinin is an inflammatory mediator that exerts its biological effects through the activation of several bradykinin receptors. The B2 receptor (B2R) is capable of coupling to different classes of G proteins in a cell specific and time-dependent manner, resulting in simultaneous or consecutive initiation of different signaling chains that may crosstalk. Blaukat et al. have shown that bradykinin activates both $G\alpha_q$ and $G\alpha_i$ pathways simultaneously and cooperative signaling between these two activated G protein pathways is required for a synergistic stimulation of ERK 1/2 *(95)*. Other studies have shown that the activated bradykinin receptor coupled to $G\alpha_q$ can activate $G\alpha_i$ and subsequently adenylate cyclase and cAMP. This activation leads to differential regulation of PLC preventing multiple stimulation of MAPK *(96)*. Bradykinin modulates α_{1b}-adrenoreceptor phosphorylation in rat-1 fibroblasts *(97)*. The B2R also has been shown to crosstalk with nucleotide receptors, such as P2Y, which are also coupled to G_q *(98,99)*.

Schindelholz et al. report growth cone collapse of neuronal growth factor (NGF)-differentiated PC12 cells evoked by bradykinin, mediated by c-Src and paxillin, revealing a crosstalk between bradykinin and growth factor receptors, such as the NGF receptor *(100)*. Bradykinin-induced transactivation of the KDR/Flk-1 (VEGF receptor 2) receptors associated with endothelial nitric oxide synthase production has also been shown in endothelial cells *(101,102)*. Work done in several systems have shown that bradykinin induces transactivation of the EGFR via both PKC-dependent and PKC-independent mechanisms, which leads to phosphorylation of downstream molecules, such as ERK 1/2, AMP responsive element-binding protein (CREB), nuclear factor (NF)-κB, and E2F *(103–105)*. EGFR transactivation by bradykinin also induces desensitization of EGFRs by a process associated with the loss of cell-surface EGFRs through clathrin-mediated endocytosis via β-arrestin and dynamin *(104)*. Whether calcium and calmodulin are required for EGFR transactivation by bradykinin remains a matter of controversy *(106–108)*. Finally, novel findings by Graness et al. show bradykinin-mediated "transinactivation" of EGFR by stimulation of a protein tyrosine phosphatase *(109)*.

2.5. Sphinosine 1-Phosphate

Sphinosine 1-phosphate (S1P) is a bioactive lipid released by activated platelets that induces cell processes, such as migration and proliferation by binding the Edg family of GPCRs. S1P induces transactivation of the vascular EGFR (VEGFR) in human umbilical vein endothelial cells, followed by Src activation and phosphorylation of the adaptor protein CrkII, to induce membrane ruffling *(110)*. In other studies, transactivation of the VEGFR by S1P is independent of ROS and is mediated by Ca^{2+} and Src, leading to the activation of the PI3K/Akt/endothelial nitric oxide synthase pathway *(111)*. S1P also stimulates Akt phosphorylation via G_i-dependent PDGFRβ transactivation *(112)*. Transactivation of EGFR by S1P has also been reported through a PKC-dependent pathway that results in the activation of the Ras–MEK–ERK pathway *(113)*.

2.6. Thrombin

Thrombin is a procoagulant protease that signals through the protease-activated receptor family that are coupled to G proteins. Transactivation of the EGFR on thrombin stimulation has been shown in a number of systems through multiple mechanisms *(114)*. Several groups also showed that thrombin transactivates the EGFR via HB-EGF, Src, and PYK2 followed by increased ERK 1/2 and p38 MAPK activation, leading to an increase in CREB activation DNA synthesis and interleukin 6 gene expression *(115–119)*. In rat VSMCs, thrombin induces the release of basic FGF that results in FGF receptor transactivation-mediated cell proliferation *(120)*. Thrombin also induces IGF-1R transactivation in rat VSMCs *(121)*.

2.7. Adrenoreceptor Agonists

AngII stimulates the release of norepinephrine from the sympathetic nerves that is a ligand for the α_1-adrenergic receptor. In carotid injury models, Majesky et al. showed that α_1-adrenergic stimulation caused PDGF-A expression, suggesting crosstalk between AngII and PDGF signaling *(122)*. Luttrell et al. also have demonstrated EGFR transactivation by G_i coupled-α-adrenergic receptors followed by tyrosine phosphorylation of the Shc adapter protein *(12)*. In addition, PDGFRs reduce actions of α_{1B}-adrenergic receptors by phosphorylating the receptors and decreasing their association with their G proteins *(93)*.

3. Growth Factor-Initiated Crosstalk Via G Proteins

3.1. Epidermal Growth Factor

Upon EGFR activation and autophosphorylation, numerous phosphotyrosines are generated that serve as docking sites for proteins, such as PLCγ, Shc, Gab1, and Grb2, which in turn activate downstream pathways. However, the EGFR also uses components involved in G protein signaling and bidirectionally interacts with GPCRs. EGF stimulation leads to increased association of $G\alpha_{12}$ with EGFR, which leads to the activation of PLCγ, ERK 1/2, and increased DNA synthesis *(123–125)*. EGFR interaction with $G\alpha_i$ inhibits $G\alpha_i$. EGFR kinase phosphorylates and associates with $G\alpha_s$ leads to the activation of $G\alpha_s$ and in the heart this mechanism leads to increased cAMP accumulation via activation of adenylate cyclase *(126–128)*.

Direct activation of EGFR also induces α_{1B}-adrenergic receptor phosphorylation by PKC via activation of PI3K *(93)*. Also, Maudley et al. reported that the EGFR exists in a preformed complex with β2-adrenergic receptor *(129,130)*. Transactivation of EGFR by GPCR agonists leads to the β-arrestin and Gβγ-mediated internalization of this complex, which is necessary for the activation of MAPK. However, EGF itself can stimulate the recruitment of β-arrestin to the EGFR, suggesting downstream interaction between the GPCR and EGFR pathways *(130)*. EGF is also known to regulate other GPCR signaling component associations, such as that between GRK2 and PDEγ, thereby regulating MAPK activation and EGF-mediated phosphorylation of RGS increases GTPase activating protein activity *(131,132)*.

3.2. Platelet-Derived Growth Factor

There is substantial evidence showing a requirement for G proteins in platelet-derived growth factor (PDGF)-stimulated pathways. Several studies have shown that activation of c-Src and ERK 1/2 downstream of PDGF stimulation is sensitive to pertussis toxin *(63,133)*. In addition, Freedman et al. showed that GTPγS binding to $G\alpha_i$ increases on PDGF stimulation *(134)*. PDGF induction of ROS also seems to require coupling of $G\alpha_{i1}$ and $G\alpha_{i2}$ to the PDGFR

(135). PDGF-induced cell migration requires the presence of EDG-1 a GPCR for S1P that activates Rac-dependent pathways *(136)*.

PDGFβ receptor signals through an endocytic pathway as well via GPCR-dependent machinery. The GRK2/β-arrestin complex constitutively associates with the PDGFR and is recruited via its association with the GPCR. On stimulation with PDGF, c-Src is recruited to the PDFGR–GPCR complex leading to β-arrestin-mediated signaling and ERK 1/2 activation *(134,137)*. RGS proteins, such as RGS2, that are GAPs involved in terminating GPCR signaling, are also recruited to the plasma membrane after PDGF stimulation, suggesting another component of GPCR signaling is involved in PDGFR signaling *(138)*.

3.3. Neuronal Growth Factor

NGF promotes the survival and differentiation of neurons and signals through its receptor TrkA, The TrkA receptor is constitutively bound to GRK2 and stimulation with NGF promotes binding of β-arrestin to this complex in a $G\alpha_{i/o}$-dependent manner. This initiates an integrative activation of the ERK 1/2 pathway via a process that involves β-arrestin 1 and clathrin-mediated endocytosis of the TrkA–GPCR/B-raf/MEK-1 signal complex. NGF also reduces cAMP levels in PC12 cells via a G protein-dependent mechanism *(139)*. Another level of GPCR crosstalk is with tyrosine kinase receptors through RGS proteins, where the RGS serves as a scaffold bridging together GPCRs and RTKs. Lou et al. were the first to show suppression of GPCR signaling by Trk, which is dependent on a PDZ domain in the RGS protein GIPC *(140)*.

3.4. Fibroblast Growth Factor

Fibroblast growth factors (FGFs) are members of a family of polypeptides synthesized by a variety of cell types that signal through one of four FGF receptors, i.e., FGFR1–4. Similar to other RTKs, FGFR stimulation with FGF results in receptor dimerization, phosphorylation, and activation of the Ras–Raf–MEK–MAPK pathway through either the Crk/FGFR substrate 2 (FRS2)/Grb2/Sos or Shc/Grb2/Sos complex. Fedorov et al. have shown that that $G_i\beta\gamma$ are involved in FGF-2 mediated activation of ERK 1/2 that promotes skeletal muscle differentiation *(141)*. Also, FGF-2 induces S1P-coupled G_i receptors by activating sphingosine kinase-1, the enzyme that converts sphingosine to S1P *(142)*. It has also been demonstrated that FGF-2 promotes dissociation of the $G_s\beta\gamma$ heterotrimer, leading to $G\alpha_s$ stimulation of adenylyl cyclase and $G\beta\gamma$ inhibition of NADPH oxidase *(143)*.

3.5. Vascular Endothelial Growth Factor

VEGF is a cytokine that is essential for angiogenesis and endothelial cell differentiation (vasculogenesis) during development *(144,145)*. VEGF regulates multiple biological functions through three major types of receptors—the

RTKs Flt1 (VEGFR-1), KDR/Flk1 (VEGFR-2), and Flt-4 (VEGFR-3), a nontyrosine kinase transmembrane protein Neuropilin-1 and heparan sulfate proteoglycans *(146–151)*. Zeng et al. have demonstrated that VEGFR-2 (KDR) stimulates MAPK activation, migration, and proliferation via $G\alpha_q$ and $G\beta\gamma$ subunits *(152,153)*. Also, KDR signaling is downregulated by VEGFR-1 (Flt-1)/ G_i/$G\beta\gamma$-mediated activation of cdc42 and Rho, demonstrating opposing effects of the two VEGFRs *(154)*.

3.6. Insulin and IGF

Insulin receptors have been shown to associate with and tyrosine phosphorylate Gi and Gs in several studies *(155,156)*. Also, insulin phosphorylates the β2 adrenergic receptor (β2-AR), leading to increased Grb2/β2-AR interaction. Grb2 inturn binds PI3K and dynamin, and this leads to the internalization of β2-AR.

IGF-1 is a 12-kDa mitogenic and survival factor hormone peptide secreted by multiple cells that interacts with its own receptor, as well as the insulin receptor. IGF-1 preferentially interacts with and uses the G_i-dependent signaling pathway by promoting $G_i\beta\gamma$ dissociation to lower cAMP levels and activate ERK 1/2 and DNA synthesis in muscle cells and fibroblasts *(157–159)*.

4. Growth Factor-Initiated RTK–RTK Crosstalk

Finally, EGFR and PDGFβ-R interact physically forming heterodimers and stimulation by EGF has been shown to increase the tyrosine phosphorylation of the PDGFβ-R leading to the recruitment of PI3K to the PDGFR *(160,161)*. Bagowski et al. also provided evidence for the negative regulation of EGF-induced *c-jun* transcription by PDGF-mediated phosphorylation of the EGFR, demonstrating crosstalk between different members of the RTK family *(162)*. Insulin receptors that are hormone-stimulated transactivate IGF-1 receptors *(163)*. Recently, Roudabush et al. showed that ERK 1/2 activation downstream of IGF-1R stimulation is mediated by transactivation of the EGFR in Cos7 cells proposing an IGF-1R–EGFR crosstalk pathway based on metalloprotease-induced shedding of pro-HB-EGF *(164)*.

5. Other Ligand-Induced Receptor Crosstalk

5.1. Integrins

Integrins, which are the primary link between extracellular matrix ligands and cytoskeletal structures, are a complex family of noncovalently associated heterodimeric transmembrane receptors composed of α and β subunits. They serve as both adhesive receptors and intracellular signaling mediators *(165,166)*. In addition to transmitting signals from the extracellular matrix to the intracellular environment ("outside-in" signaling), integrins can be modified by agonists that bind nonintegrin cellular receptors like growth factor receptors. This concept of "inside-out" signaling in turn regulates integrin activation and function. In addition, it has been shown that integrin activation of growth factor receptors can occur even in the absence of the growth factor *(167–169)*.

RTKs and growth factors interact spatially at multiple levels. At the plasma membrane, specific direct associations between integrins and RTKs, such as the PDGFR, EGFR, the insulin receptor, the IGF-1R and the VEGFR2, have been identified *(170–172)*. Another level of interaction between growth factor receptors and integrins is at the level of plasma membrane lipid rafts as shown with PDGFR by Baron et al. *(173,174)*. A third level of intersection between the growth factor and integrin pathways are at more downstream signaling molecules, such as focal adhesion kinase (FAK), and activation of a particular signaling cascade directly by integrins could lead to growth factor dimerization and phosphorylation/activation ultimately influencing MAPK activation *(175,176)*.

In addition to interacting with growth factor receptors, integrins also interact with GPCRs, such as the LPA receptor 3. Studies by Sengupta et al. show that laminin-induced cell migration in ovarian cancer cells is mediated by LPA via PLA2 and PI3K, revealing a new mechanism of crosstalk between a β1 integrin receptor and a GPCR *(177)*.

6. Conclusion

Signaling cascades often were considered to be discrete signaling cassettes that linked activation of a receptor to gene transcription and physiological function in a linear manner. Recent insights have broadened this view to encompass a complex network that allows multiple levels of crosstalk between the individual signaling units (stimulated by GPCR and RTK), leading to signal integration. This selective crosscommunication between different receptor classes generates common signals, including the stimulation of Ras GTPases and MAPKs, that control cell proliferation, differentiation, growth, and survival.

References

1. Bunemann, M. and Hosey, M. M. (1999) G-protein coupled receptor kinases as modulators of G-protein signalling. *J. Physiol.* **517,** 5–23.
2. McCormick, F. (1993) Signal transduction. How receptors turn Ras on. *Nature* **363,** 15–16.
3. Pierce, K. L., Luttrell, L. M., and Lefkowitz, R. J. (2001) New mechanisms in heptahelical receptor signaling to mitogen activated protein kinase cascades. *Oncogene* **20,** 1532–1539.
4. Waters, C., Pyne, S., and Pyne, N. J. (2004) The role of G-protein coupled receptors and associated proteins in receptor tyrosine kinase signal transduction. *Semin. Cell Dev. Biol.* **15,** 309–323.
5. Lowes, V. L., Ip, N. Y., and Wong, Y. H. (2002) Integration of signals from receptor tyrosine kinases and g protein-coupled receptors. *Neurosignals* **11,** 5–19.

6. Winitz, S., Russell, M., Qian, N. X., Gardner, A., Dwyer, L., and Johnson, G. L. (1993) Involvement of Ras and Raf in the Gi-coupled acetylcholine muscarinic m2 receptor activation of mitogen-activated protein (MAP) kinase kinase and MAP kinase. *J. Biol. Chem.* **268,** 19,196–19,199.
7. van Biesen, T., Hawes, B. E., Luttrell, D. K., et al. (1995) Receptor-tyrosine-kinase- and G beta gamma-mediated MAP kinase activation by a common signalling pathway. *Nature* **376,** 781–784.
8. Chen, Y., Grall, D., Salcini, A. E., Pelicci, P. G., Pouyssegur, J., and Van Obberghen-Schilling, E. (1996) Shc adaptor proteins are key transducers of mitogenic signaling mediated by the G protein-coupled thrombin receptor. *Embo J.* **15,** 1037–1044.
9. Bouvier, M. (2001) Oligomerization of G-protein-coupled transmitter receptors. *Nat. Rev. Neurosci.* **2,** 274–286.
10. Angers, S., Salahpour, A., Joly, E., et al. (2000) Detection of beta 2-adrenergic receptor dimerization in living cells using bioluminescence resonance energy transfer (BRET) *Proc. Natl. Acad. Sci. USA* **97,** 3684–3689.
11. Jones, K. A., Borowsky, B., Tamm, J. A., et al. (1998) GABA(B) receptors function as a heteromeric assembly of the subunits GABA(B)R1 and GABA(B)R2. *Nature* **396,** 674–679.
12. Luttrell, L. M., Della Rocca, G. J., van Biesen, T., Luttrell, D. K., and Lefkowitz, R. J. (1997) Gbetagamma subunits mediate Src-dependent phosphorylation of the epidermal growth factor receptor. A scaffold for G protein-coupled receptor-mediated Ras activation. *J. Biol. Chem.* **272,** 4637–4644.
13. Daub, H., Weiss, F. U., Wallasch, C., and Ullrich, A. (1996) Role of transactivation of the EGF receptor in signalling by G-protein-coupled receptors. *Nature* **379,** 557–560.
14. Gelband, C. H., Zhu, M., Lu, D., et al. (1997) Functional interactions between neuronal AT1 and AT2 receptors. *Endocrinology* **138,** 2195–2198.
15. Tanaka, M., Tsuchida, S., Imai, T., et al. (1999) Vascular response to angiotensin II is exaggerated through an upregulation of AT1 receptor in AT2 knockout mice. *Biochem. Biophys. Res. Commun.* **258,** 194–198.
16. Cui, T., Nakagami, H., Iwai, M., et al. (2001) Pivotal role of tyrosine phosphatase SHP-1 in AT2 receptor-mediated apoptosis in rat fetal vascular smooth muscle cell. *Cardiovasc. Res.* **49,** 863–871.
17. Sano, M., Fukuda, K., Sato, T., et al. (2001) ERK and p38 MAPK, but not NF-kappaB, are critically involved in reactive oxygen species-mediated induction of IL-6 by angiotensin II in cardiac fibroblasts. *Circ. Res.* **89,** 661–669.
18. Kim, H. E., Dalal, S. S., Young, E., Legato, M. J., Weisfeldt, M. L., and D'Armiento, J. (2000) Disruption of the myocardial extracellular matrix leads to cardiac dysfunction. *J. Clin. Invest.* **106,** 857–866.
19. Booz, G. W., Day, J. N., and Baker, K. M. (2002) Interplay between the cardiac renin angiotensin system and JAK-STAT signaling: role in cardiac hypertrophy, ischemia/reperfusion dysfunction, and heart failure. *J. Mol. Cell Cardiol.* **34,** 1443–1453.
20. Ushio-Fukai, M., Hilenski, L., Santanam, N., et al. (2001) Cholesterol depletion inhibits epidermal growth factor receptor transactivation by angiotensin II in vas-

cular smooth muscle cells: role of cholesterol-rich microdomains and focal adhesions in angiotensin II signaling. *J. Biol. Chem.* **276,** 48,269–48,275.

21. Eguchi, S., and Inagami, T. (2000) Signal transduction of angiotensin II type 1 receptor through receptor tyrosine kinase. *Regul. Pept.* **91,** 13–20.

22. Saito, Y., and Berk, B. C. (2001) Transactivation: a novel signaling pathway from angiotensin II to tyrosine kinase receptors. *J. Mol. Cell Cardiol.* **33,** 3–7.

23. Eguchi, S., Dempsey, P. J., Frank, G. D., Motley, E. D., and Inagami, T. (2001) Activation of MAPKs by angiotensin II in vascular smooth muscle cells. Metalloprotease-dependent EGF receptor activation is required for activation of ERK and p38 MAPK but not for JNK. *J. Biol. Chem.* **276,** 7957–7962.

24. Eguchi, S., Numaguchi, K., Iwasaki, H., et al. (1998) Calcium-dependent epidermal growth factor receptor transactivation mediates the angiotensin II-induced mitogen-activated protein kinase activation in vascular smooth muscle cells. *J. Biol. Chem.* **273,** 8890–8896.

25. Voisin, L., Foisy, S., Giasson, E., Lambert, C., Moreau, P., and Meloche, S. (2002) EGF receptor transactivation is obligatory for protein synthesis stimulation by G protein-coupled receptors. *Am. J. Physiol. Cell Physiol.* **283,** C446–C455.

26. Shah, B. H., and Catt, K. J. (2002) Calcium-independent activation of extracellularly regulated kinases 1 and 2 by angiotensin II in hepatic C9 cells: roles of protein kinase Cdelta, Src/proline-rich tyrosine kinase **2,** and epidermal growth receptor trans-activation. *Mol. Pharmacol.* **61,** 343–351.

27. Murasawa, S., Mori, Y., Nozawa, Y., et al. (1998) Angiotensin II type 1 receptor-induced extracellular signal-regulated protein kinase activation is mediated by Ca2+/calmodulin-dependent transactivation of epidermal growth factor receptor. *Circ. Res.* **82,** 1338–1348.

28. Wang, D., Yu, X., Cohen, R. A., and Brecher, P. (2000) Distinct effects of N-acetylcysteine and nitric oxide on angiotensin II-induced epidermal growth factor receptor phosphorylation and intracellular Ca(2+) levels. *J. Biol. Chem.* **275,** 12,223–12,230.

29. Konishi, A. and Berk, B. C. (2003) Epidermal growth factor receptor transactivation is regulated by glucose in vascular smooth muscle cells. *J. Biol. Chem.* **278,** 35,049–35,056.

30. Ishida, M., Ishida, T., Thomas, S. M., and Berk, B. C. (1998) Activation of extracellular signal-regulated kinases (ERK1/2) by angiotensin II is dependent on c-Src in vascular smooth muscle cells. *Circ. Res.* **82,** 7–12.

31. Sadoshima, J. and Izumo, S. (1996) The heterotrimeric G q protein-coupled angiotensin II receptor activates p21 ras via the tyrosine kinase-Shc-Grb2-Sos pathway in cardiac myocytes. *Embo J.* **15,** 775–787.

32. Suarez, C., Diaz-Torga, G., Gonzalez-Iglesias, A., et al. (2003) Angiotensin II phosphorylation of extracellular signal-regulated kinases in rat anterior pituitary cells. *Am. J. Physiol. Endocrinol. Metab.* **285,** E645–E653.

33. Li, X., Lee, J. W., Graves, L. M., and Earp, H. S. (1998) Angiotensin II stimulates ERK via two pathways in epithelial cells: protein kinase C suppresses a G-protein coupled receptor-EGF receptor transactivation pathway. *EMBO J.* **17,** 2574–2583.

34. Eguchi, S., Iwasaki, H., Ueno, H., et al. (1999) Intracellular signaling of angiotensin II-induced p70 S6 kinase phosphorylation at Ser(411) in vascular smooth muscle cells. Possible requirement of epidermal growth factor receptor, Ras, extracellular signal-regulated kinase, and Akt. *J. Biol. Chem.* **274,** 36,843–36,851.

35. Keely, S. J., Calandrella, S. O., and Barrett, K. E. (2000) Carbachol-stimulated transactivation of epidermal growth factor receptor and mitogen-activated protein kinase in T(84) cells is mediated by intracellular ca(2+), PYK-2, and p60(src) *J. Biol. Chem.* **275,** 12,619–12,625.

36. Soltoff, S. P. (1998) Related adhesion focal tyrosine kinase and the epidermal growth factor receptor mediate the stimulation of mitogen-activated protein kinase by the G-protein-coupled P2Y2 receptor. Phorbol ester or [Ca2+]i elevation can substitute for receptor activation. *J. Biol. Chem.* **273,** 23,110–23,117.

37. Eguchi, S., Iwasaki, H., Inagami, T., et al. (1999) Involvement of PYK2 in angiotensin II signaling of vascular smooth muscle cells. *Hypertension* **33,** 201–206.

38. Luttrell, L. M., Ferguson, S. S., Daaka, Y., et al. (1999) Beta-arrestin-dependent formation of beta2 adrenergic receptor-Src protein kinase complexes. *Science* **283,** 655–661.

39. Kodama, H., Fukuda, K., Pan, J., et al. (1998) Biphasic activation of the JAK/STAT pathway by angiotensin II in rat cardiomyocytes. *Circ. Res.* **82,** 244–250.

40. Marrero, M. B., Schieffer, B., Li, B., Sun, J., Harp, J. B., and Ling, B. N. (1997) Role of Janus kinase/signal transducer and activator of transcription and mitogen-activated protein kinase cascades in angiotensin II- and platelet-derived growth factor-induced vascular smooth muscle cell proliferation. *J. Biol. Chem.* **272,** 24,684–24,690.

41. Marrero, M. B., Schieffer, B., Paxton, W. G., et al. (1995) Direct stimulation of Jak/STAT pathway by the angiotensin II AT1 receptor. *Nature* **375,** 247–250.

42. Yamauchi, T., Ueki, K., Tobe, K., et al. (1997) Tyrosine phosphorylation of the EGF receptor by the kinase Jak2 is induced by growth hormone. *Nature* **390,** 91–96.

43. Berry, C., Hamilton, C. A., Brosnan, M. J., et al. (2000) Investigation into the sources of superoxide in human blood vessels: angiotensin II increases superoxide production in human internal mammary arteries. *Circulation* **101,** 2206–2212.

44. Touyz, R. M. and Schiffrin, E. L. (1999) Ang II-stimulated superoxide production is mediated via phospholipase D in human vascular smooth muscle cells. *Hypertension* **34,** 976–982.

45. Frank, G. D., Eguchi, S., Inagami, T., and Motley, E. D. (2001) *N*-acetylcysteine inhibits angiotensin ii-mediated activation of extracellular signal-regulated kinase and epidermal growth factor receptor. *Biochem. Biophys. Res. Commun.* **280,** 1116–1119.

46. Griendling, K. K., Sorescu, D., and Ushio-Fukai, M. (2000) NAD(P)H oxidase: role in cardiovascular biology and disease. *Circ. Res.* **86,** 494–501.

47. Ushio-Fukai, M., Griendling, K. K., Becker, P. L., Hilenski, L., Halleran, S., and Alexander, R. W. (2001) Epidermal growth factor receptor transactivation by angiotensin II requires reactive oxygen species in vascular smooth muscle cells. *Arterioscler. Thromb. Vasc. Biol.* **21,** 489–495.

48. Rao, G. N. (1996) Hydrogen peroxide induces complex formation of SHC-Grb2-SOS with receptor tyrosine kinase and activates Ras and extracellular signal-regulated protein kinases group of mitogen-activated protein kinases. *Oncogene* **13,** 713–719.

49. Touyz, R. M., Cruzado, M., Tabet, F., Yao, G., Salomon, S., and Schiffrin, E. L. (2003) Redox-dependent MAP kinase signaling by Ang II in vascular smooth muscle cells: role of receptor tyrosine kinase transactivation. *Can. J. Physiol. Pharmacol.* **81,** 159–167.

50. Seta, K. and Sadoshima, J. (2003) Phosphorylation of tyrosine 319 of the angiotensin II type 1 receptor mediates angiotensin II-induced trans-activation of the epidermal growth factor receptor. *J. Biol. Chem.* **278,** 9019–9026.

51. Prenzel, N., Zwick, E., Daub, H., et al. (1999) EGF receptor transactivation by G-protein-coupled receptors requires metalloproteinase cleavage of proHB-EGF. *Nature* **402,** 884–888.

52. Saito, S., Frank, G. D., Motley, E. D., et al. (2002) Metalloprotease inhibitor blocks angiotensin II-induced migration through inhibition of epidermal growth factor receptor transactivation. *Biochem. Biophys. Res. Commun.* **294,** 1023–1029.

53. Rouet-Benzineb, P., Gontero, B., Dreyfus, P., and Lafuma, C. (2000) Angiotensin II induces nuclear factor-kappa B activation in cultured neonatal rat cardiomyocytes through protein kinase C signaling pathway. *J. Mol. Cell Cardiol.* **32,** 1767–1778.

54. Suzuki, M., Raab, G., Moses, M. A., Fernandez, C. A., and Klagsbrun, M. (1997) Matrix metalloproteinase-3 releases active heparin-binding EGF-like growth factor by cleavage at a specific juxtamembrane site. *J. Biol. Chem.* **272,** 31,730–31,737.

55. Asakura, M., Kitakaze, M., Takashima, S., et al. (2002) Cardiac hypertrophy is inhibited by antagonism of ADAM12 processing of HB-EGF: metalloproteinase inhibitors as a new therapy. *Nat. Med.* **8,** 35–40.

56. Hao, L., Du, M., Lopez-Campistrous, A., and Fernandez-Patron, C. (2004) Agonist-induced activation of matrix metalloproteinase-7 promotes vasoconstriction through the epidermal growth factor-receptor pathway. *Circ. Res.* **94,** 68–76.

57. Frank, G. D., Mifune, M., Inagami, T., et al. (2003) Distinct mechanisms of receptor and nonreceptor tyrosine kinase activation by reactive oxygen species in vascular smooth muscle cells: role of metalloprotease and protein kinase C-delta. *Mol. Cell Biol.* **23,** 1581–1589.

58. Lin, J. and Freeman, M. R. (2003) Transactivation of ErbB1 and ErbB2 receptors by angiotensin II in normal human prostate stromal cells. *Prostate* **54,** 1–7.

59. Heeneman, S., Haendeler, J., Saito, Y., Ishida, M., and Berk, B. C. (2000) Angiotensin II induces transactivation of two different populations of the platelet-derived growth factor beta receptor. Key role for the p66 adaptor protein Shc. *J. Biol. Chem.* **275,** 15,926–15,932.

60. Linseman, D. A., Benjamin, C. W., and Jones, D. A. (1995) Convergence of angiotensin II and platelet-derived growth factor receptor signaling cascades in vascular smooth muscle cells. *J. Biol. Chem.* **270,** 12,563–12,568.

61. Abe, J., Deguchi, J., Matsumoto, T., et al. (1997) Stimulated activation of platelet-derived growth factor receptor in vivo in balloon-injured arteries: a link between angiotensin II and intimal thickening. *Circulation* **96,** 1906–1913.

62. Mondorf, U. F., Geiger, H., Herrero, M., Zeuzem, S., and Piiper, A. (2000) Involvement of the platelet-derived growth factor receptor in angiotensin II-induced activation of extracellular regulated kinases 1 and 2 in human mesangial cells. *FEBS Lett* **472**, 129–132.

63. Conway, A. M., Rakhit, S., Pyne, S., and Pyne, N. J. (1999) Platelet-derived-growth-factor stimulation of the p42/p44 mitogen-activated protein kinase pathway in airway smooth muscle: role of pertussis-toxin-sensitive G-proteins, c-Src tyrosine kinases and phosphoinositide 3-kinase. *Biochem. J.* **337**, 171–177.

64. Li, X., Ponten, A., Aase, K., et al. (2000) PDGF-C is a new protease-activated ligand for the PDGF alpha-receptor. *Nat. Cell Biol.* **2**, 302–309.

65. Du, J., Sperling, L. S., Marrero, M. B., Phillips, L., and Delafontaine, P. (1996) G-protein and tyrosine kinase receptor cross-talk in rat aortic smooth muscle cells: thrombin- and angiotensin II-induced tyrosine phosphorylation of insulin receptor substrate-1 and insulin-like growth factor 1 receptor. *Biochem. Biophys. Res. Commun.* **218**, 934–939.

66. Zahradka, P., Litchie, B., Storie, B., and Helwer, G. (2004) Transactivation of the insulin-like growth factor-I receptor by angiotensin II mediates downstream signaling from the angiotensin II type 1 receptor to phosphatidylinositol 3-kinase. *Endocrinology* **145**, 2978–2987.

67. Velloso, L. A., Folli, F., Sun, X. J., White, M. F., Saad, M. J., and Kahn, C. R. (1996) Cross-talk between the insulin and angiotensin signaling systems. *Proc. Natl. Acad. Sci. USA* **93**, 12,490–12,495.

68. Folli, F., Kahn, C. R., Hansen, H., Bouchie, J. L., and Feener, E. P. (1997) Angiotensin II inhibits insulin signaling in aortic smooth muscle cells at multiple levels. A potential role for serine phosphorylation in insulin/angiotensin II crosstalk. *J. Clin. Invest.* **100**, 2158–2169.

69. Jalink, K., Hordijk, P. L., and Moolenaar, W. H. (1994) Growth factor-like effects of lysophosphatidic acid, a novel lipid mediator. *Biochim. Biophys. Acta.* **1198**, 185–196.

70. Kranenburg, O. and Moolenaar, W. H. (2001) Ras-MAP kinase signaling by lysophosphatidic acid and other G protein-coupled receptor agonists. *Oncogene* **20**, 1540–1546.

71. Fukushima, N. and Chun, J. (2001) The LPA receptors. *Prostaglandins Other Lipid Mediat.* **64**, 21–32.

72. Dikic, I., Tokiwa, G., Lev, S., Courtneidge, S. A., and Schlessinger, J. (1996) A role for Pyk2 and Src in linking G-protein-coupled receptors with MAP kinase activation. *Nature* **383**, 547–550.

73. Chen, Y. H., Pouyssegur, J., Courtneidge, S. A., and Van Obberghen-Schilling, E. (1994) Activation of Src family kinase activity by the G protein-coupled thrombin receptor in growth-responsive fibroblasts. *J. Biol. Chem.* **269**, 27,372–27,377.

74. Daub, H., Wallasch, C., Lankenau, A., Herrlich, A., and Ullrich, A. (1997) Signal characteristics of G protein-transactivated EGF receptor. *EMBO J.* **16**, 7032–7044.

75. Cunnick, J. M., Dorsey, J. F., Munoz-Antonia, T., Mei, L., and Wu, J. (2000) Requirement of SHP2 binding to Grb2-associated binder-1 for mitogen-activated protein kinase activation in response to lysophosphatidic acid and epidermal growth factor. *J. Biol. Chem.* **275**, 13,842–13,848.

76. Sekharam, M., Cunnick, J. M., and Wu, J. (2000) Involvement of lipoxygenase in lysophosphatidic acid-stimulated hydrogen peroxide release in human HaCaT keratinocytes. *Biochem. J.* **346 Pt 3,** 751–758.

77. Chen, Q., Olashaw, N., and Wu, J. (1995) Participation of reactive oxygen species in the lysophosphatidic acid-stimulated mitogen-activated protein kinase kinase activation pathway. *J. Biol. Chem.* **270,** 28,499–28,502.

78. Bae, Y. S., Kang, S. W., Seo, M. S., et al. (1997) Epidermal growth factor (EGF)-induced generation of hydrogen peroxide. Role in EGF receptor-mediated tyrosine phosphorylation. *J. Biol. Chem.* **272,** 217–221.

79. Cunnick, J. M., Dorsey, J. F., Standley, T., et al. (1998) Role of tyrosine kinase activity of epidermal growth factor receptor in the lysophosphatidic acid-stimulated mitogen-activated protein kinase pathway. *J. Biol. Chem.* **273,** 14,468–14,475.

80. Hirota, K., Murata, M., Itoh, T., Yodoi, J., and Fukuda, K. (2001) An endogenous redox molecule, thioredoxin, regulates transactivation of epidermal growth factor receptor and activation of NF-kappaB by lysophosphatidic acid. *FEBS Lett.* **489,** 134–138.

81. Hirata, M., Umata, T., Takahashi, T., et al. (2001) Identification of serum factor inducing ectodomain shedding of proHB-EGF and studies of noncleavable mutants of proHB-EGF. *Biochem. Biophys. Res. Commun.* **283,** 915–922.

82. Umata, T., Hirata, M., Takahashi, T., et al. (2001) A dual signaling cascade that regulates the ectodomain shedding of heparin-binding epidermal growth factor-like growth factor. *J. Biol. Chem.* **276,** 30,475–30,482.

83. Liu, Z. and Armant, D. R. (2004) Lysophosphatidic acid regulates murine blastocyst development by transactivation of receptors for heparin-binding EGF-like growth factor. *Exp. Cell Res.* **296,** 317–326.

84. Moughal, N. A., Waters, C., Sambi, B., Pyne, S., and Pyne, N. J. (2004) Nerve growth factor signaling involves interaction between the Trk A receptor and lysophosphatidate receptor 1 systems: nuclear translocation of the lysophosphatidate receptor 1 and Trk A receptors in pheochromocytoma 12 cells. *Cell Signal* **16,** 127–136.

85. Wang, L., Cummings, R., Zhao, Y., et al. (2003) Involvement of phospholipase D2 in lysophosphatidate-induced transactivation of platelet-derived growth factor receptor-beta in human bronchial epithelial cells. *J. Biol. Chem.* **278,** 39,931–39,940.

86. Casas-Gonzalez, P., Ruiz-Martinez, A., and Garcia-Sainz, J. A. (2003) Lysophosphatidic acid induces alpha1B-adrenergic receptor phosphorylation through G beta gamma, phosphoinositide 3-kinase, protein kinase C and epidermal growth factor receptor transactivation. *Biochim. Biophys. Acta.* **1633,** 75–83.

87. Mickley, E. J., Gray, G. A., and Webb, D. J. (1997) Activation of endothelin ETA receptors masks the constrictor role of endothelin ETB receptors in rat isolated small mesenteric arteries. *Br. J. Pharmacol.* **120,** 1376–1382.

88. Iwasaki, H., Eguchi, S., Marumo, F., and Hirata, Y. (1998) Endothelin-1 stimulates DNA synthesis of vascular smooth-muscle cells through transactivation of epidermal growth factor receptor. *J. Cardiovasc. Pharmacol.* **31(Suppl 1),** S182–S184.

89. Iwasaki, H., Eguchi, S., Ueno, H., Marumo, F., and Hirata, Y. (1999) Endothelin-mediated vascular growth requires p42/p44 mitogen-activated protein kinase and p70 S6 kinase cascades via transactivation of epidermal growth factor receptor. *Endocrinology* **140,** 4659–4668.

90. Hua, H., Munk, S., and Whiteside, C. I. (2003) Endothelin-1 activates mesangial cell ERK1/2 via EGF-receptor transactivation and caveolin-1 interaction. *Am. J. Physiol. Renal. Physiol.* **284,** F303–F312.

91. Sihvola, R. K., Pulkkinen, V. P., Koskinen, P. K., and Lemstrom, K. B. (2002) Crosstalk of endothelin-1 and platelet-derived growth factor in cardiac allograft arteriosclerosis. *J. Am. Coll. Cardiol.* **39,** 710–717.

92. Chu, L., Takahashi, R., Norota, I., et al. (2003) Signal transduction and Ca2+ signaling in contractile regulation induced by crosstalk between endothelin-1 and norepinephrine in dog ventricular myocardium. *Circ. Res.* **92,** 1024–1032.

93. Garcia-Sainz, J. A., Vazquez-Prado, J., and del Carmen Medina, L. (2000) Alpha 1-adrenoceptors: function and phosphorylation. *Eur. J. Pharmacol.* **389,** 1–12.

94. Vazquez-Prado, J., Medina, L. C., and Garcia-Sainz, J. A. (1997) Activation of endothelin ETA receptors induces phosphorylation of alpha1b-adrenoreceptors in Rat-1 fibroblasts. *J. Biol. Chem.* **272,** 27,330–27,337.

95. Blaukat, A., Barac, A., Cross, M. J., Offermanns, S., and Dikic, I. (2000) G protein-coupled receptor-mediated mitogen-activated protein kinase activation through cooperation of Galpha(q) and Galpha(i) signals. *Mol. Cell Biol.* **20,** 6837–6848.

96. Hanke, S., Nurnberg, B., Groll, D. H., and Liebmann, C. (2001) Cross talk between beta-adrenergic and bradykinin B(2) receptors results in cooperative regulation of cyclic AMP accumulation and mitogen-activated protein kinase activity. *Mol. Cell Biol.* **21,** 8452–8460.

97. Medina, L. C., Vazquez-Prado, J., Torres-Padilla, M. E., Mendoza-Mendoza, A., Cruz Munoz, M. E., and Garcia-Sainz, J. A. (1998) Crosstalk: phosphorylation of alpha1b-adrenoceptors induced through activation of bradykinin B2 receptors. *FEBS Lett.* **422,** 141–145.

98. Czubayko, U. and Reiser, G. (1996) Desensitization of P2U receptor in neuronal cell line. Different control by the agonists ATP and UTP, as demonstrated by single-cell Ca2+ responses. *Biochem. J.* **320,** 215–219.

99. Quitterer, U. and Lohse, M. J. (1999) Crosstalk between Galpha(i)- and Galpha(q)-coupled receptors is mediated by Gbetagamma exchange. *Proc. Natl. Acad. Sci. USA* **96,** 10,626–10,631.

100. Schindelholz, B. and Reber, B. F. (1997) Bradykinin-induced collapse of rat pheochromocytoma (PC12) cell growth cones: a role for tyrosine kinase activity. *J. Neurosci.* **17,** 8391–8401.

101. Thuringer, D., Maulon, L., and Frelin, C. (2002) Rapid transactivation of the vascular endothelial growth factor receptor KDR/Flk-1 by the bradykinin B2 receptor contributes to endothelial nitric-oxide synthase activation in cardiac capillary endothelial cells. *J. Biol. Chem.* **277,** 2028–2032.

102. Miura, S., Matsuo, Y., and Saku, K. (2003) Transactivation of KDR/Flk-1 by the B2 receptor induces tube formation in human coronary endothelial cells. *Hypertension* **41,** 1118–1123.

103. Barki-Harrington, L. and Daaka, Y. (2001) Bradykinin induced mitogenesis of androgen independent prostate cancer cells. *J. Urol.* **165**, 2121–2125.

104. Grewal, J. S., Luttrell, L. M., and Raymond, J. R. (2001) G protein-coupled receptors desensitize and down-regulate epidermal growth factor receptors in renal mesangial cells. *J. Biol. Chem.* **276**, 27,335–27,344.

105. Adomeit, A., Graness, A., Gross, S., Seedorf, K., Wetzker, R., and Liebmann, C. (1999) Bradykinin B(2) receptor-mediated mitogen-activated protein kinase activation in COS-7 cells requires dual signaling via both protein kinase C pathway and epidermal growth factor receptor transactivation. *Mol. Cell Biol.* **19**, 5289–5297.

106. Mukhin, Y. V., Garnovsky, E. A., Ullian, M. E., and Garnovskaya, M. N. (2003) Bradykinin B2 receptor activates extracellular signal-regulated protein kinase in mIMCD-3 cells via epidermal growth factor receptor transactivation. *J. Pharmacol. Exp. Ther.* **304**, 968–977.

107. Zwick, E., Wallasch, C., Daub, H., and Ullrich, A. (1999) Distinct calcium-dependent pathways of epidermal growth factor receptor transactivation and PYK2 tyrosine phosphorylation in PC12 cells. *J. Biol. Chem.* **274**, 20,989–20,996.

108. Zwick, E., Daub, H., Aoki, N., et al. (1997) Critical role of calcium- dependent epidermal growth factor receptor transactivation in PC12 cell membrane depolarization and bradykinin signaling. *J. Biol. Chem.* **272**, 24,767–24,770.

109. Graness, A., Hanke, S., Boehmer, F. D., Presek, P., and Liebmann, C. (2000) Protein-tyrosine-phosphatase-mediated epidermal growth factor (EGF) receptor transinactivation and EGF receptor-independent stimulation of mitogen-activated protein kinase by bradykinin in A431 cells. *Biochem. J.* **347**, 441–447.

110. Endo, A., Nagashima, K., Kurose, H., Mochizuki, S., Matsuda, M., and Mochizuki, N. (2002) Sphingosine 1-phosphate induces membrane ruffling and increases motility of human umbilical vein endothelial cells via vascular endothelial growth factor receptor and CrkII. *J. Biol. Chem.* **277**, 23,747–23,754.

111. Tanimoto, T., Jin, Z. G., and Berk, B. C. (2002) Transactivation of vascular endothelial growth factor (VEGF) receptor Flk-1/KDR is involved in sphingosine 1-phosphate-stimulated phosphorylation of Akt and endothelial nitric-oxide synthase (eNOS) *J. Biol. Chem.* **277**, 42,997–43,001.

112. Baudhuin, L. M., Jiang, Y., Zaslavsky, A., Ishii, I., Chun, J., and Xu, Y. (2004) S1P3-mediated Akt activation and cross-talk with platelet-derived growth factor receptor (PDGFR) *Faseb. J.* **18**, 341–343.

113. Kim, J. H., Song, W. K., and Chun, J. S. (2000) Sphingosine 1-phosphate activates Erk-1/-2 by transactivating epidermal growth factor receptor in rat-2 cells. *IUBMB Life* **50**, 119–124.

114. Chan, A. K., Kalmes, A., Hawkins, S., Daum, G., and Clowes, A. W. (2003) Blockade of the epidermal growth factor receptor decreases intimal hyperplasia in balloon-injured rat carotid artery. *J. Vasc. Surg.* **37**, 644–649.

115. Bobe, R., Yin, X., Roussanne, M. C., et al. (2003) Evidence for ERK1/2 activation by thrombin that is independent of EGFR transactivation. *Am. J. Physiol.l* **285**, H745–H754.

116. Sabri, A., Guo, J., Elouardighi, H., Darrow, A. L., Andrade-Gordon, P., and Steinberg, S. F. (2003) Mechanisms of protease-activated receptor-4 actions in cardiomyocytes. Role of Src tyrosine kinase. *J. Biol. Chem.* **278**, 11,714–11,720.

117. Sabri, A., Short, J., Guo, J., and Steinberg, S. F. (2002) Protease-activated receptor-1-mediated DNA synthesis in cardiac fibroblast is via epidermal growth factor receptor transactivation: distinct PAR-1 signaling pathways in cardiac fibroblasts and cardiomyocytes. *Circ. Res.* **91,** 532–539.

118. Tokunou, T., Ichiki, T., Takeda, K., Funakoshi, Y., Iino, N., and Takeshita, A. (2001) cAMP response element-binding protein mediates thrombin-induced proliferation of vascular smooth muscle cells. *Arterioscler. Thromb. Vasc. Biol* **21,** 1764–1769.

119. Kanda, Y., Mizuno, K., Kuroki, Y., and Watanabe, Y. (2001) Thrombin-induced p38 mitogen-activated protein kinase activation is mediated by epidermal growth factor receptor transactivation pathway. *Br. J. Pharmacol.* **132,** 1657–1664.

120. Rauch, B. H., Millette, E., Kenagy, R. D., Daum, G., and Clowes, A. W. (2004) Thrombin- and factor Xa-induced DNA synthesis is mediated by transactivation of fibroblast growth factor receptor-1 in human vascular smooth muscle cells. *Circ. Res.* **94,** 340–345.

121. Rao, G. N., Delafontaine, P., and Runge, M. S. (1995) Thrombin stimulates phosphorylation of insulin-like growth factor-1 receptor, insulin receptor substrate-1, and phospholipase C-gamma 1 in rat aortic smooth muscle cells. *J. Biol. Chem.* **270,** 27,871–27,875.

122. Majesky, M. W., Daemen, M. J., and Schwartz, S. M. (1990) Alpha 1-adrenergic stimulation of platelet-derived growth factor A-chain gene expression in rat aorta. *J. Biol. Chem.* **265,** 1082–1088.

123. Piiper, A., Stryjek-Kaminska, D., Klengel, R., and Zeuzem, S. (1997) Epidermal growth factor inhibits bombesin-induced activation of phospholipase C-beta1 in rat pancreatic acinar cells. *Gastroenterology* **113,** 1747–1755.

124. Melien, O., Sandnes, D., Johansen, E. J., and Christoffersen, T. (2000) Effects of pertussis toxin on extracellular signal-regulated kinase activation in hepatocytes by hormones and receptor-independent agents: evidence suggesting a stimulatory role of G(i) proteins at a level distal to receptor coupling. *J. Cell Physiol.* **184,** 27–36.

125. Zhang, B. H., Ho, V., and Farrell, G. C. (2001) Specific involvement of G(alphai2) with epidermal growth factor receptor signaling in rat hepatocytes, and the inhibitory effect of chronic ethanol. *Biochem. Pharmacol.* **61,** 1021–1027.

126. Poppleton, H., Sun, H., Fulgham, D., Bertics, P., and Patel, T. B. (1996) Activation of Gsalpha by the epidermal growth factor receptor involves phosphorylation. *J. Biol. Chem.* **271,** 6947–6951.

127. Sun, H., Chen, Z., Poppleton, H., et al. (1997) The juxtamembrane, cytosolic region of the epidermal growth factor receptor is involved in association with alpha-subunit of Gs. *J. Biol. Chem.* **272,** 5413–5420.

128. Nair, B. G. and Patel, T. B. (1993) Regulation of cardiac adenylyl cyclase by epidermal growth factor (EGF) Role of EGF receptor protein tyrosine kinase activity. *Biochem. Pharmacol.* **46,** 1239–1245.

129. Maudsley, S., Pierce, K. L., Zamah, A. M., et al. (2000) The beta(2)-adrenergic receptor mediates extracellular signal-regulated kinase activation via assembly of a multi-receptor complex with the epidermal growth factor receptor. *J. Biol. Chem.* **275,** 9572–9580.

130. Kim, J., Ahn, S., Guo, R., and Daaka, Y. (2003) Regulation of epidermal growth factor receptor internalization by G protein-coupled receptors. *Biochemistry* **42,** 2887–2894.
131. Derrien, A., Zheng, B., Osterhout, J. L., et al. (2003) Src-mediated RGS16 tyrosine phosphorylation promotes RGS16 stability. *J. Biol. Chem.* **278,** 16,107–16,116.
132. Wan, K. F., Sambi, B. S., Frame, M., Tate, R., and Pyne, N. J. (2001) The inhibitory gamma subunit of the type 6 retinal cyclic guanosine monophosphate phosphodiesterase is a novel intermediate regulating p42/p44 mitogen-activated protein kinase signaling in human embryonic kidney 293 cells. *J. Biol. Chem.* **276,** 37,802–37,808.
133. Rosenfeldt, H. M., Hobson, J. P., Maceyka, M., et al. (2001) EDG-1 links the PDGF receptor to Src and focal adhesion kinase activation leading to lamellipodia formation and cell migration. *FASEB J.* **15,** 2649–2659.
134. Freedman, N. J., Kim, L. K., Murray, J. P., et al. (2002) Phosphorylation of the platelet-derived growth factor receptor-beta and epidermal growth factor receptor by G protein-coupled receptor kinase-2. Mechanisms for selectivity of desensitization. *J. Biol. Chem.* **277,** 48,261–48,269.
135. Kreuzer, J., Viedt, C., Brandes, R. P., et al. (2003) Platelet-derived growth factor activates production of reactive oxygen species by NAD(P)H oxidase in smooth muscle cells through Gi1,2. *FASEB J.* **17,** 38–40.
136. Hobson, J. P., Rosenfeldt, H. M., Barak, L. S., et al. (2001) Role of the sphingosine-1-phosphate receptor EDG-1 in PDGF-induced cell motility. *Science* **291,** 1800–1803.
137. Alderton, F., Rakhit, S., Kong, K. C., et al. (2001) Tethering of the platelet-derived growth factor beta receptor to G-protein-coupled receptors. A novel platform for integrative signaling by these receptor classes in mammalian cells. *J. Biol. Chem.* **276,** 28,578–28,585.
138. Cho, H., Harrison, K., Schwartz, O., and Kehrl, J. H. (2003) The aorta and heart differentially express RGS (regulators of G-protein signalling) proteins that selectively regulate sphingosine 1-phosphate, angiotensin II and endothelin-1 signalling. *Biochem. J.* **371,** 973–980.
139. Rakhit, S., Pyne, S., and Pyne, N. J. (2001) Nerve growth factor stimulation of p42/p44 mitogen-activated protein kinase in PC12 cells: role of G(i/o), G protein-coupled receptor kinase **2,** beta-arrestin I, and endocytic processing. *Mol. Pharmacol.* **60,** 63–70.
140. Lou, X., Yano, H., Lee, F., Chao, M. V., and Farquhar, M. G. (2001) GIPC and GAIP form a complex with TrkA: a putative link between G protein and receptor tyrosine kinase pathways. *Mol. Biol. Cell* **12,** 615–627.
141. Fedorov, Y. V., Jones, N. C., and Olwin, B. B. (1998) Regulation of myogenesis by fibroblast growth factors requires beta-gamma subunits of pertussis toxin-sensitive G proteins. *Mol. Cell Biol.* **18,** 5780–5787.
142. Xu, C. B., Zhang, Y., Stenman, E., and Edvinsson, L. (2002) D-erythro-N,N-dimethylsphingosine inhibits bFGF-induced proliferation of cerebral, aortic and coronary smooth muscle cells. *Atherosclerosis* **164,** 237–243.

143. Krieger-Brauer, H. I., Medda, P., and Kather, H. (2000) Basic fibroblast growth factor utilizes both types of component subunits of Gs for dual signaling in human adipocytes. Stimulation of adenylyl cyclase via Galph(s) and inhibition of NADPH oxidase by Gbeta gamma(s) *J. Biol. Chem.* **275,** 35,920–35,925.

144. Ferrara, N. (1996) Vascular endothelial growth factor. *Eur. J. Cancer* 32A, 2413–2422.

145. Risau, W. (1997) Mechanisms of angiogenesis. *Nature* **386,** 671–674.

146. Petrova, T. V., Makinen, T., and Alitalo, K. (1999) Signaling via vascular endothelial growth factor receptors. *Exp. Cell Res.* **253,** 117–130.

147. Neufeld, G., Cohen, T., Gengrinovitch, S., and Poltorak, Z. (1999) Vascular endothelial growth factor (VEGF) and its receptors. *FASEB J.* **13,** 9–22.

148. Migdal, M., Huppertz, B., Tessler, S., et al. (1998) Neuropilin-1 is a placenta growth factor-2 receptor. *J. Biol. Chem.* **273,** 22,272–22,278.

149. Makinen, T., Olofsson, B., Karpanen, T., et al. (1999) Differential binding of vascular endothelial growth factor B splice and proteolytic isoforms to neuropilin-1. *J. Biol. Chem.* **274,** 21,217–21,222.

150. Soker, S., Takashima, S., Miao, H. Q., Neufeld, G., and Klagsbrun, M. (1998) Neuropilin-1 is expressed by endothelial and tumor cells as an isoform-specific receptor for vascular endothelial growth factor. *Cell* **92,** 735–745.

151. Cohen, T., Gitay-Goren, H., Sharon, R., et al. (1995) VEGF121, a vascular endothelial growth factor (VEGF) isoform lacking heparin binding ability, requires cell-surface heparan sulfates for efficient binding to the VEGF receptors of human melanoma cells. *J. Biol. Chem.* **270,** 11,322–11,326.

152. Zeng, H., Zhao, D., and Mukhopadhyay, D. (2002) KDR stimulates endothelial cell migration through heterotrimeric G protein Gq/11-mediated activation of a small GTPase RhoA. *J. Biol. Chem.* **277,** 46,791–46,798.

153. Zeng, H., Zhao, D., Yang, S., Datta, K., and Mukhopadhyay, D. (2003) Heterotrimeric G alpha q/G alpha 11 proteins function upstream of vascular endothelial growth factor (VEGF) receptor-2 (KDR) phosphorylation in vascular permeability factor/VEGF signaling. *J. Biol. Chem.* **278,** 20,738–20,745.

154. Zeng, H., Zhao, D., and Mukhopadhyay, D. (2002) Flt-1-mediated down-regulation of endothelial cell proliferation through pertussis toxin-sensitive G proteins, beta gamma subunits, small GTPase CDC42, and partly by Rac-1. *J. Biol. Chem.* **277,** 4003–4009.

155. Imamura, T., Vollenweider, P., Egawa, K., et al. (1999) G alpha-q/11 protein plays a key role in insulin-induced glucose transport in 3T3-L1 adipocytes. *Mol. Cell Biol.* **19,** 6765–6774.

156. Sanchez-Margalet, V., Gonzalez-Yanes, C., Santos-Alvarez, J., and Najib, S. (1999) Insulin activates G alpha IL-2 protein in rat hepatoma (HTC) cell membranes. *Cell Mol. Life Sci.* **55,** 142–147.

157. Profrock, A., Schnefel, S., and Schulz, I. (1991) Receptors for insulin interact with Gi-proteins and for epidermal growth factor with Gi- and Gs-proteins in rat pancreatic acinar cells. *Biochem. Biophys. Res. Commun.* **175,** 380–386.

158. Kuemmerle, J. F. and Murthy, K. S. (2001) Coupling of the insulin-like growth factor-I receptor tyrosine kinase to Gi2 in human intestinal smooth muscle:

Gbetagamma-dependent mitogen-activated protein kinase activation and growth. *J. Biol. Chem.* **276,** 7187–7194.

159. Hallak, H., Seiler, A. E., Green, J. S., Ross, B. N., and Rubin, R. (2000) Association of heterotrimeric G(i) with the insulin-like growth factor-I receptor. Release of G(betagamma) subunits upon receptor activation. *J. Biol. Chem.* **275,** 2255–2258.

160. Liu, P., and Anderson, R. G. (1999) Spatial organization of EGF receptor transmodulation by PDGF. *Biochem. Biophys. Res. Commun.* **261,** 695–700.

161. Habib, A. A., Hognason, T., Ren, J., Stefansson, K., and Ratan, R. R. (1998) The epidermal growth factor receptor associates with and recruits phosphatidylinositol 3-kinase to the platelet-derived growth factor beta receptor. *J. Biol. Chem.* **273,** 6885–6891.

162. Bagowski, C. P., Stein-Gerlach, M., Choidas, A., and Ullrich, A. (1999) Cell-type specific phosphorylation of threonines T654 and T669 by PKD defines the signal capacity of the EGF receptor. *EMBO J.* **18,** 5567–5576.

163. Tartare, S., Ballotti, R., and Van Obberghen, E. (1991) Interaction between heterologous receptor tyrosine kinases. Hormone-stimulated insulin receptors activate unoccupied IGF-I receptors. *FEBS Lett.* **295,** 219–222.

164. Roudabush, F. L., Pierce, K. L., Maudsley, S., Khan, K. D., and Luttrell, L. M. (2000) Transactivation of the EGF receptor mediates IGF-1-stimulated shc phosphorylation and ERK1/2 activation in COS-7 cells. *J. Biol. Chem.* **275,** 22,583–22,589.

165. Giancotti, F. G. and Ruoslahti, E. (1999) Integrin signaling. *Science* **285,** 1028–1032.

166. Humphries, M. J. (2000) Integrin structure. *Biochem. Soc. Trans.* **28,** 311–339.

167. Sundberg, C. and Rubin, K. (1996) Stimulation of beta1 integrins on fibroblasts induces PDGF independent tyrosine phosphorylation of PDGF beta-receptors. *J. Cell Biol.* **132,** 741–752.

168. Chen, K. D., Li, Y. S., Kim, M., et al. (1999) Mechanotransduction in response to shear stress. Roles of receptor tyrosine kinases, integrins, and Shc. *J. Biol. Chem.* **274,** 18,393–18,400.

169. Moro, L., Venturino, M., Bozzo, C., et al. (1998) Integrins induce activation of EGF receptor: role in MAP kinase induction and adhesion-dependent cell survival. *EMBO J.* **17,** 6622–6632.

170. Falcioni, R., Antonini, A., Nistico, P., et al. (1997) Alpha 6 beta 4 and alpha 6 beta 1 integrins associate with ErbB-2 in human carcinoma cell lines. *Exp. Cell Res.* **236,** 76–85.

171. Mariotti, A., Kedeshian, P. A., Dans, M., Curatola, A. M., Gagnoux-Palacios, L., and Giancotti, F. G. (2001) EGF-R signaling through Fyn kinase disrupts the function of integrin alpha6beta4 at hemidesmosomes: role in epithelial cell migration and carcinoma invasion. *J. Cell Biol.* **155,** 447–458.

172. Schneller, M., Vuori, K., and Ruoslahti, E. (1997) Alphavbeta3 integrin associates with activated insulin and PDGFbeta receptors and potentiates the biological activity of PDGF. *EMBO J.* **16,** 5600–5607.

173. Baron, W., Decker, L., Colognato, H., and French-Constant, C. (2003) Regulation of integrin growth factor interactions in oligodendrocytes by lipid raft microdomains. *Curr. Biol.* **13,** 151–155.

174. Leitinger, B. and Hogg, N. (2002) The involvement of lipid rafts in the regulation of integrin function. *J. Cell Sci.* **115,** 963–972.

175. Sieg, D. J., Hauck, C. R., Ilic, D., et al. (2000) FAK integrates growth-factor and integrin signals to promote cell migration. *Nat. Cell Biol.* **2,** 249–256.

176. Aplin, A. E. and Juliano, R. L. (1999) Integrin and cytoskeletal regulation of growth factor signaling to the MAP kinase pathway. *J. Cell Sci.* **112,** 695–706.

177. Sengupta, S., Xiao, Y. J., and Xu, Y. (2003) A novel laminin-induced LPA autocrine loop in the migration of ovarian cancer cells. *FASEB J.* **17,** 1570–1572.

3

Protein–Protein Interactions in Signaling Cascades

Bruce J. Mayer

Summary

The process of signal transduction is dependent on specific protein–protein interactions. In many cases, these interactions are mediated by modular protein domains that confer specific binding activity to the proteins in which they are found. Rapid progress has been made in the biochemical characterization of binding interactions, the identification of binding partners, and determination of the three-dimensional structures of binding modules and their ligands. The resulting information establishes the logical framework for our current understanding of the signal transduction machinery. In this chapter, a variety of protein interaction modules that participate in signaling are discussed, and issues relating to binding specificity and the significance of a particular interaction are considered.

Key Words: Modular domains; protein ligands; signal transduction; specificity; SH2 domain; SH3 domain; bromo domain; chromo domain; PDZ domain; ankyrin repeats.

1. Introduction

Understanding how signals from the outside environment are transmitted into the cell and how those signals are interpreted and integrated within the cytosol has been a long-standing goal of biological research. The past two decades have seen tremendous progress toward this goal and have brought to light the pivotal role played by specific protein–protein interactions (reviewed in **refs. 1** and **2**). Indeed, the way in which we regard signaling mechanisms has shifted fundamentally from an earlier emphasis on the regulation of enzymes and their substrate specificities to our current focus on the regulation and specificity of protein binding interactions. We now fully appreciate that the cell is not a simple aqueous solution but instead a dense gel of interacting proteins in which the actual activity of an enzyme is as dependent on its binding partners and subcellular localization as it is on the kinetic parameters of its catalytic activity.

From: *Methods in Molecular Biology, vol. 332: Transmembrane Signaling Protocols, Second Edition*
Edited by: H. Ali and B. Haribabu © Humana Press Inc., Totowa, NJ

Early biochemical work on metabolic pathways had emphasized the concepts of pathways and cascades, in which one step leads to subsequent steps in a relatively linear fashion, often with amplification at each step. These concepts proved inadequate, however, to fully describe signal transduction mechanisms. A good example is the case of receptor tyrosine kinases. In the early 1980s, it was discovered that the receptors for many mitogenic growth factors, such as epidermal growth factor and platelet-derived growth factor, were transmembrane protein tyrosine kinases. It seemed obvious that the key to understanding signal transmission would be to identify the substrate proteins phosphorylated by the liganded receptors, which must surely include the effectors responsible for stimulating the cell to proliferate. When lysates of growth factor-stimulated cells were analyzed with phosphotyrosine (pTyr)-specific antibodies, however, a problem arose: by far the most prominent tyrosine-phosphorylated protein was the receptor itself. Clearly this finding was inconsistent with models in which the receptor initiates and amplifies a signaling cascade by phosphorylating many substrate proteins. We now know that the key to signal transmission in this case is the creation of binding sites on the receptor, via autophosphorylation, for effector proteins containing modular protein binding domains (reviewed in **ref. 3**). From such findings a new paradigm for signaling emerged, in which an enzyme's most important function can be to modify protein-binding activities in response to ligand.

When closely examined, even "classical" signaling pathways reveal the central role of specific and regulated protein–protein interactions. Among the best-understood signaling cascades are those mediated by heterotrimeric G proteins *(4,5)*. In the β-adrenergic pathway, for example, an agonist-stimulated receptor activates many molecules of a heterotrimeric G protein, each of which can then activate a molecule of the enzyme adenyl cyclase. The resulting rise in intracellular cyclic adenosine monophosphate in turn activates many molecules of protein kinase (PK) A, which then phosphorylate many intracellular proteins on serine and threonine residues. The details of this relatively simple signaling apparatus reveal at least five critical protein–protein interactions: The heterotrimeric G protein binds to the liganded (but not the unliganded) receptor; conformational changes brought about by receptor binding and concomitant guanosine diphosphate release and guanosine triphosphate (GTP) binding induce the dissociation of the α subunit of the G protein (Gα) from its β and γ subunits (Gβγ), and from the liganded receptor; the released Gα subunit binds to and activates the cyclase. Meanwhile, Gβγ binds to and relocalizes the β-adrenergic receptor kinase, leading to receptor phosphorylation and densensitization. Finally, cyclic adenosine monophosphate binding causes the dissociation of the regulatory subunit of PKA, thereby releasing the active catalytic subunit. Although enzymes (kinases, GTPases) are involved, it is obvious that changes in protein–protein interactions play a central role in signal transmission.

Much of current signal transduction research is directly or indirectly concerned with the analysis of protein–protein interactions. Indeed, such interactions are now thought to be so important to function that one of the first experiments performed on a protein may be a search for its interaction partners, and many efforts are now underway to define the global protein interaction network ("interactome") for various model organisms *(6–10)*. This chapter briefly reviews a few examples of the many well-characterized protein–protein interaction motifs known to be involved in intracellular signaling and then discusses several experimental considerations, such as the specificity and regulation of binding interactions and how the significance of a particular interaction can be assessed.

2. Examples of Protein-Binding Modules Involved in Signal Transduction

It is now clear that not only are protein–protein interactions important for signaling but that many signaling proteins contain recognizable modular domains or motifs that confer binding specificity. This finding is indeed fortunate because it allows us in many cases to predict what type of binding interactions to expect based on simple inspection of the amino acid sequence of a protein of interest. Such a modular system makes intuitive sense from an evolutionary standpoint—domains can be shuffled and existing interaction pairs fine-tuned during evolution so that specific binding surfaces need not arise *de novo* for each pair of interacting proteins.

Protein interaction modules or motifs that are recognizable by sequence similarity fall into two classes. First there are those, such as the Src homology (SH)2 and SH3 domains, which are independently folding units that confer a characteristic and specialized type of binding interaction (to tyrosine-phosphorylated peptides, for example). In such modules, the most conserved residues are those that are directly involved in binding to ligands. The amino and carboxyl termini of such domains are located close together on the same face of the domain (away from the ligand binding site), allowing the domain to be inserted into a variety of proteins with only minimal disruption of the conformation of the "host" protein. The other broad class of motifs is those, such as ankyrin or armadillo repeats, in which the sequence similarity relates to a common folded structure but does not necessarily predict the specific type of binding interaction. These motifs may be repeated many times in proteins containing them and assemble into higher order structures (in which case they are more properly termed "repeats"). Such motifs represent an evolutionary solution to the design problem of small, stable folded structures that can easily evolve to display variable surface residues capable of mediating specific binding interactions. An example that will not be discussed further is the variety of zinc-binding "fingers" that mediate specific protein–protein and protein– DNA interactions, whose compact folded structures are stabilized by metal binding.

The number and variety of protein binding domains continues to expand as genomic and proteomic information is mined. Several online databases provide an excellent starting place for exploring the diversity of protein interaction domains. These include PFAM (http://pfam.wustl.edu/), Prosite (http://us.expasy.org/prosite/), and SMART (http://smart.embl-heidelberg.de). In the following subheadings, I consider several examples of regulated and unregulated modular domains, as well as one scaffold repeat, to provide a broad overview of the range and variety of protein interaction modules involved in signaling.

2.1. SH2 Domains: Regulation by Tyrosine Phosphorylation

The discovery that SH2 domains bind specifically to tyrosine-phosphorylated peptides but not to the corresponding unphosphorylated sites emphatically underscored the importance of regulated protein–protein interactions in signal transduction. SH2 domains consist of approx 100 amino acids and were first recognized as regions of sequence similarity between the Src tyrosine kinase and other distantly related kinases (hence the name Src Homology domain 2, or SH2 *[11]*). Their importance became apparent in light of several simultaneous discoveries: that many cytosolic proteins implicated in signaling contained SH2 domains, that these proteins could often be shown to bind tightly to ligand-activated growth-factor receptors, and that bacterially expressed SH2 domains could be shown to bind to tyrosine-phosphorylated proteins including activated receptors *(12–16)*. We now know that these domains serve a general role in signaling in multicellular organisms, mediating the relocalization of SH2-containing proteins and the assembly of protein complexes in response to changes in tyrosine phosphorylation. There are approx 115 SH2 domains in the human genome, making them one of the more common protein binding modules. Because SH2 domains and true tyrosine kinases are both lacking in yeast, they must be relatively recent evolutionary innovations, presumably enlisted to deal with the greater signaling demands of multicellular life.

Much is known about the structure and binding interactions of these domains, and only a brief summary will be given here; a number of recent reviews are available *(17–19)*. Binding of SH2 domains to tyrosine- phosphorylated sites is quite tight, with measured dissociation constants (K_Ds) in the range of 10^{-8} to 10^{-7} *M (20)*, and is absolutely dependent on phosphorylation, as binding to unphosphorylated ligands is undetectable. High-affinity binding to short synthetic phosphopeptides is observed, indicating that the interaction generally is independent of the larger protein bearing the phosphorylated site. The first X-ray crystal structures of SH2 domains bound to high-affinity ligands showed that the peptide is draped along the surface of the domain, with the phosphorylated tyrosine projecting deep into a highly conserved, positively charged pocket *(21,22)*. This mode of interaction—an extended linear peptide making multiple contacts with a binding groove on the surface, often coupled with a "plug and socket" interaction providing additional specificity—has proven to be a hallmark of modular protein interaction domains.

Considerable specificity exists among SH2 domains for different phosphorylated peptide sites, and a degenerate peptide library approach allowed the binding specificities of a number of SH2 domains to be determined *(23,24)*. Specificity was found to be dependent on the three (in rare cases, up to five or six) amino acids C-terminal to the phosphorylated tyrosine, with residues N-terminal to the pTyr having little or no effect on binding *(17)*. However, all SH2 domains have a detectable affinity for pTyr itself (indeed, this can be used as a purification scheme to isolate SH2 domains), so peptide specificity is relative rather than absolute. Which SH2 domains will bind to a particular site in vivo will depend on the local concentration, as well as the relative affinities of potential binding partners. It has also been shown that some SH2 domains also can bind with modest affinity to inositol lipids phosphorylated on the 3' position *(25)*; therefore, it is worth remembering that protein binding domains might have hitherto unappreciated activities that will affect their behavior in vivo. One recent example is the SH2 domain of the Abl tyrosine kinase, which interacts in *cis* with the catalytic domain in a pTyr-independent fashion to repress kinase activity *(26,27)*.

In addition to the SH2 domain, a second modular protein domain, the pTyr binding or PTB domain, also evolved to bind specifically to tyrosine-phosphorylated peptides. The PTB was first appreciated in the Shc adaptor, when it was found that many Shc binding sites consisted of an **NPx*po*Y** motif (where **x** can be any amino acid and ***po*Y** represents pTyr) quite different from typical SH2-binding sites. It was ultimately shown that binding to these sites mapped to an approx 160-amino acid region of Shc with no sequence homology to the SH2 domain *(28–31)*. The degree of sequence similarity among known PTB domains is weak, making identification from sequence problematic; the insulin receptor substrate 1 PTB was only identified by virtue of its binding activity *(30,32)*. Apparent affinity for NPx*po*Y peptides is approx 10^{-6} *M (31,32)*, but because few PTB domains have been identified, the range of target specificities and affinities is not known. In several examples, in fact, tyrosine phosphorylation of the binding site is not required and may in fact inhibit binding *(33,34)*. The tertiary structures of the Shc and insulin receptor substrate 1 PTB domains are virtually identical to that of another protein module, the pleckstrin homology (PH) domain *(35,36)*, many of which specifically bind phosphoinositol lipids *(37)*, so one could regard PTB domains as a specialized subset of the larger PH domain family.

In addition to these PTB motifs, several other protein domains have been implicated in binding specifically to serine- or threonine-phosphorylated peptides (reviewed in **ref. 38**). These domains include 14-3-3 proteins, some WW domains, and some WD-40 repeat proteins. Unlike the case of pTyr, it appears there is no widely distributed protein module that is specifically dedicated to binding phosphoserine or phosphothreonine. The 14-3-3 proteins are a class of small, abundant proteins that generally bind to phosphorylated partners but because they are not embedded in other larger proteins, they cannot be considered modular protein-binding domains. As for the examples of other domains implicated in phosphoserine- and phosphothreonine-binding, these may be considered evolutionary specializations of larger classes of protein modules whose binding does not generally depend on phosphorylation.

2.2. Bromo and Chromo Domains: The Histone Code

Chromatin is the site of numerous regulated protein–protein interactions that serve to modulate its transcriptional activity *(39)*. These modifications include acetylation, methylation, phosphorylation, and ubiquitination. In 1999 Strahl and Allis proposed the "histone code" hypothesis, in which the pattern of covalent modification of histones dictated the recruitment and assembly of specific protein complexes, which in turn regulated chromatin activity *(40)*. This premise was soon bolstered by the discovery that a small modular protein domain, the bromo domain, bound specifically to peptides derived from histone H3 or H4 tails only when a specific lysine residue was acetylated *(41)*. The bromo domain is found in many histone acetyltransferases and transcriptional activators, and the extent of histone acetylation generally correlates with transcriptional activity. Like other modular protein binding domains, the bromo domain is often found in multiple copies and/or in association with other functional domains in proteins. Binding affinities of a single bromo domain for acetylated peptide ligands in the range of approx 40 to 400 μ*M* have been reported *(41,42)*, with higher affinities observed for binding of doubly acetylated peptides to tandem domains *(42)*.

The first structural studies of the approx 110-amino-acid domain revealed that it consists of a compact antiparallel four-helix bundle, with a deep hydrophobic pocket at the base of the bundle providing an apparent acetyl-lysine (*ac*Lys) binding site *(41,42)*. Subsequent structural studies of bromodomains complexed with acetylated peptides demonstrated that the peptide binds in extended conformation, with the *ac*Lys sidechain projecting deep into a cleft *(43,44)*. The cleft is lined by conserved hydrophobic residues that interact with the aliphatic chain of the *ac*Lys, whereas a network of hydrogen bonds are formed between the *N*-acetyl group, backbone atoms of the domain, and ordered water molecules. The positive charge and lack of hydrophobic methyl group of unacetylated lysine make its binding much less favorable, providing the basis for specificity for acetylated ligands. In addition to *ac*Lys itself, adjacent N-terminal and C-terminal ligand residues contact the domain and provide specificity for particular sites. Thus, like SH2 domains, overall specificity is determined by a pocket that recognizes a specific modified amino acid plus an extended groove contacting adjacent residues.

More recently, a second modular protein domain, the chromo domain, was found to play a similar role in specific recognition of modified histones. In this case, the domain recognizes histone tails in which specific lysine residues have been *N*-methylated, a modification associated with epigenetic silencing *(39)*. In the heterochromatin binding protein HP1 and a number of other examples, a second related domain termed the chromo shadow domain is also found. Even before its function was known, structural studies on the approx 50-residue chromo domain revealed a compact fold consisting of an antiparallel β sheet and a single α helix *(45)*. Analogy with the (structurally unrelated) bromodomain inspired several groups to test whether chromo domains specifically recognize methylated histones. It was found that full-length HP1 and its isolated chromo domain bound with high affinity to methylated histone tail peptides, but not to the unmethylated counterparts *(46,47)*. The chromo shadow domain did not bind detectably, but because this domain can mediate HP1 dimerization *(48)*, it is likely to play a role in increasing the avidity of HP1 binding to chromatin containing multiple binding sites.

Subsequent structural studies showed that trimethylated lysine (me_3Lys) is recognized via an "aromatic cage" of chromodomain residues that provides a hydrophobic environment for the methylated groups of the me_3Lys head group, as well as providing favorable amino-aromatic interactions with its nitrogen atom *(49,50)*. The depth of the methyl binding pocket also provides specificity, as no other amino acid sidechain is long enough to bind favorably. A second interesting aspect of binding is that the ligand backbone adopts a β strand conformation, and its binding induces the ordering of N-terminal chromo domain residues to complete a β-sandwich structure. This "β-augmentation" strategy is seen in a number of other peptide–domain interactions *(51)*. In the case of HP1, binding affinity for methylated peptides is quite high (approx 4 μ*M*), with virtually no affinity for unmethylated peptides *(49,50,52)*. Sequence comparison showed that among all chromo domains, the aromatic cage residues and other residues that contact ligand are conserved in only a subset of these *(49,50)*, consistent with experimental data that not all chromo domains recognize methylated substrates. Finally, the structure of the chromo domain of Polycomb, which binds a different methylated histone peptide than HP1, revealed that the ligand-binding pocket of this domain is partially formed by a dimer interface. This suggests that Polycomb dimers may serve to coordinately bind the histone tails from adjacent nucleosomes, consistent with a role in chromatin condensation and inactivation *(53)*.

2.3. SH3 Domains: Recognizing Proline

One of the best-studied examples of a modular protein domain whose binding activity is NOT regulated by covalent modification is the SH3 domain. Like the SH2 domain, this small interaction module was first identified as a region with sequence similarity to the Src kinase. SH3 domains have subsequently been found in a wide range of proteins in all eukaryotes (reviewed in **ref. 54**). Although they are often found in the same proteins as SH2 domains, this location does not reflect any structural or functional similarity in the domains themselves, but is more likely related to the frequent involvement of these domains in signal transduction complexes. Indeed, there is a class of proteins that consist entirely of these two domains, termed the SH2/SH3 adaptors, which serve as molecular "crosslinkers" to nucleate the formation of signaling complexes.

SH3 domains bind to short, proline-rich binding sites in proteins *(55–57)*. From structural studies and peptide library screens, it is known that peptide ligands consist of two turns of a left-handed polyproline-2 (PPII) helix. Surprisingly, ligands can bind in either an N-to-C- or a C-to-N-terminal orientation as a result of the pseudosymmetry of the PPII helix *(58,59)*. Most SH3 domains bind to motifs with the consensus +xΦPxΦP (class 1) or ΦPxΦPx+ (class 2), where + represents a basic residue, **x** can be any amino acid, and Φ represents a hydrophobic residue. PxxP is considered the minimal core SH3 binding site, but a few exceptions exist that do not conform to this consensus. Although considerable specificity exists among different SH3 domains for their binding sites, the differences in affinity between high- and low-affinity sites can be quite small, so it is difficult to predict *a priori* which specific sites might bind in vivo (*see* discussion of specificity in **refs. 54** and **60**). Affinities generally are quite modest, with K_D values in the range of 10^{-6} to 10^{-5} M for specific SH3–peptide interactions. However, in a number of cases, significantly higher affinities have been documented between an SH3 domain and a longer peptide or protein target, demonstrating that interactions outside the PxxP binding groove can be exploited to increase specificity.

The first and perhaps most celebrated role ascribed to the SH3 domain is in recruiting Sos, a guanine nucleotide exchange factor for Ras, to the membrane leading to Ras activation *(61)*. This activation is mediated by the SH2/SH3 adaptor Grb2, which contains one SH2 and two SH3 domains. Sos and Grb2 exist as a preformed SH3-mediated complex in the cytosol, and this complex is recruited to the membrane by binding of the Grb2 SH2 domain to tyrosine-phosphorylated sites generated by activated growth factor receptors. As in this example, SH3-mediated binding generally is not thought to be regulated directly; in most cases, these domains bind their ligands constitutively, thus functioning as an intermolecular adhesive rather than a switch. However, there are examples where phosphorylation of a protein allosterically regulates the availability of its SH3 domain and/or proline-rich target peptides. Similarly, direct phosphorylation of SH3 binding sites (such as the binding site on Pak for the Nck SH3 domain [61]) has been shown to prevent binding in some cases.

Several other protein modules also have evolved to bind specifically to proline-rich ligands, including the WW and EVH1 domains (reviewed in **ref. 62**). These domains are completely unrelated structurally to each other or to SH3 domains, yet they exploit similar features of proline-rich peptides, such as the PPII helix conformation, to drive binding. The relative hydrophobicity of proline coupled with its propensity to be found on the surface of proteins, and the relatively low entropic cost of binding owing to its lack of rotational freedom, make proline-rich peptides well suited to mediate protein–protein interactions *(63)*.

2.4. PDZ Domains

The PDZ domain is a second example of a modular domain whose binding does not depend on covalent modification of the ligand. This module is found in a wide variety of proteins, many of which localize to specialized submembranous structures such as tight junctions and postsynaptic densities, and is thought to perform a scaffolding function in assembling these structures (reviewed in **ref. 64**). The PDZ domain is unusual in that it is found not only in eukaryotes but in several prokaryotic proteins as well, suggesting it was one of the first protein interaction modules to evolve. The domain is also one of the most abundant in many eukaryotes, being found in hundreds of human proteins, and often multiple copies of the domain (in some cases more than 10) are found in a single protein. In the vast majority of cases, the binding sites for PDZ domains are at the extreme C-terminus of the target protein, with several broad classes of sites currently known: $(S/T)x\Phi$-COO$^-$ (class 1), $\Phi x\Phi$-COO$^-$ (class 2), and $(D/E)x\Phi$-COO$^-$ (class 3), where x can be any amino acid, Φ represents any hydrophobic amino acid, and COO$^-$ represents the C-terminus *(65,66)*. The terminal carboxyl group of the ligand is bound by a loop whose sequence is highly conserved among PDZ domains, which is consistent with its role as an important determinant of binding *(67)*. However, in at least one case, a PDZ domain has been shown to bind an internal site *(68)*, and sequence divergence in the carboxylate-binding loop of some PDZ domains suggests that the C-terminal carboxyl group will not be required in all cases. As previously discussed for the chromo domain, the peptide ligand binds via β-augmentation, adding a strand to an existing β sheet in the domain *(51)*. Measured affinities for ligands are quite high, with dissociation constants in the 10^{-7} *M* range *(65,69)*, comparable to SH2 domains and tighter than for most SH3 domains. PDZ domains also have been shown to interact specifically with other PDZ domains *(70)*, which is likely to be important for assembling large supramolecular structures by proteins containing these domains.

PDZ domains are thought to provide a mechanism for assembling large submembranous complexes that perform a variety of roles, such as coupling transmembrane receptors to multiple intracellular effector proteins or clustering transmembrane proteins such as ion channels into highly concentrated patches. A well-characterized example is the *Drosophila* InaD protein, which contains five PDZ domains. InaD is found in the rhabdomeres of the retina, where it serves as a scaffold to assemble proteins involved in the light response. Phospholipase C-β, eye-PKC, and the Trp calcium channel have each been shown to bind specifically to a different PDZ domain in InaD; mutation of the PDZ domains resulted in mislocalization of the corresponding binding proteins and defective light response *(71)*. Presumably, the PDZ-mediated clustering of signal response proteins is important for the very rapid and amplified response to activation of rhodopsin by a single photon.

2.5. Ankyrin Repeats: A Versatile Scaffold

Only one example will be considered of the large and diverse class of protein repeats that have been implicated in protein–protein interaction (reviewed in **ref. 72**). Such domains are widespread and are likely to have evolved rapidly via intragenic duplication and recombination. As mentioned earlier, they differ from the modular protein binding domains in that they do not function autonomously, but must assemble in multiple copies, and their presence does not predict binding to any specific class of peptide ligand. In evolution they have been widely used as a structural framework or scaffold, on which specific protein–protein interactions can evolve in an *ad hoc* fashion.

Ankyrin repeats were first identified as a motif in the membrane matrix protein ankyrin and have subsequently been identified by sequence similarity in a wide variety of proteins, including a few prokaryotic and extracellular examples *(73)*. The repeat itself consists of approx 33 residues and is always present in at least 4 (and as many as 24) tandem-repeat copies. This small size and the presence of multiple copies suggested that individual repeats are relatively unstable and that multiple repeats fold into a more stable higher order structure; this theory was confirmed by the first X-ray crystal structures of ankyrin repeat-containing proteins *(74,75)*. An individual repeat consists of a β-hairpin flanked by two antiparallel α helices oriented perpendicular to the plane of the β-hairpin. The helices of adjacent repeats organize into bundles, generating a structure reminiscent of a hand with the fingers (the β-hairpins) at right angles to the palm (the helical bundles). The structures indicate that protein–protein association is mediated by the residues in the loops at the tips of the β-hairpins and in the concave surface between the fingers and palm. This type of interaction surface is quite different from the well-defined peptide binding groove or pocket found in independently functioning modular domains, such as the SH2, SH3, or PDZ.

Many ankyrin-repeat proteins are known to participate in protein–protein interactions, with perhaps some of the best examples being ankyrin itself (which binds to the anion transporter, Na/K adenosine triphosphatase, tubulin, and the sodium channel) and the inhibitory subunits of the nuclear factor-κB family of transcription factors that bind to and inhibit the activity of the DNA-binding subunits. Other repeat structures, such as WD-40 repeats, armadillo repeats, and many others, similarly can mediate interaction with widely divergent classes of binding proteins.

3. Specificity

Two of the defining parameters of protein–protein interactions are binding specificity and whether that specificity can be regulated. Specificity is of course a function of both the affinity for target sites and the affinity for "nonspecific" sites. In cases in which specificity is very high for a single target molecule (for example, the regulatory subunit of PKA for its catalytic subunit), we might term the two proteins subunits of a holoenzyme, but clearly there is no fundamental difference between such an interaction and one that is somewhat less specific, for example the binding of the same heterotrimeric G protein β subunit to several different α and γ subunits, or one that is much less specific, for instance, the binding of an SH3 domain to proline-rich sites in tens or hundreds of different proteins.

Specificity is usually defined either in terms of dissociation constants or in a more practical sense of signal-to-background (e.g., a specific association gives a dark plaque or a blue colony in a sea of light plaques or colonies). It is worth thinking of specificity a bit more carefully in terms of concentrations of proteins in a cell. A protein that represents 1/10,000 of total cell protein is present in the cytosol at a concentration on the order of 10^{-7} M; simplistically, for two interacting proteins at this level of abundance, the K_D for the complex would have to be submicromolar for a significant percentage to associate in vivo. K_D values for known biologically relevant interactions are generally in this range, for example, 10^{-9} M for the association of the regulatory and catalytic subunits of PKA, and 10^{-8} to 10^{-7} M for complexes of SH2 domains with tyrosine-phosphorylated targets. Significant interactions can certainly have less impressive K_D values; however, as mentioned previously, individual SH3 domain–peptide interactions usually have affinities in the range of 10^{-5} to 10^{-6} M, but factors such as additional contacts in actual binding complexes (e.g., multiple binding sites and multiple SH3 domains) can raise the overall affinity. An extreme example is actin, where the K_D for binding of monomers to the end of a filament is approx 10^{-5} M, but complex formation (polymerization) is favored because the total intracellular concentration of actin is very high.

The actual behavior of protein complexes in the cell is also highly dependent on other factors, such as the average lifetime of those complexes and the local concentrations of potential binding partners. By definition, the dissociation constant is the equal to the off-rate, k_{off}, divided by the on-rate, or k_{on}. Because the on-rate cannot be faster than the diffusion-limited rate of collision (on the order of 10^8 or $10^9\ M^{-1}\ s^{-1}$), this means that for moderate K_D values the off-rate must be relatively fast, and the half-life of complexes ranges from seconds to a few minutes. The practical implication is that what we may think of as "stable" complexes in the cell are constantly breathing apart and reassociating on a relatively rapid time scale. If truly stable, long-term binding is important, then either the affinity of the interaction must be very high, or the complex must be held together by multiple independent interactions, thus minimizing the likelihood of dissociation should one of the interactions temporarily unbind. Furthermore, the relatively short time scale of biological interactions highlights the importance of local concentrations: because the composition of complexes necessarily reflects the local availability of binding partners, it follows that the relocalization of a protein within the cell can rapidly change its spectrum of interactions.

Regulation of specificity is often (but not always) critical if the complexes are to be important to signaling. Although specific, unregulated protein complexes might be important for function, and are certainly worth knowing about, it is changes in binding that drive the transduction of biological signals. This is of practical importance because it can provide an experimental handle to identify interactions involved in signaling (for example, proteins that associate with a G protein only when it is bound to GTP and not GDP, are likely to be important effector molecules). Changes in binding specificity can be the result of allosteric alterations in one of the binding partners, depending, for example, on what nucleotide is bound to a G protein or to direct covalent modification of the binding site (e.g., tyrosine phosphorylation to create an SH2 domain binding site). Obviously there are many cases in which such distinctions are blurred; in just one example, dephosphorylation of a key tyrosine residue in the Src family of tyrosine kinases destabilizes both direct (SH2–pTyr mediated) and indirect intramolecular interactions, leading to a more open conformation, activation of the catalytic domain, and increased interactions with other proteins in *trans* (*76,77*).

4. Is a Binding Interaction Significant?

Perhaps the most vexing question facing those working on signaling pathways is whether a potential interaction is functionally significant. Because sequence inspection leads to predictions about potential interaction partners, and because the techniques for detecting potential interactions are so efficient, there are often not one or two but tens or hundreds of candidate binding proteins for any given protein of interest. Of course in some cases this may actually reflect the messy reality that the protein partitions among many different complexes in the cell, each of which might be important to some aspect of cellular behavior. But how can we experimentally evaluate the significance of any single proposed interaction?

The problem is one of establishing the relationship between in vitro (or otherwise experimentally manipulated) binding data and the actual biological properties of the proteins in their normal cellular environment. A detailed discussion of specific controls for the various methods used to identify binding partners is well beyond the scope of this chapter, but it is worth considering some general criteria. At the very least, the two proteins should be present in the cell in the same subcellular compartment at a suitable concentration for the interaction to occur. Although this would seem to require some detailed knowledge of the dissociation constant and the concentration in various compartments, in fact it can be quite easy to get the rough estimates of these parameters needed to evaluate an interaction. For example, if two interacting proteins are of very low abundance (a few thousand molecules per cell) and the apparent K_D from simple in vitro binding studies using recombinant proteins is greater than 10^{-6} M, the interaction is unlikely to be significant in vivo. However, if immunofluorescence studies suggest that these same two proteins are colocalized in a small fraction of the total volume of the cell (at the plasma membrane, for example, or at focal adhesions), it still might be possible for the interaction to be favored because of high local concentration. In vivo imaging approaches, such as fluorescence resonance energy transfer, can now be used to probe the extent and localization of pairwise interactions as they occur in living cells (78).

Coimmunoprecipitation of two proteins is often taken as strong evidence of in vivo binding, because the increase in volume after cell lysis and during antibody binding, and the repeated washing of the immune complexes, would seem to eliminate all but the tightest interactions. However, several caveats must be kept in mind. First, it is important to know what fraction of the total coimmunoprecipitating protein is associated with the complex, because the high sensitivity of detection (usually by immunoblotting of the immune complexes) means that a tiny fraction of the total pool of protein can be detected. It should also be kept in mind that by lysing the cell in detergent-containing buffers, proteins that are normally in different subcellular compartments are mixed, and furthermore, that detergent can change binding properties relative to the intracellular environment. Finally, the k_{off}s of many biologically significant interactions are sufficiently rapid so that it may be difficult to detect even a biologically meaningful interaction by coimmunoprecipitation.

Concentrations of the protein of interest vs concentrations of other competing cellular components also must be considered. Many studies present convincing evidence of association between two proteins when one or both is highly overexpressed, as in transiently transfected tissue culture cells. However, the extremely high levels of expression and concomitantly high intracellular concentrations mean that interactions might be favored that would not be observed at endogenous levels of abundance. The most extreme example of this type of bias is when two purified proteins are shown to interact in vitro. The "sticky" nature of proteins in general, especially in purified form in which some may be partially denatured and aggregated, means that such results are meaningful only when carefully controlled, for example, where binding does not occur under the same conditions with a point mutant predicted from genetics or structural studies to abolish binding activity. Of course, the availability of highly purified proteins allows detailed analysis of the kinetic parameters of binding by methods such as surface plasmon resonance, fluorescence quenching, and isothermal titration calorimetry.

Experimental approaches in which both partners are overexpressed or purified are complemented by those that score binding to a candidate protein under conditions in which all potential binding partners are present at their in vivo levels of relative abundance. For example, in pulldown assays, an immobilized purified protein is used to "fish" a total cell lysate for binding partners, and the bound proteins then are displayed by silver staining, metabolic labeling, or immunoblotting. Another example would be "far-Western" filter-binding assays, in which total cell lysates are separated on sodium dodecyl sulfate polyacrylamide gels, transferred to membranes, and probed with purified proteins. In such experiments, all proteins in the cell lysate compete for binding at their natural relative level of abundance. The observation that this type of assay is often quite "dirty" owing to nonspecific binding to highly abundant proteins is sobering. Furthermore, high concentrations of the probe domain tend to blunt specificity, as all high affinity binding sites in the sample can become saturated, thus driving binding of the probe to lower affinity sites (79). In practice, many nonspecific background bands can be eliminated by appropriate controls, and a protein that specifically binds to the protein of interest in such an assay is likely to be meaningful.

One recent approach that has been used successfully to detect relatively stable complexes existing in the cell at endogenous levels of expression is the tandem affinity purification tag approach (80). In this method a protein of interest is expressed in cells at relatively low levels as a fusion with a tandem affinity tag. This tandem affinity purification tag permits two different rounds of affinity purification of the tagged protein and any associated binding partners, greatly reducing background nonspecific binding compared with a single-step purification. Specifically bound proteins can then be identified by mass spectrometry.

In a somewhat different way, interactions identified by screening expression libraries or via yeast two-hybrid screens have a good chance of being significant because many thousands of plaques or colonies score negative for each that scores positive. It must be remembered, however, that potential binding proteins are being highly overexpressed, either in yeast or more extremely in phage plaques; this allows detection of interactions that might not occur at in vivo levels of abundance. These assays therefore are biased toward identification of proteins with relatively high affinity, whereas those discussed in the previous paragraphs are biased toward high-abundance proteins.

Genetics is perhaps the most unambiguous and unbiased test of significance, for instance, when mutation or deletion of the gene for one protein can be shown to have phenotypic effects that are dependent on the putative interacting partner. Genetic studies first suggested the importance of Grb2 and Sos in activating Ras, and lent credence to the subsequent biochemical data showing that Grb2 and Sos could bind to each other. Unfortunately, it often is difficult, if not impossible, to test the importance of a proposed interaction genetically, so pseudogenetic approaches using dominant inhibitory mutants have been commonly used. These experiments are based on the principle that if an exogenous protein is highly overexpressed, it will compete with its endogenous counterpart for binding to other cellular proteins. If the exogenously expressed protein has an intact binding domain, but is mutated such that other important functions (e.g., catalytic activity, other binding domains) are impaired, then normal signaling through proteins that bind to the mutant will be blocked. These approaches can be informative but must be interpreted carefully. One potential pitfall is illustrated by a situation in which the overexpressed protein competes away not only its endogenous counterpart, but also other more important endogenous proteins that might bind to the same site. For this reason, the effects of dominant-negative mutants are best interpreted in comparison with overexpression of the wild-type protein at comparable levels.

Even genetic approaches may ultimately be unable to address whether a particular protein interaction is actually relevant to a biological output of interest. Consider that, in yeast and mice, it is possible to "knock in" a mutant gene encoding a protein lacking a supposed interaction motif or binding site. Although any biological effects of such a manipulation implicate the mutated region in an essential interaction, the identity of the actual interacting protein that is important for the biological activity remains unknown. In a few rare instances, however, it has been possible to demonstrate directly the role of a specific protein–protein interaction. For example, expression of a chimera in which the SH2 domain of the Grb2 adaptor was fused to Sos could rescue developmental phenotypes in Grb2 knockout embryonic stem cells, strongly suggesting that it is the SH3-mediated association of Sos and Grb2 that is critical for normal development *(81)*.

It is clear that novel experimental approaches are needed if we are to address the functional consequences of specific protein–protein interactions in a systematic way. One such approach is the Functional Interaction Trap *(82–84)*. In this method, a relatively nonspecific interaction, for example, between an SH3 domain and its binding site, is replaced with an engineered, highly specific protein binding interface (e.g., two amphipathic coiled-coil segments that heterodimerize with high affinity). Thus, when two such engineered proteins are expressed in the cell, they are forced to interact specifically in a pairwise fashion, in the absence of interactions with other potential binding partners. This allows the functional consequence of that specific interaction to be assessed in vivo. Once libraries of full-length complementary DNAs are available, such a strategy could serve as the basis for proteomic screens to determine the functional consequences of the pairwise interaction of any protein of choice with all other proteins in the proteome.

5. Prospects

The advent of powerful approaches to isolate binding partners for interacting proteins, and the availability of high-resolution three-dimensional structures of such interacting proteins, has opened a fruitful avenue to analyze and understand signaling pathways. Clearly, one of the most efficient ways to work up or down signaling pathways is through the identification of binding partners for the participants. Global proteome-wide approaches are now being used to define not only all of the possible interactions that can occur in the cell, but also those stable interactions that actually do occur at endogenous levels of expression. Thus, we have already entered a new era in which the most fundamental roadblock to understanding signaling pathways is not lack of information, but rather too much information.

The greatest strides over the next few years will likely be in developing methods to interpret and make use of this flood of information. Toward this end, rapid advances in imaging are beginning to allow the quantitative analysis of binding interactions and their subcellular localization over time in living cells. Coupled with biophysical parameters of binding (k_{on}, k_{off}, rate of diffusion), such data will allow quantitative models of signal transduction networks to be constructed and validated via computer-based simulations (85). In the future such quantitative models will serve not only as repositories for organizing vast amounts of physical information about proteins and their interaction partners, but will also serve to test and refine hypotheses through the comparison of actual experimental data and model predictions. A truly comprehensive understanding of the role of protein–protein interactions in signaling is now within striking distance, setting the stage for the rational, targeted manipulation of these interactions as a means to modify biological activities.

References

1. Pawson, T. and Nash, P. (2003) Assembly of cell regulatory systems through protein interaction domains. *Science* **300,** 445–452.

2. Pawson, T. and Nash, P. (2000) Protein-protein interactions define specificity in signal transduction. *Genes Dev.* **14,** 1027–1047.
3. Schlessinger, J. (2000) Cell signaling by receptor tyrosine kinases. *Cell* **103,** 211–225.
4. Neves, S. R., Ram, P. T., and Iyengar, R. (2002) G protein pathways. *Science* **296,** 1636–1639.
5. Neer, E. J. (1995) Heterotrimeric G proteins: organizers of transmembrane signals. *Cell* **80,** 249–257.
6. Giot, L., Bader, J. S., Brouwer, C., et al. (2003) A protein interaction map of Drosophila melanogaster. *Science* **302,** 1727–1736.
7. Ito, T., Tashiro, K., Muta, S., et al. (2000) Toward a protein-protein interaction map of the budding yeast: a comprehensive system to examine two-hybrid interactions in all possible combinations between the yeast proteins. *Proc. Natl. Acad. Sci. USA* **97,** 1143–1147.
8. Li, S., Armstrong, C. M., Bertin, N., et al. (2004) A map of the interactome network of the metazoan C. elegans. *Science* **303,** 540–543.
9. Schwikowski, B., Uetz, P., and Fields, S. (2000) A network of protein–protein interactions in yeast. *Nat. Biotechnol.* **18,** 1257–1261.
10. Uetz, P., Giot, L., Cagney, G., et al. (2000) A comprehensive analysis of protein-protein interactions in Saccharomyces cerevisiae. *Nature* **403,** 623–627.
11. Sadowski, I., Stone, J. C., and Pawson, T. (1986) A noncatalytic domain conserved among cytoplasmic protein-tyrosine kinases modifies the kinase function and transforming activity of fujinami sarcoma virus P130$^{gag\text{-}fps}$. *Mol. Cell. Biol.* **6,** 4396–4408.
12. Anderson, D., Koch, C. A., Grey, L., Ellis, C., Moran, M. F., and Pawson, T. (1990) Binding of SH2 domains of phospholipase C_g1, GAP, and src to activated growth factor receptors. *Science* **250,** 979–982.
13. Margolis, B., Li, N., Koch, A., et al. (1990) The tyrosine-phosphorylated carboxyterminus of the EGF receptor is a binding site for GAP and PLC-γ. *EMBO J.* **9,** 4375–4380.
14. Matsuda, M., Mayer, B. J., Fukui, Y., and Hanafusa, H. (1990) Binding of transforming protein, P47$^{gag\text{-}crk}$, to a broad range of phosphotyrosine-containing proteins. *Science* **248,** 1537–1539.
15. Mayer, B. J., Jackson, P. K., and Baltimore, D. (1991) The noncatalytic *src* homology region 2 segment of *abl* tyrosine kinase binds to tyrosine-phosphorylated cellular proteins with high affinity. *Proc. Natl. Acad. Sci. USA* **88,** 627–631.
16. Moran, M. F., Koch, C. A., Anderson, D., et al. (1990) Src homology region 2 domains direct protein-protein interactions in signal transduction. *Proc. Natl. Acad. Sci. USA* **87,** 8622–8626.
17. Bradshaw, J. M. and Waksman, G. (2002) Molecular recognition by SH2 domains. *Adv. Protein Chem.* **61,** 161–210.
18. Pawson, T., Gish, G. D., and Nash, P. (2001) SH2 domains, interaction modules and cellular wiring. *Trends Cell Biol.* **11,** 504–511.
19. Pawson, T. (2004) Specificity in signal transduction: from phosphotyrosine-SH2 domain interactions to complex cellular systems. *Cell* **116,** 191–203.
20. Ladbury, J. E., Lemmon, M. A., Zhou, M., Green, J., Botfield, M. C., and Schlessinger, J. (1995) Measurement of binding of tyrosyl phosphopeptides to SH2 domains: a reappraisal. *Proc. Natl. Acad. Sci. USA* **92,** 3199–3202.

21. Eck, M. J., Shoelson, S. E., and Harrison, S. C. (1993) Recognition of a high-affinity phosphotyrosyl peptide by the Src homology 2 domain of p59[lck]. *Nature* **362,** 87–91.

22. Waksman, G., Shoelson, S. E., Pant, N., Cowburn, D., and Kuriyan, D. (1993) Binding of a high affinity phosphotyrosyl psptide to the Src SH2 domain: crystal structures of the complexed and peptide-free forms. *Cell* **72,** 779–790.

23. Songyang, Z., Shoelson, S. E., Chaudhuri, M., et al. (1993) SH2 domains recognize specific phosphopeptide sequences. *Cell* **72,** 767–778.

24. Songyang, Z., Shoelson, S. E., McGlade, J., et al. (1994) Specific motifs recognized by the SH2 domains of Csk, 3BP2, fps/fes, GRB-2, HCP, SHC, Syk, and Vav. *Mol. Cell. Biol.* **14,** 2777–2785.

25. Rameh, L. E., Chen, C.-S., and Cantley, L. C. (1995) Phosphatidylinositol $(3,4,5)P_3$ interacts with SH2 domains and modulates PI 3-kinase association with tyrosine-phosphorylated proteins. *Cell* **83,** 821–830.

26. Nagar, B., Hantschel, O., Young, M. A., et al. (2003) Structural basis for the autoinhibition of c-Abl tyrosine kinase. *Cell* **112,** 859–871.

27. Hantschel, O., Nagar, B., Guettler, S., et al. (2003) A myristoyl/phosphotyrosine switch regulates c-Abl. *Cell* **112,** 845–857.

28. Kavanaugh, W. M., and Williams, L. T. (1994) An alternative to SH2 domains for binding tyrosine-phosphorylated proteins. *Science* **266,** 1862–1865.

29. Blaikie, P., Immanuel, D., Wu, J., Li, N., Yajnik, V., and Margolis, B. (1994) A region in Shc distict from the SH2 domain can bind tyrosine-phosphorylated growth factor receptors. *J. Biol. Chem.* **269,** 32,031–32,034.

30. Gustafson, T. A., He, W., Craparo, A., Schaub, C. D., and O'Neill, T. J. (1995) Phosphotyrosine-dependent interaction of SHC and insulin receptor substrate 1 with the NPEY motif of the insulin receptor via a novel non-SH2 domain. *Mol. Cell. Biol.* **15,** 2500–2508.

31. van der Geer, P., Wiley, S., Ka-Man Lai, V., et al. (1995) A conserved amino-terminal Shc domain binds to phosphotyrosine motifs in activated receptors and phosphopeptides. *Curr. Biol.* **5,** 404–412.

32. Wolf, G., Trub, T., Ottinger, E., et al. (1995) PTB domains of IRS-1 and Shc have distinct but overlapping binding specificities. *J. Biol. Chem.* **270,** 27,407–27,410.

33. Dho, S. E., Jacob, S., Wolting, C. D., French, M. B., Rohrschneider, L. R., and McGlade, C. J. (1998) The mammalian numb phosphotyrosine-binding domain: Characterization of binding specificity and identification of a novel PDZ domain-containing numb binding protein, LNX. *J. Biol. Chem.* **273,** 9179–9187.

34. Zhang, Z., Lee, C.-H., Mandiyan, V., et al. (1997) Sequence-specific recognition of the internalization motif of the Alzheimer's amyloid precursor protein by the X11 PTB domain. *EMBO J.* **16,** 6141–6150.

35. Eck, M. J., Dhe-Paganon, S., Trub, T., Nolte, R. T., and Shoelson, S. E. (1996) Structure of the IRS-1 PTP domain bound to the juxtamembrane region of the insulin receptor. *Cell* **85,** 695–705.

36. Zhou, M.-M., Ravichandran, K. S., Olejniczak, E. T., et al. (1995) Structure and ligand recognition of the phosphotyrosine binding domain of Shc. *Nature* **92,** 7784–7788.

37. Lemmon, M. A. (2003) Phosphoinositide recognition domains. *Traffic* **4**, 201–213.
38. Yaffe, M. B. and Elia, A. E. (2001) Phosphoserine/threonine-binding domains. *Curr. Opin. Cell Biol.* **13**, 131–138.
39. Khorasanizadeh, S. (2004) The nucleosome: from genomic organization to genomic regulation. *Cell* **116**, 259–272.
40. Strahl, B. D. and Allis, C. D. (2000) The language of covalent histone modifications. *Nature* **403**, 41–45.
41. Dhalluin, C., Carlson, J. E., Zeng, L., He, C., Aggarwal, A. K., and Zhou, M. M. (1999) Structure and ligand of a histone acetyltransferase bromodomain. *Nature* **399**, 491–496.
42. Jacobson, R. H., Ladurner, A. G., King, D. S., and Tjian, R. (2000) Structure and function of a human TAFII250 double bromodomain module. *Science* **288**, 1422–1425.
43. Owen, D. J., Ornaghi, P., Yang, J. C., et al. (2000) The structural basis for the recognition of acetylated histone H4 by the bromodomain of histone acetyltransferase gcn5p. *EMBO J.* **19**, 6141–6149.
44. Mujtaba, S., He, Y., Zeng, L., et al. (2002) Structural basis of lysine-acetylated HIV-1 Tat recognition by PCAF bromodomain. *Mol. Cell* **9**, 575–586.
45. Ball, L. J., Murzina, N. V., Broadhurst, R. W., et al. (1997) Structure of the chromatin binding (chromo) domain from mouse modifier protein 1. *EMBO J.* **16**, 2473–2481.
46. Bannister, A. J., Zegerman, P., Partridge, J. F., et al. (2001) Selective recognition of methylated lysine 9 on histone H3 by the HP1 chromo domain. *Nature* **410**, 120–124.
47. Lachner, M., O'Carroll, D., Rea, S., Mechtler, K., and Jenuwein, T. (2001) Methylation of histone H3 lysine 9 creates a binding site for HP1 proteins. *Nature* **410**, 116–120.
48. Brasher, S. V., Smith, B. O., Fogh, R. H., et al. (2000) The structure of mouse HP1 suggests a unique mode of single peptide recognition by the shadow chromo domain dimer. *EMBO J.* **19**, 1587–1597.
49. Nielsen, P. R., Nietlispach, D., Mott, H. R., et al. (2002) Structure of the HP1 chromodomain bound to histone H3 methylated at lysine 9. *Nature* **416**, 103–107.
50. Jacobs, S. A. and Khorasanizadeh, S. (2002) Structure of HP1 chromodomain bound to a lysine 9-methylated histone H3 tail. *Science* **295**, 2080–2083.
51. Harrison, S. C. (1996) Peptide-surface association: The case of PDZ and PTB domains. *Cell* **86**, 341–343.
52. Jacobs, S. A., Taverna, S. D., Zhang, Y., et al. (2001) Specificity of the HP1 chromo domain for the methylated N-terminus of histone H3. *EMBO J.* **20**, 5232–5241.
53. Min, J., Zhang, Y., and Xu, R. M. (2003) Structural basis for specific binding of Polycomb chromodomain to histone H3 methylated at Lys 27. *Genes Dev.* **17**, 1823–1828.
54. Mayer, B. J. (2001) SH3 domains: complexity in moderation. *J. Cell Sci.* **114**, 1253–1263.
55. Gout, I., Dhand, R., Hiles, I. D., et al. (1993) The GTPase dynamin binds to and is activated by a subset of SH3 domains. *Cell* **75**, 25–36.

56. Ren, R., Mayer, B. J., Cicchetti, P., and Baltimore, D. (1993) Identification of a 10-amino acid proline-rich SH3 binding site. *Science* **259,** 1157–1161.
57. Yu, H., Chen, J. K., Feng, S., Dalgarno, D. C., Brauer, A. W., and Schreiber, S. L. (1994) Structural basis for the binding of proline-rich peptides to SH3 domains. *Cell* **76,** 933–945.
58. Feng, S., Chen, J. K., Yu, H., Simon, J. A., and Schreiber, S. L. (1994) Two binding orientations for peptides to the Src SH3 domain: development of a general model for SH3-ligand interactions. *Science* **266,** 1241–1247.
59. Lim, W. A., Richards, F. M., and Fox, R. O. (1994) Structural determinants of peptide-binding orientation and of sequence specificity in SH3 domains. *Nature* **372,** 375–379.
60. Ladbury, J. E., and Arold, S. (2000) Searching for specificity in SH domains. *Chem. Biol.* **7,** R3–R8.
61. McCormick, F. (1993) How receptors turn ras on. *Nature* **363,** 15–16.
62. Zarrinpar, A., Bhattacharyya, R. P., and Lim, W. A. (2003) The structure and function of proline recognition domains. *Sci STKE* **2003(179),** RE8.
63. Kay, B. K., Williamson, M. P., and Sudol, M. (2000) The importance of being proline: the interaction of proline-rich motifs in signaling proteins with their cognate domains. *FASEB J.* **14,** 231–241.
64. Nourry, C., Grant, S. G., and Borg, J. P. (2003) PDZ domain proteins: plug and play! *Sci STKE* **2003(179),** RE7.
65. Songyang, Z., Fanning, A. S., Fu, C., et al. (1997) Recognition of unique carboxyl-terminal motifs by distinct PDZ domains. *Science* **275,** 73–77.
66. Stricker, N. L., Christopherson, K. S., Yi, B. A., et al. (1997) PDZ domain of neuronal nitric oxide synthase recognizes novel C-terminal peptide sequences. *Nat. Biotechnol.* **15,** 336–342.
67. Doyle, D. A., Lee, A., Lewis, J., Kim, E., Sheng, M., and MacKinnon, R. (1996) Crystal structures of a complexed and peptide-free membrane protein-binding domain: molecular basis of peptide recognition by PDZ. *Cell* **85,** 1067–1076.
68. Shieh, B. H. and Zhu, M. Y. (1996) Regulation of the TRP channel by INAD in *Drosophila* photoreceptors. *Neuron* **16,** 991–998.
69. Marfatia, S. M., Morais Cabral, J. H., Lin, L., et al. (1996) Modular organization of the PDZ domains in the human discs-large protein suggests a mechanism for coupling PDZ domain-binding proteins to ATP and the membrane cytoskeleton. *J. Cell Biol.* **135,** 753–766.
70. Brenman, J. E., Chao, D. S., Gee, S. H., et al. (1996) Interaction of nitric oxide synthase with the postsynaptic density protein PSD-95 and α1-syntrophin mediated by PDZ domains. *Cell* **84,** 757–767.
71. Tsunoda, S., Sierralta, J., Sun, Y., et al. (1997) A multivalent PDZ-domain protein assembles signalling complexes in a G-protein-coupled cascade. *Nature* **388,** 243–249.
72. Andrade, M. A., Perez-Iratxeta, C., and Ponting, C. P. (2001) Protein repeats: structures, functions, and evolution. *J. Struct. Biol.* **134,** 117–131.
73. Sedgwick, S. G. and Smerdon, S. J. (1999) The ankyrin repeat: a diversity of interactions on a common structural framework. *Trends Biochem. Sci.* **24,** 311–316.

74. Gorina, S. and Pavletich, N. P. (1996) Structure of the p53 tumor suppressor bound to the ankyrin and SH3 domains of 53BP2. *Science* **274,** 1001–1005.

75. Batchelor, A. H., Piper, D. E., de la Brousse, F. C., McKnight, S. L., and Wolberger, C. (1998) The structure of GABPα/β: An ETS domain-ankyrin repeat heterodimer bound to DNA. *Science* **279,** 1037–1041.

76. Sicheri, F., Moarefi, I., and Kuriyan, J. (1997) Crystal structure of the Src family tyrosine kinase Hck. *Nature* **385,** 602–609.

77. Xu, W., Harrison, S. C., and Eck, M. J. (1997) Three-dimensional structure of the tyrosine kinase c-Src. *Nature* **385,** 595–602.

78. Sekar, R. B. and Periasamy, A. (2003) Fluorescence resonance energy transfer (FRET) microscopy imaging of live cell protein localizations. *J. Cell Biol.* **160,** 629–633.

79. Nollau, P. and Mayer, B. J. (2001) Profiling the global tyrosine phosphorylation state by Src Homology 2 domain binding. *Proc. Natl. Acad. Sci. USA* **98,** 13,531–13,536.

80. Rigaut, G., Shevchenko, A., Rutz, B., Wilm, M., Mann, M., and Seraphin, B. (1999) A generic protein purification method for protein complex characterization and proteome exploration. *Nat. Biotechnol.* **17,** 1030–1032.

81. Cheng, A. M., Saxton, T. M., Sakai, R., et al. (1998) Mammalian Grb2 regulates multiple steps in embryonic development and malignant transformation. *Cell* **95,** 793–803.

82. Fujiwara, K., Poikonen, K., Aleman, L., Valtavaara, M., Saksela, K., and Mayer, B. J. (2002) A single-chain antibody/epitope system for functional analysis of protein–protein interactions. *Biochemistry* **41,** 12,729–12,738.

83. Sharma, A., Antoku, S., Fujiwara, K., and Mayer, B. J. (2003) Functional Interaction Trap: a strategy for validating the functional consequences of tyrosine phosphorylation of specific substrates *in vivo*. *Mol. Cell. Proteomics* **2,** 1217–1224.

84. Sharma, A., Antoku, S., and Mayer, B. J. (2004) The Functional Interaction Trap: A novel strategy to study specific protein-protein interactions, in *Methods in Proteome and Protein Analysis (MPSA 2002)* (Kamp, R. M., Calvete, J., and Choli-Papadopoulou, T., eds.). Springer-Verlag, Berlin, pp. 165–181.

85. Slepchenko, B. M., Schaff, J. C., Macara, I., and Loew, L. M. (2003) Quantitative cell biology with the *virtual cell*. *Trends Cell Biol.* **13,** 570–576.

II

SPECIFIC TOPICS

A. Membrane Receptors and Signaling

4

Biological Role of the CXCR4–SDF-1 Axis in Normal Human Hematopoietic Cells

Marcin Majka and Mariusz Z. Ratajczak

Summary

Stromal-derived factor (SDF)-1, an α-chemokine that binds to G protein-coupled seven transmembrane-spanning receptor, CXCR4, plays an important and unique role in regulating the trafficking of normal hematopoietic stem/progenitor cells and their homing/retention in bone marrow. The same axis also modulates several biological processes in more differentiated cells from the granulocyte-monocytic, erythroid, and megakaryocytic lineages. In this chapter, experimental details are described for the isolation of early human hematopoietic cells, such as CD34+ mononuclear cells, myeloblasts, erythroblasts, and megakaryoblasts. These cells can be used routinely for studying the role of the CXCR4–SDF-1 axis in normal human hematopoiesis.

Key Words: Hematopoietic stem/progenitor cells; myeloblasts; erythroblasts; megakaryoblasts; ex vivo expansion; CXCR4–SDF-1 axis.

1. Introduction

Chemokines, small proinflammatory chemoattractive cytokines that bind to specific G protein-coupled seven transmembrane-spanning receptors that are present on the plasma membrane of target cells, are the major regulators of cell trafficking and adhesion *(1–4)*. Some of them have also been reported to modulate cell growth and survival *(5–7)*.

There are more than 50 different chemokines and 20 different chemokine receptors cloned *(8–10)*. Most chemokines bind to multiple receptors, and the same receptor may bind more than one chemokine; however, stromal-derived factor (SDF)-1 is the one exception to this rule. This chemokine binds only to CXCR4, and CXCR4 is its only receptor *(1)*. This fact already suggests that the SDF-1–CXCR4 axis plays an important and unique biological role. In support

From: *Methods in Molecular Biology, vol. 332: Transmembrane Signaling Protocols, Second Edition*
Edited by: H. Ali and B. Haribabu © Humana Press Inc., Totowa, NJ

of this notion, murine "knockout" data have provided strong evidence that SDF-1 secreted by bone marrow (BM) stroma cells is important during embryogenesis for the colonization of BM by fetal liver-derived hematopoietic stem/progenitor cells (HSPCs *[11,12]*). Also, later in adult life it plays an essential role in the retention/homing of these cells in the BM microenviron-ment *(13,14)*. Thus, it is not surprising that perturbation of the CXCR4–SDF-1 axis during mobilization plays a pivotal role in the egress of HSPC from BM into peripheral blood (PB *[15]*). In contrast, proper functioning of the CXCR4–SDF-1 axis is crucial in directing the engraftment of HSPC into BM during hematopoietic transplantation *(16–19)*.

The biological effects and signaling pathways of the CXCR4–SDF-1 axis have been extensively studied for hematopoietic cells. The major biological effects of SDF-1 are related to the ability of this chemokine to induce motility, chemotactic responses, adhesion, and secretion of matrix metalloproteinases and angiopoietic factors (e.g., vascular endothelial growth factor [VEGF]) in the cells bearing cognate CXCR4 *(20– 22)*. SDF-1 increases the adhesion of early hematopoietic cells to vascular cell adhesion molecule-1, intercellular adhesion molecule-1, fibronectin, and fibrinogen by activating/modulating the function of several cell surface integrins *(23,24)*. There are contradictory observations regarding whether SDF-1, in addition to regulating the traffick-ing of cells, also may directly affect their proliferation and survival *(25)*. It has been suggested that SDF-1 may be secreted by HSPCs and play an important autocrine/paracrine role in their development and survival *(26)*. However, sur-prisingly, SDF-1 was found to inhibit growth and induce apoptosis in T-lym-phocytic Jurkat cells *(27)*. In our hands, SDF-1 did not affect the survival/ proliferation of other primary hematopoietic progenitor cells *(21,28)* or any of several established hematolymphopoietic cells lines *(22,29)*.

The main focus of our laboratory is to study the role of the CXCR4–SDF-1 axis in normal human hematopoietic cells. These cells can be isolated from BM, PB, and cord blood (CB). The highest percentage of HSPCs is present in BM, and very few of these cells can be isolated from steady-state PB. How-ever, the number of early hematopoietic cells can be increased in PB during so-called mobilization, during which a patient is treated with mobilizing agents (e.g., granulocyte colony stimulating factor [CSF] or cyclophosphamide), and these cells egress from BM to PB. After this treatment, PB is enriched in circu-lating HSPCs and is then referred to as mobilized peripheral blood (mPB). A relatively high number of early hematopoietic cells are also present in CB, which can be envisioned as fetal/neonatal blood mobilized by the stress related to the delivery process *(30)*.

Nevertheless, the cells isolated from BM, mPB, or CB are heterogeneous. To obtain a fraction of mononuclear cells (MNCs), BM aspirate, mPB, or CB samples have to be depleted of erythrocytes and mature granulocytes. Among the remaining BM MNCs after depletion, approx 0.5 to 4% express the CD34 antigen, a marker of HSPCs. The percentage of these cells is a magnitude lower among mPB- and CB-derived MNCs. There are strategies described to isolate these rare cells from the suspension of BM-, mPB-, or CB-MNCs based on immunomagnetic separation by using α-CD34 antibodies conjugated with paramagnetic beads. CD34$^+$ cells isolated by employing these strategies could be subsequently used for experiments.

Because BM aspirates contain a wide spectrum of hematopoietic cells from different lineages at different levels of maturation/differentiation, separation of a homogenous cell fraction for experiments (e.g., by using immunomagnetic methods, a fluorescence-activated cell sorter, or elutriation procedures) is difficult and does not guarantee a sufficient yield of these cells. Thus, we developed a well-controlled in vitro strategy in which CD34$^+$ HSPCs are expanded toward granulocyto-monocytic, erythroid, and megakaryocytic lineages in quantities that allow for a sufficient number of these lineage-expanded primary cells for several experimental procedures. To obtain lineage-specific expansion, CD34$^+$ progenitors are expanded in serum-free media (to avoid the influence of unspecific stimulants that are present in animal or human sera) that is supplemented with combinations of the appropriate lineage-specific recombinant growth factors/cytokines.

This chapter describes the experimental details for isolating human CD34$^+$ MNCs that are enriched in HSPCs, as well as procedures for expanding these cells toward the granulocyto-monocytic, erythroid, and megakaryocytic lineages. These cells may be subsequently employed for studying the role of the CXCR4–SDF-1 axis in normal human hematopoiesis.

2. Materials

2.1. Isolation of BM-, mPB-, or CB-Derived CD34$^+$ MNCs Using Immunomagnetic Beads

1. Ficoll-Histopaque (Amersham Biosciences, AB, Uppsala, Sweden).
2. Commercially available kit for isolating CD34 cells (e.g., Miltenyi Biotec, Auburn, CA).
3. Magnetic MiniMacs kit from Miltenyi.
4. Isolation buffer: phosphate-buffered saline (PBS) supplemented with 2 mM ethylene diamine tetraacetic acid and 0.5% albumin, fraction V.
6. Isove's Modification of Dulbecco's Medium (IMDM).

2.2. Procedure for Isolating a Population of BM-Derived CD34$^+$, c-kit$^+$ Rhodamine 123dim MNC

1. Heat-inactivated fetal bovine serum (FBS).
2. 100-mm Petri dishes.
3. Monoclonal antibodies anti-CD34 phycoerithin (PE), anti-c-kit Cy5 and fluorochrome dye—Rhodamine 123 (Becton Dickenson, Mountain View, CA).

2.3. Long-Term Culture-Initiating Cells, Cobblestone Area-Forming Cells, and DELTA Assays to Study the Most Primitive Hematopoietic Progenitor Cells

1. BM aspiration kit.
2. Horse serum.
3. Hydrocortisone.

2.4. Colony-Forming Unit Tests to Evaluate More Differentiated Progenitor Cells

1. CB or mPB.
2. MethoCult H4230 methylcellulose (Stem Cell Technologies Inc., Vancouver, Canada).
3. Human recombinant growth factors: granulocyto-monocyte (GM)-CSF, erythro-poietin (EPO), kit ligand (KL), thrombopoietin (TPO), and interleukin (IL)-3 (all R&D Systems, Minneapolis, MN).

2.5. Expansion of Colony-Forming Unit–GM-Derived Myeloblasts, Colony-Forming Unit–Megakaryocyte-Derived Megakaryoblasts, and Burst-Forming Unit of Erythrocyte-Derived Erythroblasts in Liquid Serum-Free Cultures

All required reagents are described in **Subeadings 2.1.** through **2.4.**

3. Methods

3.1. Isolation of BM-, mPB-, or CB-Derived CD34+ MNC Using Immunomagnetic Beads

1. Dilute aliquots of 10 to 20 mL of BM-, mPB-, or CB-derived suspension, 1:1 or 1:2, with PBS at room temperature (RT).
2. Layer the cell suspension over Ficoll-Histopaque gradient (1.077 g/dm³) (*see* **Note 1**) and separate MNCs from granulocytes and erythrocytes by spinning for 30 min at 400g at RT with low or no brake.
3. Collect the interface containing BM-, mPB-, or CB-derived MNCs and wash them twice with IMDM or PBS at 4°C.
4. Resuspend the BM-, mPB, or CB-derived MNC (10^7) in 300 μL of isolation buffer; add 100 μL of blocking solution A1 (containing antibodies against the Fc receptor) and 100 μL of solution A2 (containing anti-CD34 antigen antibodies). Mix well and incubate for 15 min at 4 to 6°C.
5. Wash cells by diluting them 20-fold in isolation buffer and spinning for 5 min at 200 to 250g at 4°C).
6. Resuspend pellet in 400 μL of isolation buffer and add 100 μL of solution B (contains anti-mouse antibodies conjugated with paramagnetic particles). Incubate cells for 15 min at 4 to 6°C.
7. Wash as in **step 2**. During this wash, prepare column by washing/priming it with 500 μL of isolation buffer.

8. Resuspend pellet in 500 µL of isolation buffer and apply to the column. After cell suspension runs through the column, wash the column three times with 500 µL of isolation buffer.
9. After the last wash, take the column from the magnet and wash the cells out with 500 to 1000 µL of isolation buffer into a sterile 15-mL tube.
10. Wash, count cells, check their viability by a 0.5% Trypan blue exclusion test, and resuspend them in growth medium (*see* **Note 2**). The purity of BM-derived CD34⁺ cells isolated by immunomagnetic beads is shown in **Fig. 1A**.

3.2. Procedure for Isolating a Population of BM-Derived CD34⁺, c-kit⁺ Rhodamine 123dim MNCs

1. Aspirate 10 to 20 mL of bone marrow using an aspiration kit.
2. Dilute bone marrow 1:1 or 1:2 with PBS at RT.
3. Layer the cell suspension over a Ficoll-Histopaque gradient (*see* **Note 1**) (1.077 g/dm³) and separate MNCs from granulocytes and erythrocytes by spinning for 30 min at $400g$ at RT with low or no brake.
5. Collect the interface containing MNCs and wash them twice with IMDM or PBS at 4°C.
6. Resuspend cell pellet in ice-cold PBS with 2% FBS at a concentration of no greater than 10^8/mL.
7. Add anti-CD34 and anti-c-kit (CD117) antibodies according to the manufacturer's protocol and Rhodamine 123 to a final concentration of 5 mM and incubate for 30 min at 4°C.
8. After staining, wash cells twice with ice-cold medium or PBS.
9. Resuspend pellet with PBS containing 2% FBS and sort cells positive for CD34 and c-kit antigens and dim for Rhodamine 123 by using a cell sorter.

3.3. Long-Term Culture-Initiating Cells, Cobblestone Area-Forming Cells, and DELTA Assays to Study the Most Primitive Hematopoietic Progenitor Cells

3.3.1. Growing a Feeder Layer of Cells for Long-Term Culture-Initiating Cells and Cobblestone Area-Forming Cell Assays

1. MNCs obtained after Ficoll-Histopaque separation are seeded at 2×10^6/mL in 24- or 96-well plates in IMDM supplemented with 12.5% FBS, 12.5% horse serum (HS), and $10^{-6} M$ hydrocortisone.
2. Change half of the culture medium once a week and add fresh medium as described previously in **steps 20** to **22**.
3. After 3 to 4 wk, the feeder layer should be irradiated with 15 Gy to eliminate the remaining hematopoietic cells among stromal cells and to stop the proliferation of the stromal cells.
4. BM-derived fibroblasts/stromal cells derived from these cultures could be subsequently used as a feeder layer for long-term culture-initiating cell (LTCIC) and cobblestone area-forming cell (CAFC) assays. Alternatively, the primary stromal cell feeder layer can be replaced with the established cell lines, e.g., M2–10B4.

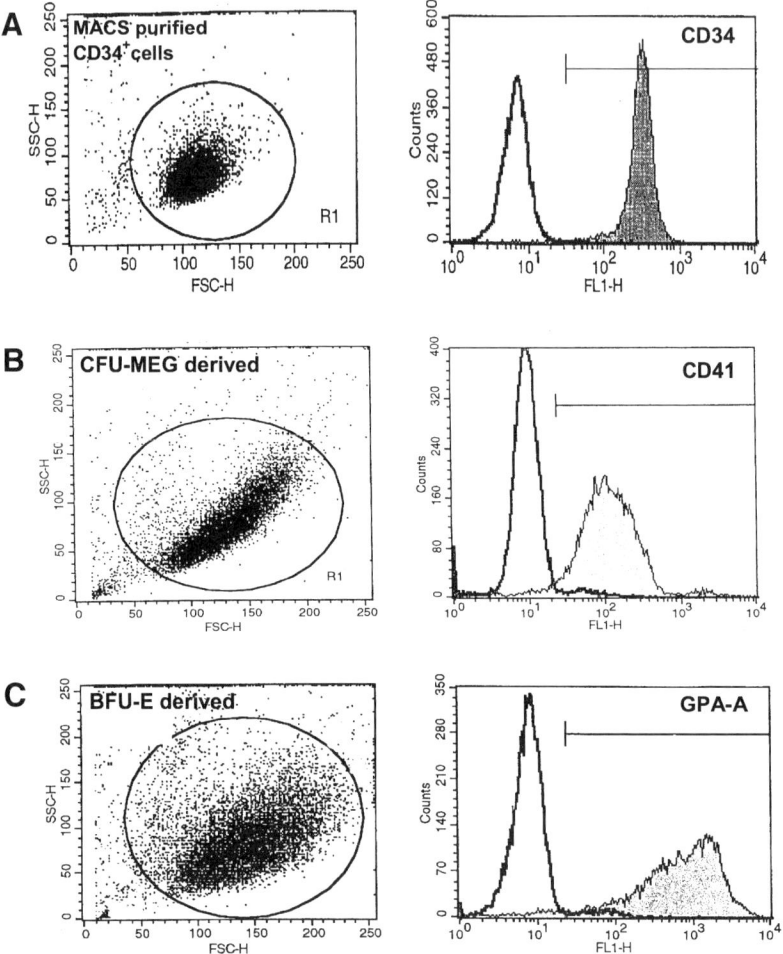

Fig 1. **(A)** Purity percentage of CD34+ cells isolated by immunomagnetic beads human bone marrow-derived CD34+ cells. **(B)** Purity of expanded ex vivo human megakaryoblasts (CD41+ cells). **(C)** Purity percentage of GPA-A+ cells expanded ex vivo human megakaryoblasts (GPA-A+ cells). Representative staining is shown.

3.3.2. Procedure for Performing LTCIC Assay

1. Seed CD34+ cells at 1×10^4/mL over an irradiated fibroblast layer in IMDM supplemented with 12.5% FBS, 12.5% HS, and 10^{-6} mol/L hydrocortisone.
2. Change half the medium once or twice weekly.
3. During every medium exchange, the cells are collected from the medium and plated in secondary methylcellulose cultures to grow colony-forming unit (CFU)-GM colonies.

4. Score CFU-GM colonies after 11 d by employing an inverted microscope.
5. LTCIC assay can be carried out for 2 to 6 mo.

3.3.3. Procedure for Performing CAFC Assay

1. Seed CD34+ cells (*see* **Note 2**) at 1×10^4/mL over an irradiated fibroblast layer in IMDM supplemented with 12.5% FBS, 12.5% HS, and $10^{-6} M$ hydrocortisone.
2. Change half the medium once or twice weekly.
3. After 1 to 2 mo, count the groups of 15 or more closely attached cells that form a cobblestone area on the fibroblast feeder layer by employing an inverted microscope.

3.3.4. Procedure for Performing DELTA Assay

1. Seed CD34+ cells (*see* **Note 2**) at 1×10^3/mL per well in a 96-well plate with addition of IL-3 (20 ng/mL), KL (50 ng/mL), and FLT-3L (100 ng/mL).
2. Replace old medium with fresh medium supplemented with the same combination of growth factors every week.
3. Cells collected during medium change are plated in secondary methylcellulose cultures to grow CFU-GM colonies.
4. Score CFU-GM colonies after 11 d by using an inverted microscope.

3.4. CFU Tests to Evaluate More Differentiated Progenitor Cells

3.4.1. CFU of Mixed Lineages Assay

1. Resuspend 10^5 MNC cells or 2×10^4 CD34+ cells in 0.4 mL of IMDM.
2. Add GM-CSF (5 ng/mL), EPO (2 ng/mL), KL (10 ng/mL), TPO (50 ng/mL), and IL-3 (5 ng/mL; *see* **Note 3**). Mix and add 1.8 mL of MethoCult H4230 methylcellulose. Final volume is 2.2 mL. Mix very well by slowly pipetting up and down several times avoiding the generation of air bubbles.
3. Using a glass 2-mL pipet, apply cell mixture into two 3-mm Petri dishes (1 mL into each dish). Place two small dishes in one big dish (100 mm) and add one more small dish (bottom only) with sterile water (to prevent drying out of the methylcellulose).
4. Grow colonies in a humidified incubator (37°C, 5% CO_2, 95% humidity). Count CFUs of mixed lineages colonies using an inverted light microscope after 14 d. CFUs of mixed lineages colonies should contain cells mainly from the erythroid-, GM-, and megakaryocytic lineages.

3.4.2. CFUs of CFU-GM Assay

1. Resuspend 10^5 of MNC cells or 2×10^4 CD34+ cells in 0.4 mL of IMDM.
2. Add IL-3 (10 ng/mL) and GM-CSF (5 ng/mL; *see* **Note 3**), mix, and add 1.8 mL of MethoCult H4230 methylcellulose. Final volume is 2.2 mL. Mix the cell suspension very well by slowly pipetting up and down several times without creating air bubbles.

3. Using a glass 2-mL pipet, apply cell mixture into two 30-mm Petri dishes (1 mL into each dish). Place two small dishes in one big dish (100 mm) and add one more small dish (bottom only) with sterile water (to prevent drying out of the methylcellulose).

4. Grow colonies in humidified incubator (37°C, 5% CO_2, 95% humidity). Count CFU-GM colonies using inverted light microscope after 11 d. CFU-GM colony should contain at least 50 cells (*see* **Note 4**).

3.4.3. Burst-Forming Unit of Erythrocytes

1. Resuspend 10^5 MNC cells or 2×10^4 CD34+ cells in 0.4 mL IMDM.
2. Add 2 U/mL of EPO and 10 ng/mL of KL (*see* **Note 3**). Mix and add 1.8 mL of MethoCult H4230 metylcellulose. Final volume is 2.2 mL. Mix very well by slowly pipetting up and down several times avoiding the creation of air bubbles.
3. Using a glass 2-mL pipet, apply cells into two 30-mm Petri dishes (1 mL into each). Place two small dishes in one big dish (100 mm) and add one more small dish (bottom only) with sterile water (to prevent the drying out of the methylcellulose).
4. Grow colonies in a humidified incubator (37°C, 5% CO_2, 95% humidity). Count burst-forming unit of erythrocyte (BFU-E) colonies using an inverted light microscope after 11 d. BFU-E colonies should contain at least 200 cells (*see* **Note 4**).

3.4.4. CFU of Megakaryocytes Assay

1. Resuspend 10^5 of MNC cells or 2×10^4 CD34+ cells in 0.4 mL of IMDM. Add 50 ng/mL of TPO and 10 ng/mL of IL-3 (*see* **Note 3**), 1 mL of CFU of megakaryocytes (CFU-Meg) growth medium, 0.6 mL of horse serum, and 0.2 mL of bovine plasma. For 20 mL of CFU-Meg growth medium, mix 16 mL of IMDM, 4 mL of 10% bovine serum albumin in PBS, 500 µL of L-aspargine, 40 µL of α-thioglicerol, and 100 µL of 100 mM $CaCl_2$ (in water).
2. Mix very well by slowly pipetting up and down several times avoiding the creation of air bubbles.
3. Using a glass 2-mL pipet, apply cell mixture into two 30-mm Petri dishes (1 mL into each). Observe formation of the plasma clot. Place two small dishes in one big dish (100 mm) and add one more small dish (bottom only) with sterile water (to prevent the drying out of the plasma clot).
4. Grow colonies in a humidified incubator (37°C, 5% CO_2, 95% humidity). Count CFU-Meg colonies using a fluorescent inverted microscope after 11 d after staining with CD41-fluorescein isothiocyanate (FITC) antibody. CFU-Meg colonies should contain at least 10 to 20 green fluorescent cells.

3.4.4.1. STAINING OF CFU-MEG COLONIES

1. Fix dishes containing CFU-Meg plasma clot cultures in a mixture of acetone and methanol (1:3) twice for 10 min.
2. Wash dishes once with PBS for 5 min, with water for 5 min, and once more shortly with water.

3. Add 350 to 450 µL of PBS containing primary antibody against human CD41 (1:200 dilution). Incubate 1 h in dark in the incubator (37°C).
4. Wash dishes three times for 5 min in PBS.
5. Add 350 to 450 µL of PBS containing secondary FITC conjugated anti- mouse antibody (1:80 dilution). Incubate 1 h in a dark incubator (37°C).
6. Wash once in PBS for 5 min, subsequently in water for 5 min and once shortly with water.
7. Add a few drops of glicerol-barbitol and cover with a round cover glass. Store covered dishes at 4°C in the dark. Count colonies by employing an inverted immunofluorescence microscope.

3.5. Expansion of CFU-GM-Derived Myeloblasts, CFU-Meg-Derived Megakaryoblasts, and BFU-E-Derived Erythroblasts in Liquid Serum-Free Cultures

1. Seed 5×10^4 CD34$^+$ cells/well in 1 mL of serum-free medium (*see* **Note 5**; Stem Cell Technologies) in a 24-well plate.
2. For expansion of erythroblasts, supplement medium with 2 U/mL of EPO and 10 ng/mL KL (*see* **Note 3**). For expansion of megakaryoblasts, add 50 ng/mL TPO and 10 ng/mL IL-3 (*see* **Note 3**). For expansion of myeloblasts, add to the liquid cultures GM-CSF (5 ng/mL) + IL-3 (10 ng/mL; *see* **Note 3**). Grow cells for 8 to 11 d, changing medium every 4 to 7 d or if needed.
3. At the end of the expansion, the purity of erythroblasts is checked by fluorescent-activated cell sorting (FACS) after staining with anti-glycophorine-A antibodies (**Fig. 1B**), purity of megakaryoblasts after staining with anti-CD41 antibodies (**Fig. 1C**), and purity of myeloblasts after staining with anti-CD33 antibodies.

These lineage-expanded myeloblasts, erythroblast, and megakaryoblasts may be used as a population of target cells to study the role of SDF-1 in chemotaxis, adhesion, proliferation, and survival of normal hematopoietic cells. Both messenger RNA and proteins isolated from these cells stimulated by SDF-1 may be used for the molecular analysis of gene expression. Finally, these normal human hematopoietic lineage-expanded cells may be employed to study various aspects of SDF-1-signaling, such as calcium flux studies (**Fig. 2**), phosphorylation of signal transduction proteins, and actin polymerization.

4. Notes

1. Ficoll-Histopaque should be warmed to 17°C.
2. Check purity of isolated CD34$^+$ cells by FACS.
3. Use recombinant growth factors and cytokines only.
4. Colonies could be aspirated with Pasteur pipet from methylocellulose cultures, solubilized in PBS and the purity of erythroblasts could be checked by FACS after staining with anti-glycophorine-A antibodies (**Fig. 1B**), purity of megakaryoblasts after staining with anti-CD41 antibodies (**Fig. 1C**), and purity of myeloblasts after staining with anti-CD33 antibodies.

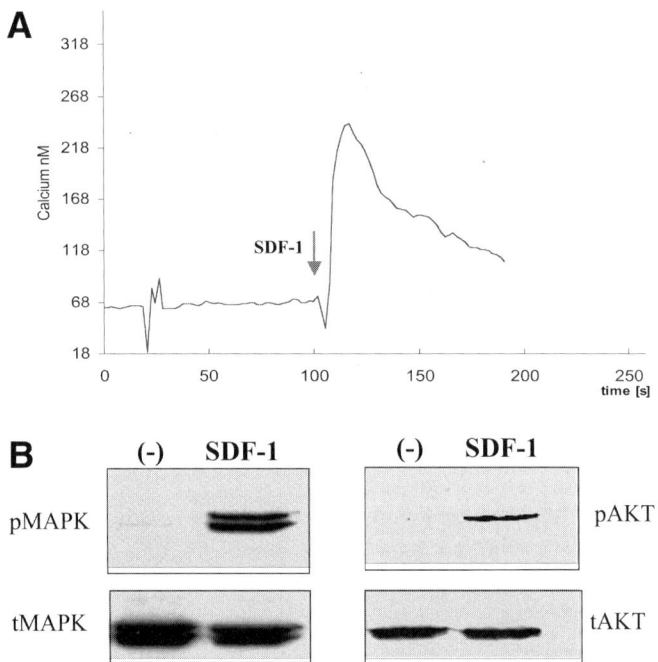

Fig 2. (**A**) Calcium flux studies on megakaryoblasts derived from serum free expansion cultures. Cells were loaded with Fura-2 and stimulated by stromal-derived factor (SDF)-1 (500 ng/mL). Calcium flux was evaluated by spectrophotofluorimeter. (**B**) Signal transduction studies on normal human megakaryoblasts derived from serum free expansion cultures. Cells were made quiescent by serum and growth factors deprivation (–) and subsequently stimulated by SDF-1 (300 ng/mL) for 1 min (SDF-1). Left panel, phosphorylation of mitogen-activated protein kinase (MAPK) p42/44 and total level of MAPK p42/44. Right panel, phosphorylation of AKT and total level of AKT. Representative studies are shown.

5. Initial expansion should be is most efficient at initial density 5×10^4 CD34+ cells/ well in 1 mL of serum-free medium.

References

1. Kucia, M., Jankowski, K., Reca, R., et al. (2004) CXCR4-SDF-1 signalling, locomotion, chemotaxis and adhesion. *J. Mol. Histol.* **35,** 233–245.
2. Melchers, F., Rolink, A. G., and Schaniel, C. (1999) The role of chemokines in regulating cell migration during humoral immune responses. *Cell* **99,** 351–354.
3. Rossi, D. and Zlotnik, A. (2000) The biology of chemokines and their receptors. *Annu. Rev. Immunol.* **18,** 217–242.
4. Rollins, B. J. (1997) Chemokines. *Blood* **90,** 909–928.

5. Maze, R., Sherry, B., Kwon, B. S., Cerami, A., and Broxmeyer, H. E. (1992) Myelosuppressive effects in vivo of purified recombinant murine macrophage inflammatory protein-1 alpha. *J. Immunol.* **149,** 1004–1009.

6. Broxmeyer, H. E. (2001) Regulation of hematopoiesis by chemokine family members. *Int. J. Hematol.***74,** 9–17.

7. Youn, B. S. Mantel, C., and Broxmeyer, H. E. (2000) Chemokines, chemokine receptors and hematopoiesis. *Immunol. Rev.* **177,** 150–174.

8. Zlotnik, A. and Yoshie, O. (2000) Chemokines: a new classification system and their role in immunity. *Immunity* **12,** 121–127.

9. Cyster, J. G. (2005) Chemokines, sphingosine-1-phosphate, and cell migration in secondary lymphoid organs. *Annu. Rev. Immunol.* **23,** 127–159.

10. Rot, A. and von Andrian, U. H. (2004) Chemokines in innate and adaptive host defense: basic chemokinese grammar for immune cells. *Annu. Rev. Immunol.* **22,** 891–928.

11. Nagasawa, T., Hirota, S., Tachibana, K., et al. (1996) Defects of B–cell lymphopoiesis and bone–marrow myelopoiesis in mice lacking the CXC chemokine PBSF/SDF–1. *Nature* **382,** 635–638.

12. Zou, Y., Kottmann, A. H., Kuroda, M., Taniuchi, I., and Littman, D. R. Function of the chemokine receptor CXCR4 in haematopoiesis and in cerebellar development. *Nature* **393,** 595–599.

13. Peled, A., Grabovsky, V., Habler, L., et al. (1997) The chemokine SDF–1 stimulates integrin–mediated arrest of CD34$^+$ cells on vascular endothelium under shear flow. *J. Clin. Invest.* **104,** 1199–1211.

14. Wysoczynski, M., Reca, R., Ratajczak, J., et al. (2005) Incorporation of CXCR4 into membrane lipid rafts primes homing-related responses of hematopoietic stem/progenitor cells to an SDF-1 gradient. *Blood* **105,** 40–48.

15. Liles, W. C., Broxmeyer, H. E., Rodger, E., et al. (2003) Mobilization of hematopoietic progenitor cells in healthy volunteers by AMD3100, a CXCR4 antagonist. *Blood* **102,** 2728–2730.

16. Peled, A., Petit, I., Kollet, O., et al. (1999) Dependence of human stem cells engraftment and reopulation of NOD/SCID mice on CXCR4. *Science* **283,** 845–848.

17. Lapidot, T. (2001) Mechanism of human stem cell migration and repopulation of NOD/SCID and B2mnull NOD/SCID mice. The role of SDF-1/CXCR4 interactions. *Ann. NY Acad. Sci.* **938,** 83–95.

18. Lapidot, T. and Kollet, O. (2002) The essential roles of the chemokine SDF-1 and its receptor CXCR4 in human stem cell homing and repopulation of transplanted immune-deficient NOD/SCID and NOD/SCID/B2m(null) mice. *Leukemia* **16,** 1992–2003.

19. Kahn, J., Byk, T., Jansson-Sjostrand, L., et al. (2004) Overexpression of CXCR4 on human CD34+ progenitors increases their proliferation, migration, and NOD/SCID repopulation. *Blood* **103,** 2942–2949.

20. Janowska–Wieczorek, A., Marquez, L. A., Dobrowsky, A., Ratajczak, M. Z., and Cabuhat, M. L. (2000) Differential MMP and TIMP production by human marrow and blood CD34$^+$ cells in response to chemokines. *Exp. Hematol.* **28,** 1274–1285.

21. Majka, M., Janowska–Wieczorek, A., Ratajczak, J., et al. (2000) Stromal derived factor–1 and thrombopoietin regulate distinct aspects of human megakaryopoiesis. _Blood_ **96,** 4142–4151.
22. Kijowski, J., Baj-Krzyworzeka, M., Majka, M., et al. (2001) The SDF-1-CXCR4 axis stimulates VEGF secretion and activates integrins but does not affect proliferation and survival in lymphohematopoietic cells. _Stem Cells_ **19,** 453–466.
23. Peled, A., Kollet, O., Ponomaryov, T., et al. (2000) The chemokine SDF–1 activates the integrins LFA–1, VLA–4, and VLA–5 on immature human CD34⁺ cells: role in transendothelial/stromal migration and engraftment of NOD/SCID mice. _Blood_ **95,** 3289–3296.
24. Avigdor, A., Goichberg, P., Shivtiel, S., et al. (2004) CD44 and hyaluronic acid cooperate with SDF-1 in the trafficking of human CD34+ stem/progenitor cells to bone marrow. _Blood_ **103,** 2981–2989.
25. Lataillade, J. J., Domenech, J., and Le Bousse-Kerdiles, M. C. (2004) Stromal cell-derived factor-1 (SDF-1)\CXCR4 couple plays multiple roles on haematopoietic progenitors at the border between the old cytokine and new chemokine worlds: survival, cell cycling and trafficking. _Eur Cytokine Netw._ **15,** 177–188.
26. Broxmeyer, H. E., Cooper, S., Kohli, L., et al. (2003) Transgenic expression of stromal cell – derived factor – 1/cxc chemokine lingad 12 enhances myeloid progenitor cell survival/antiapoptosis in vitro in response to growth factor withdrawal and enhances myelopoiesis in vivo. _J. Immunol._ **170,** 421–429.
27. Colamussi, M. L., Secchiero, P., Gonelli, A., Marchisio, M., Zauli, G., and Capitani, S. (2001) Stromal derived factor-1 alpha (SDF-1 alpha) induces CD4+ T cell apoptosis via the functional up-regulation of the Fas (CD95)/Fas ligand (CD95L) pathway. _J. Leukoc. Biol._ **69,** 263–270.
28. Majka, M., Ratajczak, J., Lee, B., et al. (2000) The role of HIV related chemokine receptors and chemokines in human erythropoiesis in vitro. _Stem Cells_ **18,** 128–138.
29. Majka, M., Rozmyslowicz, T., Honczarenko, M., et al. (2000) Biological significance of the expression of HIV-related chemokine coreceptors (CCR5 and CXCR4) and their ligands by human hematopoietic cell lines. _Leukemia_ **14,** 1821–1832.
30. Schmitz, N. and Barrett, J. (2002) Optimizing engraftment—source and dose of stem cells. _Semin. Hematol._ **39,** 3–14.

5

Functional Expression of CXCR4 in *Saccharomyces cerevisiae* in the Development of Powerful Tools for the Pharmacological Characterization of CXCR4

Zi-xuan Wang, James R. Broach, and Stephen C. Peiper

Summary

CXCR4, the receptor for stromal cell-derived factor (SDF)-1, was expressed in *Saccharomyces cerevisiae*, coupled to the pheromone response pathway via a chimeric G_α subunit. Engagement of CXCR4 by SDF-1 resulted in expression of reporter genes, *HIS3* or *lacZ*, under the transcriptional control of a *FUS1* promoter, which is pheromone-responsive. CXCR4 mutants with constitutive signaling activity were generated by random mutagenesis of receptor coding sequences and selection for complementation of histidine auxotrophy in the yeast strain by autonomous expression of the *FUS1-HIS3* reporter gene. Linkage of CXCR4 to the pheromone response pathway in yeast provides a system that lends itself to screening of receptor antagonists. The use of constitutively active mutants to screen for inhibitors of the weak partial agonist and inverse agonist pharmacologic types offers a sensitive, efficient approach that is independent of ligand.

Key Words: CXCR4; *Saccharomyces cerevisiae*; GPCR; constitutively active mutant (CAM); antagonists; *FUS1-HIS3, FUS1-lacZ*.

1. Introduction

G protein-coupled receptors (GPCRs) are the largest gene family, accounting for approx 5 to 10% of the human genome *(1)*, and are prominent drug targets; antagonists represent approximately one in three of front-line pharmaceutical agents. GPCRs are the phylogenetically oldest type of receptor, which function in *Saccharomyces cerevisiae* to transduce signals of mating factor triggering responses *(2)* that prepare cells for mating, including cell cycle arrest, transcriptional activation of certain genes, and the development of cell surface projections and, more recently described, to sense glucose *(3)*, activating path-

From: *Methods in Molecular Biology, vol. 332: Transmembrane Signaling Protocols, Second Edition*
Edited by: H. Ali and B. Haribabu © Humana Press Inc., Totowa, NJ

ways that control filamentous growth. The existence of GPCR-coupled signal transduction pathways in yeast opens avenues to access the power of yeast genetics and ease of growth to study human GPCRs.

CXCR4 is a GPCR for stromal cell-derived factor (SDF)-1 (also designated CXCL12), a CXC chemokine involved in migration of leukocytes, endothelial cells, stem cells, and primordial germ cells. CXCR4 plays a role in pathological physiology as an entry coreceptor for T-tropic strains of human immune deficiency virus-1 and in programming the metastatic spread of cancers that express the receptor to organs that secrete SDF-1. This chapter presents the technical approaches used to express CXCR4 in *S. cerevisiae* coupled to the pheromone response signaling pathway, thereby gaining access to the genetic power and ease of growth of yeast to approach this molecular target. Specifically, this system was used to characterize activation of CXCR4 by SDF-1, to generate CXCR4 mutants with constitutive signaling activity (constitutively active mutants [CAMs]), and to characterize antagonists of this receptor *(4)*. Methods used for this approach are described herein.

2. Materials

The yeast strains and vectors used in this system were genetically engineered at Cadus Pharmarceuticals, Inc. *(5)* and unpublished data.

1. Yeast strain CY12946.
2. *FUS1-lacZ* reporter plasmid Cp1584.
3. Expression vector Cp4258.
4. Human CXCR4 complementary DNA.
5. Yeast medium components:
   a. Yeast nitrogen base (YNB) without amino acids (Sigma; cat. no. Y 0626).
   b. Dextrose.
   c. Yeast synthetic drop-out medium supplement (Sigma; cat. no. Y 2001).
   d. Leucine, histidine, tryptophan, uracil.
   e. HEPES sodium salt (Sigma, cat. no. H 3784).
   f. SDF-1 (Leinco; cat. no.S111).
6. Flat-bottom 96-well plates and 384-well black plates.
7. Oligonucleotide primers.
8. Restriction enzymes, Taq DNA polymerase, and T4 DNA ligase.
9. Agarose gels and DNA sequencing facility.
10. Fluorescein di-β-D-galactopyranoside (FDG): fluorogenic substrate of β-galactosidase (Molecular Probes; cat. no. F1179).
11. QIAprep Spin Miniprep Kit, Qiagene (cat. no. 27106).
12. Zymoprep Yeast Plasmid Miniprep™ (ZYMO Research; cat. no. D2001).
13. NovaBlue Competent Cells: Novagen (cat. no. 69825-4).
12. Frozen-EZ Yeast Transformation II™ (ZYMO Research; cat. no. T2001).
13. Zymoclean™ Gel DNA Recovery Kit (ZYMO Research; cat. no. D4002).

14. DNA Clean and Concentrator™-5 (ZYMO Research; cat. no. D4004).
15. 9E10 antibody, anti-C-Myc (Santa Cruz Biotechnology, Inc.; cat. no. sc-40).
16. Fusionα™ (Packard) was used to detect fluorescent intensity in 384-well plates for *lacZ* assay.
17. Plate reader, read absorbance optical density at 600 n*M* (OD$_{600}$) of yeast in 96-well plates.

2.1. Yeast Medium Preparation

2.1.1. Medium Stocks

1. 10X YNB (500 mL): weigh 33.5 g of YNB without amino acids (Sigma; cat. no. Y-0626); add water to 500 mL to dissolve. If needed, incubate in a 65°C water bath to help dissolve.
2. 10X of dextrose (500 mL) (20%): weigh 100 g of dextrose (Fisher Chemicals; cat. no. D14-500), add water to 500 mL, and dissolve.
3. 10X Dropout (500 mL): weigh 7 g of yeast synthetic drop-out medium supplement (Sigma; cat. no. Y-2001), add water to 500 mL, and dissolve.
4. 100X Leucine (6 mg/mL): weigh 300 mg and dissolve in 50 mL of water.
5. 100X Histidine (2 mg/mL): weigh 100 mg and dissolve in 50 mL of water.
6. 100X Tryptophan (4 mg/mL): weigh 200 mg and dissolve in 50 mL of water.
7. 100X Uracil (2 mg/mL): weigh 100 mg and dissolve in 50 mL of water.
8. 0.5 *M* HEPES, pH 6.8: weigh 23.8 g of HEPES sodium salts (molecular weight 238.8) and dissolve in 180 mL of water, adjust the pH to 6.8 with 10 *N* NaOH, and adjust the final volume to 200 mL.
9. All the solutions should be sterilized by filtration (0.2-μm pore size).

All liquid synthetic medium for yeast culture should contain 1X YNB, 1X Dropout, 1X dextrose, and the proper content of 1X amino acids.

2.2. Solid Medium for Plates

1. To make 200 mL of solid medium for plates, measure following items:
   a. Agar: 4 g (2%).
   b. 10X YNB: 20 mL.
   c. H$_2$O: 132–138 mL (depending on how many kinds of amino acids are needed).
   d. 10X NaOH: 50 μL.

Autoclave. Allow to cool to approx 80°C, then add:

   a. 10X Dropout: 20 mL.
   b. 10X Dextrose: 20 mL.
   c. 100X AA: 2 mL each as needed and pour the plates.

2.3. Preparation of FDG Substrate Solution Components

1. FDG: dissolve 5 mg of FDG with 769 μL of dimethylsulfoxide to reach 10 m*M* and make aliquots. Tubes are wrapped with foil to avoid light exposure and stored at –20°C.

2. 1 *M* 1,4-Piperazinediethanesulfonic acid (PIPES), pH 7.2: weigh 12.1 g of PIPES, dissolve it in 25 mL of H_2O, adjust pH with 10 *N* NaOH (approx 5 mL of 10 *N* NaOH), and adjust the final volume to 40 mL.

   a. FDG substrate solution (100 μL).

   b. 10 m*M* FDG: 10 μL.

   c. 20% Triton X-100: 12 μL.

   d. 1 *M* PIPES: 13 μL.

   e. H_2O: 75 μL.

3. Methods

The methods presented in detail in this section describe yeast strains and reporter gene plasmids (**Subheading 3.1.**), construction of the receptor expression vector (**Subheading 3.2.**), transformation of yeast cells (**Subheading 3.3.**), growth of yeast (**Subheading 3.4.**), reporter gene assays (**Subheading 3.5.**), generation of receptor CAMs (**Subheading 3.6.**), screening for GPCR antagonists (**Subheading 3.7.**), and a summary of important concepts (**Subheading 3.8.**). More complementary information can be found in two previous publications from this laboratory *(4)* and others *(6)*.

3.1. Yeast Strains and Reporter Gene Plasmids

The *S. cerevisiae* strain CY12946 (*FUS1p-HIS3 gpa1*::GPA1-Gαi2(5) far1 1442 sst2 2 ste14 ::trp1::LYS2 ste3 1156 tbt1-1 his3 leu2 lys2 trp1 ura3 can1) was used to functionally express CXCR4 and to generate CXCR4 CAMs. This strain demonstrates histidine auxotrophy that is complemented by a *his3* gene under the transcriptional control of a (pheromone-responsive) *FUS1* promoter. It expresses a chimeric G_α subunit that is encoded by the *gpa1* gene (yeast G α-subunit) with the last five codons replaced by those from the rat gene encoding the $G_{\alpha i}$ subunit. Genes responsible for cell cycle arrest after mating factor activation have been deleted *(5)*, as has the gene encoding the endogenous GPCR for the alpha mating factor. This strain requires the supplementation of leucine, histidine, tryptophan, and uracil in synthetic medium unless a compensatory expression of these synthetic enzymes encoded by the plasmid vector or the activation of the receptor to synthesize His3. His growth assay can be readily done with this strain with the synthetic medium lacking histidine to determine whether CXCR4 is activated via the hybrid G protein.

In addition to the *FUS1-HIS3* reporter gene system, *lacZ* also can be used as a reporter gene that can be quantitated by enzymatic assay for its product, β galactosidase. The plasmid Cp1584 encoding the *FUS1-lacZ* reporter gene uses auxotrophic selection for tryptophan, as has been previously described *(5)*. Yeast strain CY12946 can be transformed with Cp1584 and grow on/in the medium lacking tryptophan to maintain the plasmid. This strain then can be prepared as the competent cells for the *lacZ* reporter gene assay.

3.2. Construction of the CXCR4 Expression Vectors

The yeast LEU2 2-μ plasmid Cp4258 was used as an expression vector for CXCR4. This plasmid contains the phosphoglycerate kinase 1 promoter, which directs constitutive expression. The remainder of the expression cassette includes the prepro domain of α factor (codon 1-89), composed of the signal peptide and the propeptide that is proteolytically cleaved by KEX2, a yeast host protease, a multiple cloning site (*Nco*I and *Xba*I), and the transcriptional termination region from *PHO5* (repressible acid phosphatase). The open reading frame (ORF) encoding CXCR4 was engineered using polymerase chain reaction to contain Myc and His epitope tags at the N-terminus and C-terminus, respectively, flanked by *Nco*I and *Xba*I sites compatible with directional cloning into corresponding restriction sites in the Cp4258 vector. The expression construct was transformed into competent NovaBlue cells. Plasmid DNA from colonies were extracted from *Escherichia coli* using Qiagen miniprep kits and recombinant Cp4258-CXCR4 construct was confirmed by DNA sequencing with Cp4258F (5'-CGATGTTGCTGTTTTGCC-3') and Cp4258R (5'-CGTATCTGACGTAGGTGTCG-3').

3.3. Transformation of Yeast Cells

Constructs were introduced into yeast cells using the Frozen-EZ Yeast Transformation-II kit (Zymo Research). This is a simple procedure that is similar to methods that require lithium cations but probably involves some metabolic pathways that are not fully understood at this time. It does not involve a spheroplast step and can be used to produce competent cells from *Candida albicans*, *Schizosaccharomyces pombe*, or *Pichia pastoris*, as well as *S. cerevisiae*. A detailed procedure can be found in the manufacturer's protocol. The following represent the critical steps:

1. Cells are grown in yeast synthetic broth with the proper selective pressure at 30°C until mid-log phase ($OD_{600} = 0.5$ to 1).
2. The cells of 10-mL culture are then pelleted at $600g$ using a bench-top centrifuge.
3. Pellets are resuspended in 10 mL of solution 1 and centrifuged as in **step 2** to wash.
4. The pellet is resuspended in 1 mL of solution 2.
5. At this point, the competent cells can be dispensed into 50-μL aliquots and used for transformation immediately or stored at –70°C after gradual freezing.
6. For transformation, 0.2 to 1.0 μg of DNA is added to a 50-μL aliquot of competent cells, and then 500 μL of solution 3 is added and mixed thoroughly. The transformation reaction is incubated at 30°C water bath for 45 min.
7. The mixture is spread on a plate containing appropriate selective medium and incubated in 30°C for 2 to 3 d.

More than one construct can be cotransformed into the host yeast by incubating with multiple plasmids and plating out the reactions on the proper selective medium. Colonies are then picked, expanded, and subjected to functional tests.

The presence of the ORF encoding the receptor is confirmed by polymerase chain reaction amplification of yeast DNA and expression of the GPCR protein verified by Western blotting using a monoclonal antibody to the N-terminal Myc epitope tag. Cell surface expression and functional coupling to the hybrid G protein in the yeast need to be tested with the reporter gene assays with the corresponding ligand, SDF-1/CXCL12. Although actual cell surface expression experiments are difficult to perform in yeast, the detection of activated signaling provides indirect evidence for proper trafficking in these cells.

3.4. Growth of Yeast

A variety of experiments require yeast cells expressing CXCR4. Cultures may be grown on solid medium in plates or in liquid medium in a fashion very similar to that used for bacteria. Cultures grown on plates are prepared by streaking stocks onto synthetic medium agar plates that have deletion of the appropriate nutrients for selection according to the strain and enzymes encoded by the plasmids. Typically, liquid cultures are initiated by inoculating single colonies from a plate into 100 µL of broth in wells of flat-bottom 96-well plate or 3 mL in bacteria culture tubes and growing these cultures in 30°C incubator or shaker overnight or up to 2 d, depending on the viability of the stock. Although an exhaustive survey of growth conditions has not been performed, excellent results in reporter gene assay experiments have been obtained using yeast grown to log phase, with a density reflected by OD_{600} between 1 and 2. This density roughly corresponds to 1 to 2×10^7 cells per milliliter. Absorbance reading can be obtained directly for the yeast grown in wells of a microplate, but conversion to values obtained by standard analysis in cuvets is required (*see* **Note 1**).

3.5. Reporter Gene Assays

Two reporter genes (*His3* and *lacZ*) are in standard use for determining GPCR-dependent activation of the pheromone responsive *FUS1* promoter.

3.5.1. Histidine-Independent Growth Assay

The *FUS1-HIS3* reporter gene complements histidine auxotrophy of the host strain through induction of histidine synthesis. Thus, cells from the parental strain that lack the ability to grow in medium deficient in histidine can proliferate without supplementation of this amino acid when CXCR4 activation triggers the pheromone response pathway. The method for measuring expression of this reporter gene is a simple growth assay of (histidine auxotrophic) cells in medium lacking histidine as follows:

1. Inoculate a colony of CY12946 yeast tranducing the Cp4258 CXCR4 construct in leucine-deficient medium to maintain expression of this plasmid and grow

overnight at 30°C. Because yeast do not grow rapidly, the density of these overnight cultures varies, but is typically 1 to 2 OD.

2. The overnight cultures are pelleted and resuspended in leucine and histidine-deficient medium to wash yeast cells. Cells are then pelleted in a microcentrifuge and reconstituted to a proper volume of leucine- and histine-deficient medium to a final reading of OD_{600} approx 0.01.

3. A 90-μL aliquot of the diluted culture is then transferred to a 96-well plate for the growth experiment. The final volume of the cultures is adjusted to 100 μL, allowing for the addition of, for instance, chemokines and antagonist candidates. Typically, all experimental conditions are determined in duplicate in wells of a flat-bottomed microtiter plate. Absorbance at 600 nm is determined in an ELISA reader every 2 h during daytime for 36 to 48 h, beginning after growth for approx 18 h. This approach can be adapted to accommodate shorter initial growth intervals by increasing the OD_{600} of yeast cells in the culture reaction (*see* **Note 2**). Plotting of absorbance vs time gives a growth curve.

4. For experiments with single end points, one reading at a proper time can be used to determine the activation of CXCR4 instead of multiple readings. Functional expression of the GPCR is determined in dose response experiments with single end points by demonstrating ligand-dependant growth in the absence of histidine, indicating activation of the *FUS1-HIS3* reporter gene.

5. The positive control for the histidine-independent growth assay is with WT-CXCR4 with 1 to 2 μ*M* SDF-1 in the growth medium or a CXCR4 mutant with constitutively autonomous signaling activity. The specificity of signaling can be demonstrated by exposing yeast strains expressing other chemokine receptors (we have used CCR5) to SDF-1/CXCL12, which uniquely activates CXCR4.

6. A typical experiment demonstrating activation of the *FUS1-HIS3* reporter gene is shown in **Fig. 1**. Cells of the CY12946 yeast strain expressing CXCR4, or other chemokine receptors, such as CCR5, do not grow in medium lacking histidine. Addition of SDF-1/CXCL12 (1–3 μ*M*) to strains expressing CXCR4, but not control cells or strains expressing other chemokine receptors, results in growth similar to that observed in medium containing histidine. Expression of a CXCR4 mutant with substitution of Ser for Asn-119 in transmembrane helix 3 confers expression of the *FUS1-HIS3* reporter gene without addition of SDF-1/CXCL12, indicative of autonomous activation of the pheromone response pathway.

The second reporter gene commonly used is *FUS1-lacZ*, which encodes β-galactosidase. Expression of this reporter gene by ligand-induced GPCR activation of the pheromone response pathway is determined by enzymatic analysis of β-galactosidase activity.

1. Yeast CY12946 with Cp4258-CXCR4 and Cp1584 *FUS1-lacZ* plasmids are cultured overnight in 96-well plates in synthetic medium lacking leucine and tryptophan (OD_{600} ~1–2).

2. To set up the *lacZ* assay, the cultures for analysis are prepared by calculating the volume of cells needed to obtain a 100-μL culture OD_{600} approx 0.15. The proce-

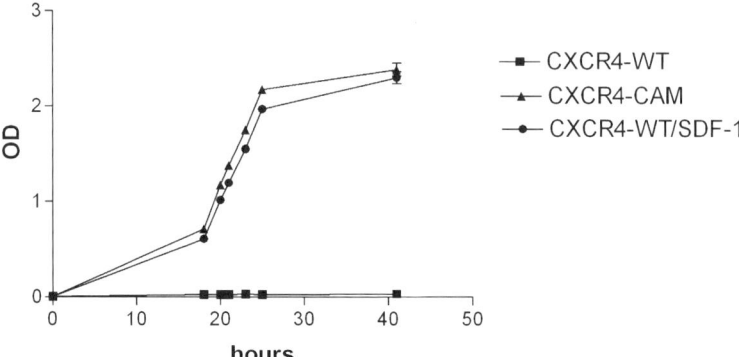

Fig. 1. Growth of histidine auxotrophic yeast strains expressing chemokine receptors coupled to a pheromone-responsive *his3* reporter gene. This figure illustrates the induction of growth of yeast strains deficient in histidine synthesis by activation of a pheromone-responsive *HIS3* reporter gene through CXCR4 signaling. Yeast strains expressed CXCR4-wild-type (WT) or CXCR4-constitutively active mutant (CAM). Whereas yeast-expressing CXCR4-WT grew only when exposed to SDF-1/CXCL12, the autonomous signaling of the CXCR4-CAM resulted in activation of the *FUS1-HIS3* reporter gene independent of the physiological ligand.

dure is as described previously, except that the medium used is adjusted to pH 6.8 with 0.5 *M* HEPES buffer and lacks both leucine and tryptophan to maintain plasmids encoding CXCR4 and the *lacZ* reporter gene. The cells are then grown to mid-log phase (OD_{600} ~0.5–0.8), typically around 6 h. The initial density of the yeast cell cultures may be adapted to increase the initial incubation to permit overnight growth (*see* **Note 2**).

3. Before the culture is ready, prepare FDG substrate solution (as described in **Subheading 2.3.**).
4. A 10-µL aliquot of each well of the yeast cultures is transferred to a 384-well black microtiter plate and 2 µL of the FDG substrate solution is added, pipetting several times to mix thoroughly. This manipulation must be performed in the absence of direct light. The plate is then covered with plastic wrap and foil and incubated at 37°C for 30 to 60 min (in the dark), depending on the strength of the signal. Cleavage of the substrate by β-galactosidase (encoded by the pheromone-responsive *lacZ* reporter gene) to produce the fluorescent product is determined in a universal plate reader, Fusionα (Packard). The data are transferred to Microsoft Excel for analysis.

3.6. Generation of CAMs

The generation of CAMs is performed using the *FUS1-HIS3* reporter gene to select for receptor mutants that autonomously trigger the pheromone response pathway. Pools of GPCR mutants are developed by random mutagenesis of the ORF. Because autonomous signaling results in constitutive expression of *HIS3*, pools are screened for growth of histidine auxotrophic strains in medium lacking both the GPCR ligand, SDF-1/CXCL12 in the case of CXCR4, and histidine.

The key to generating GPCR CAMs is the development of a pool of GPCR ORFs containing random mutations at an appropriate frequency. A rate of 0.2 to 0.4% is ideal for CXCR4, because it results in two to four nucleotide substitutions in the 1000-base pair ORF, thereby programming approximately one to two amino acid changes in the receptor. Random mutagenesis may be performed using PCR strategies with inclusion of deoxyinosine 5'-triphosphate (dITP) or Mn^{2+} in the amplification reaction or mismatch of the dNTP concentrations (conditions we have used are detailed under **Note 3**).

1. Amplification reactions contain primers that flank the 5' and 3' ends of the ORF and include the Myc and His epitope tags, respectively.
2. The amplification products are subjected to electrophoresis in a 1% agarose gel and the band corresponding to the full-length ORF is excised from the gel and eluted by centrifugation using the Zymoclean™ Gel DNA Recovery Kit.
3. This pool is then digested with *Nco*I and *Xba*I to generate cohesive ends compatible with those in the Cp4258 vector.
4. After ligation of the pool of ORF random mutants into Cp4258 and transformation of competent *E. coli*, approx 10 colonies are selected for DNA extraction and nucleotide sequence analysis. This enables confirmation of the rate of random mutation in the GPCR ORF.
5. If the rate is appropriate, the entire ligation reaction is transformed into competent *E. coli*, scraping of the individual colonies, and extraction of plasmid DNA from the pool.
6. This plasmid pool is then used to transform yeast cells. An aliquot of the transformation reaction is plated on solid medium supplemented with histidine to determine the efficiency of transformation and total number of events.
7. The majority of the transformation reaction is plated on medium lacking histidine and grown for 2 to 3 d. Yeast colonies that grow in the absence of histidine are candidates for containing CAMs.
8. Plasmids are extracted from the individual yeast colonies using Zymoprep Yeast Plasmid Miniprep™ and transformed into NovaBlue competent cells.
9. The individual plasmids are retransformed into the CY12946 yeast strain and tested for the ability to autonomously activate the pheromone-responsive *HIS3* reporter gene.
10. If this is confirmed, the individual candidate plasmid is transformed into the CY12946 yeast strain with *FUS1-lacZ* reporter gene plasmid Cp1584, and tested for induction of β-galactosidase activity.

With this approach, we obtained a CXCR4 mutant with constitutively active signaling conferred by conversion of Asn-119 in the third transmembrane domain to Ser.

3.7. Screening for GPCR Antagonists

The simplicity of growth of yeast cells makes this GPCR expression system well suited for high-throughput screening of antagonists. Two strategies are possible using this system: (1) inhibition of ligand-induced GPCR signal transduction, and (2) alteration of GPCR CAM signaling. Whereas the former approach is capable of detecting all pharmacological classes of antagonist, weak partial agonists, and inverse agonists, the latter approach is capable of detecting neutral antagonists, frequently with greater sensitivity.

The approach involving inhibition of ligand-induced GPCR signal transduction is straightforward. CY12946 yeast cells cotransformed with Cp1584 and Cp4258-CXCR4-WT are incubated with SDF-1 at a final concentration of 1 to 2 μM and 10 μL of each candidate antagonist to make a final concentration of 10 μM (screening may be performed at 0.1, 1, or 10 μM) for approx 6 h, as described in **Subheading 3.6**. As shown in **Fig. 2A**, β-galactosidase assays are performed to determine whether CXCR4 antagonist candidates FC001 to FC072 *(7)*, a combination of T140 conformation- and sequenced-based molecular-size reduction library, block SDF-1 induction of the *lacZ* reporter gene. Although this assay is able to detect any type of antagonist, it has the disadvantage of being reliant on access to significant quantities of ligand, which, albeit not be a problem for many GPCRs, represents a major obstacle for CXCR4. Whereas activation of CXCR4 signaling in mammalian cells requires concentrations of approx 1 nM, triggering of the pheromone response pathway in yeast requires concentrations of approx 1 to 5 μM to reach a reliable readout. This dramatic difference in ligand concentration may result from a barrier effect on SDF-1 binding to CXCR4 by the yeast cell wall or the absence of pathways to sulfate critical tyrosine residues in the CXCR4 N-terminal extracellular domain in yeast. Thus, high-throughput screening of CXCR4 antagonists with ligand would require prohibitive amounts of SDF-1. It is possible to use an autocrine system in which yeast cells are programmed to express both CXCR4 and SDF-1/CXCL12. This approach has potential pitfalls that include the preferential interaction of ligand and receptor in intracellular compartments prior to access to the antagonist candidate and variation in expression levels and stability of the chemokine ligand (which we observed for SDF-1/CXCL12).

In contrast, the yeast signaling system driven by the GPCR CAM does not require ligand and is highly suited for screening for antagonists of the weak partial agonist and inverse agonist pharmacological types, as shown in **Fig. 2B**. The former class of agents further activates GPCR CAM signaling and the latter extinguish GPCR CAM signaling by shifting the conformational equilibrium to the inactive state. The use of GPCR CAMs for the detection of weak partial agonists and inverse agonists requires only minor modification of the

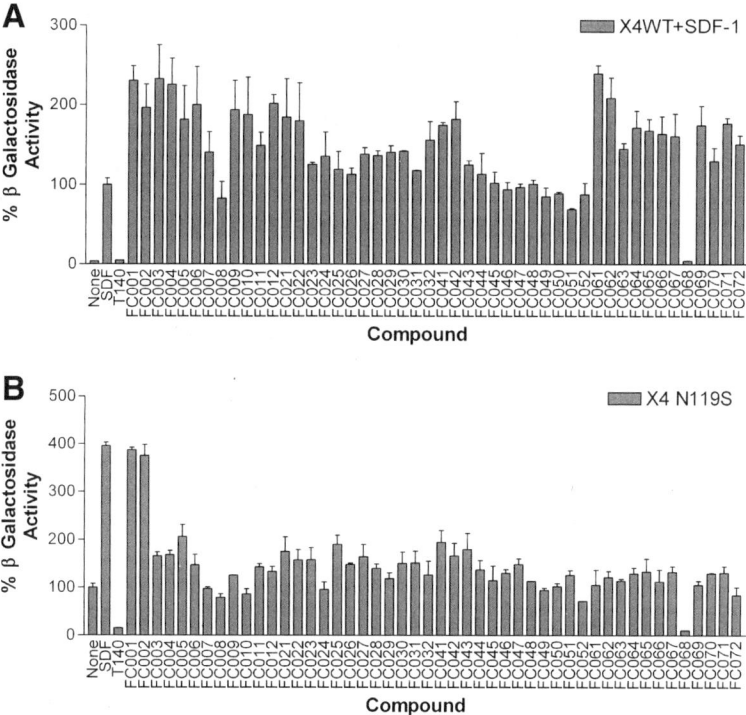

Fig. 2. Identification of antagonists using yeast strains expressing CXCR4 coupled to the pheromone response pathway. Yeast strains expressing native CXCR4 or CXCR4(N119S), a constitutively active mutant, were used to screen a combinatorial library of cyclic pentapeptides, as shown in (**A**) and (**B**), respectively. The effects of these compounds on the activation of native CXCR4 by SDF-1/CXCL12 are demonstrated in (**A**). Changes in the levels of activation of the pheromone responsive *lacZ* reporter gene, reflected by β-galactosidase activity, driven by the CXCR4-CAM in response to SDF-1, the natural ligand, T140, a 14-residue inverse agonist, FC001 and FC002, weak partial agonists, and FC68, an inverse agonist, are illustrated in (**B**).

assay for the *FUS1-lacZ* reporter gene. Screening is performed in cultures containing 10 μ*M* of candidate compounds with yeast containing CXCR4-N119S during growth to mid-log phase. β-galactosidase activity is then determined as described in **Subheading 3.5.**

3.8. Summary

CXCR4 was expressed in yeast functionally coupled to the pheromone response pathway. Exposure to SDF-1-activated signaling mediated by CXCR4 via a hybrid G_a subunit that induced the expression of two different reporter genes. This approach also was applied successfully to CCR5, but several other receptors for CC chemokines were not functional in this yeast signaling system. Both CXCR4 and CCR5 required high concentrations of SDF-1/CXCL12 and RANTES/CCL5 to trigger signaling. To adapt this system for studying mechanisms for receptor activation and screening and characterization of antagonists, constitutively active mutants were genetically engineered. These variants represent powerful tools for studying chemokine receptor signaling in mammalian cells, for screening libraries for antagonists, and for the pharmacological characterization of candidates. This system also can be applied to identify and discover natural ligands for orphan receptors.

4. Notes

1. To determine the cell density of yeast cultures grown in 96-well plates, we obtain OD of 100 μL yeast in 96-well plate at 600 nM with a plate reader, then multiply by 3.3 to convert the reading to the OD obtained in standard disposable cuvets with a 1-cm path length ($A_{600\,nM}$ in microtiter plate × 3.3   OD).
2. A doubling time of 2 to 2.5 h is used to estimate proliferation of yeast during the initial time interval for analyzing time course growth experiments and single-end point assays.
3. Conditions for random mutagenesis PCR used:

Components	Condition 1	Condition 2	Condition 3
H_2O	21 μL	15.5 μL	19 μL
10X buffer	5 μL	5 μL	5 μL
dNTP	2.5 mM each, 4 μL	2 mM dATP, 10 mM dG, dT, and dC, 5 μL	2.5 mM each, 4 μL
dITP	—	—	400 μM, 5 μL
5' and 3' primers	10 μM, 4 μL	10 μM, 5 μL	10 μM, 5 μL
$MgCl_2$	20 mM, 5 μL	20 mM, 7.5 μL	25 mM, 5 μL
$MnSO_4$	1 mM, 5 μL	6 mM, 5 μL	—
Template	100 ng/μL, 1 μL	100 ng/μL, 1 μL	100 ng/μL, 1 μL
Taq	1 μL	1 μL	1 μL

Acknowledgments

Research described in this chapter was supported in part by Philip Morris USA Inc. and by Philip Morris International.

References

1. Malbon, C. C. (2004) Frizzleds: new members of the superfamily of G-protein-coupled receptors. *Front. Biosci.* **9,** 1048–1058.
2. Bardwell, L., Cook, J. G., Inouye, C. J., and Thorner, J. (1994) Signal propagation and regulation in the mating pheromone response pathway of the yeast *Saccharomyces cerevisiae. Dev Biol* **166,** 363–379.

3. Ozcan, S., Dover, J., and Johnston, M. (1998) Glucose sensing and signaling by two glucose receptors in the yeast *Saccharomyces cerevisiae. EMBO J.* **17,** 2566–2573.

4. Zhang, W. B., Navenot, J. M., Haribabu, B., et al. (2002) A point mutation that confers constitutive activity to CXCR4 reveals that T140 is an inverse agonist and that AMD3100 and ALX40–4C are weak partial agonists. *J. Biol. Chem.* **277,** 24,515–24,521.

5. Klein, C., Paul, J. I., Sauve, K., et al. (1998) Identification of surrogate agonists for the human FPRL-1 receptor by autocrine selection in yeast. *Nat. Biotechnol.* **16,** 1334–1337.

6. Zhang, W. B., Wang, Z. X., Murray, J. L., Fujii, N., Broach, J., and Peiper, S. C. (2004) Functional expression of CXCR4 in *S. cerevisiae*: development of tools for mechanistic and pharmacologic studies. *Ernst. Schering. Res. Found Workshop*, 125–152.

7. Fujii, N., Oishi, S., Hiramatsu, K., et al. (2003) Molecular-size reduction of a potent CXCR4-chemokine antagonist using orthogonal combination of conformation- and sequence-based libraries. *Angew. Chem. Int. Ed. Engl.* **42,** 3251–3253.

6

Characterization of Constitutively Active Mutants of G Protein-Coupled Receptors

Jean-Marc Navenot, Zi-xuan Wang, and Stephen C. Peiper

Summary

The ability of G protein-coupled receptors to transduce signaling typically is induced by the binding of an appropriate ligand (agonist), resulting in a conformational change of the receptor and the subsequent interaction with the G protein heterotrimer. Some mutants of G protein-coupled receptors, known as constitutively active mutants, have the capacity to activate the G protein-signaling cascade even in the absence of ligand. In this chapter, we describe three methods that most directly allow characterization of constitutively active mutants and discriminate them from the wild-type receptors. All methods are based on the spontaneous signaling function in the absence of ligand and its consequences on the receptor.

Key Words: G protein-coupled receptor; signaling; constitutive activity; phosphorylation; trafficking.

1. Introduction

With approx 1000 known members, G protein-coupled receptors (GPCRs) represent the largest family of membrane receptors. Their functions are as diverse as their physiological ligands, which range from photons activating retinal to large polypeptides. Despite this extreme diversity of nature of ligands and function of the receptors, GPCRs signal through a limited number of G proteins, although it recently has become evident that part of the signaling is independent of G proteins. Furthermore, all GPCRs share a similar structure, which consists of seven transmembrane (TM) helices connected by three extracellular domains and three intracellular domains, an extracellular N-terminal fragment, and a cytoplasmic C-terminal fragment. Because of their complexity, limited information is available about the precise structure of GPCRs and

From: *Methods in Molecular Biology, vol. 332: Transmembrane Signaling Protocols, Second Edition*
Edited by: H. Ali and B. Haribabu © Humana Press Inc., Totowa, NJ

their mechanism of activation. The canonical GPCR rhodopsin has been crystallized, and most of structural information obtained on other receptors is based on their analogy with rhodopsin *(1)*.

However, there is strong evidence that the mechanism of activation of GPCRs involves relative movements of rotation between TM3 and TM6 and conformational changes of TM6 *(2,3)*. To gain insight into the structural features of GPCR activation, constitutively active mutants (CAMs), which can spontaneously transduce a signal in the absence of ligand, constitute an interesting model. Numerous CAMs have been described, either naturally occurring or generated in laboratories by mutagenesis. In humans, CAMs are associated with a number of diseases *(4)*. Some of these CAMs are mutants of human receptors, whereas others have a viral origin *(5–8)*. For instance, the Kaposi's sarcoma-associated herpesvirus encodes for a CAM that, when expressed in infected cells, is responsible for at least part of the physiopathology *(9,10)*. The biological relevance of these in vitro observations was confirmed in a transgenic model developed in mice *(11,12)*. The constitutive activity of Kaposi's sarcoma-associated herpesvirus-GPCR and other CAMs has been shown to be modulated by human ligands *(13–15)*. We focused on the chemokine receptors CXCR4 and CCR5, which are the main co-receptors necessary for infection by HIV-1 *(16,17)*. Whereas CCR5 is quite dispensable (individuals with a genetic defect in CCR5 expression are healthy and are highly resistant to HIV-1 infection), the critical biological role of CXCR4 has been emphasized by the fact that the knock-out of the receptor itself, as well as the knockout of its ligand stromal cell-derived factor (SDF)-1 (CXCL12), has proven lethal in mice where embryos die *in utero* from major cardiac, neurological, and vascular development defects *(18–22)*. Because of its multiple and critical biological functions, the investigation of the structure–function relationships involved in CXCR4 signal transduction is highly valuable. Because CAMs are a promising tool to approaching the mechanistic aspects of GPCR activation, CAMs of CXCR4 and CCR5, as well as CCR2, have been obtained and characterized *(16,17)*. All result from single-point mutations in the TM domains of the receptors. The constitutive activity of the CAMs of CXCR4 can be modulated by the binding of SDF-1 or synthetic ligands, which could be characterized as agonists or inverse agonists of the receptor.

This chapter describes a strategy to characterize CAMs of GPCRs in mammalian cells using CXCR4 as an example to illustrate autonomous and residual ligand-induced signaling and desensitization. Because the signaling of GPCRs normally is induced by the binding of the ligand and the signaling of CAMs is independent from the ligand, CAMs typically will have a reduced ability to respond to their ligand even though the binding is preserved. As a consequence, the signaling assays frequently performed to assess the transient response of a receptor to its ligand (e.g., calcium flux) usually will show a reduced response or even an absence of response, which is why CAMs can easily be mistaken with nonfunctional receptors. This chapter describes three methods that can discriminate CAMs from normal or nonfunctional receptors and that are either a direct measurement of the receptor activity or a direct consequence of this activity. First, we will describe the activation of G proteins and its evaluation by the γ-S-guanosine triphosphate (GTP) assay. Then, we will study a method to evaluate the phosphorylation of the receptor resulting from its constitutive activation of GPCR kinases (GRKs) and, finally, we will describe the effect of that constitutive phosphorylation of the receptor on its trafficking inside the cells by immunofluorescence and confocal microscopy.

1.1. Evaluation of Constitutive Activity of CXCR4 by γ-S-GTP Binding Assay

This assay can be considered as the most direct evidence of the constitutive activity of GPCRs in mammalian cells. This assay is based on the mechanism of activation of G-proteins. GTP binding proteins are composed of three subunits (α, β, γ), the α subunit being the one that binds guanosine diphosphate (GDP)/GTP. In a resting state, the GDP-bound α subunit makes a stable complex with the $\beta\gamma$ subunits. Upon activation by an agonist-activated GPCR, GDP is replaced by GTP on the α subunit, which results in the dissociation of the $\beta\gamma$ complex. This dissociation confers a GTPase activity to the α subunit, which then can hydrolyze GTP into GDP + Pi. The GDP-bound form of the α subunit can then reassociate with a $\beta\gamma$ complex, restoring a heterotrimer that can be activated again by a GPCR. This cycle can be interrupted by a nonhydrolyzable form of GTP (γ-S-GTP) in which the double-bound oxygen on the third (or γ) phosphate is replaced by a sulfur, hence preventing the hydrolysis of that phosphate. If a radioactive form (γ-[^{35}S]-GTP) is used, the activation of a GPCR will result in the accumulation on the cytoplasmic face of the plasma membrane of G proteins bound to radioactive GTP, accumulation that can be assessed by measuring the radioactivity bound to the membrane by liquid scintillation.

1.2. Evaluation of Constitutive Phosphorylation of GPCR CAMs

Activation of GPCRs by their ligand leads to the phosphorylation of cytoplasmic residues of the receptor by GRKs. This phosphorylation is responsible for the internalization of the receptors and the termination of the signal but also is involved directly in part of the signaling. This phosphorylation can be investigated by loading the cells with [^{32}P] orthophosphate, which will be metabolically integrated into adenosine triphosphate. Upon activation of the cells, the labeled phosphates will be used to phosphorylate the molecules involved in signal transduction, including the receptor itself. Immunoprecititation of the receptor with an antibody specific for the molecule itself or for an epitope tag expressed at the N-terminus (or the C-terminus) followed by autoradiography will reveal the level of phosphorylation. CAMs of GPCR will activate GRKs even in the absence of ligand and are expected to display a high level of constitutive phosphorylation.

2. Materials

2.1. Constitutive Activity of CXCR4 by γ-S-GTP Binding Assay

1. Polypropylene 5-mL round bottom tubes.
2. 1.5-mL microcentrifuge tubes.
3. Chinese hamster ovary (CHO) cells.
4. Citrate buffer: 15 mM Na citrate, 135 mM KCl.
5. Lysis buffer: 5 mM Tris-HCl, pH 7.5; 5 mM ethylene diamine tetraacetic acid (EDTA); 5 mM ethylenebis(oxyethylenenitrilo)tetraacetic acid (EGTA).
6. Reaction buffer: 50 mM Tris-HCl, pH 7.5, 5 mM MgCl$_2$, 1 mM EGTA, 100 mM NaCl.
7. Syringes with 28.5-gage needle (insulin syringe).
8. Protein-assay kit (Bio-Rad).
9. 96-Well plates with flat bottom for colorimetric assays.
10. Bovine serum albumin (BSA).
11. SDF-1.
12. γ-[^{35}S]-GTP.
13. γ-S-GTP.
14. GDP.
15. Phosphate-buffered saline (PBS).
16. GF/C filters (Whatman).
17. Vacuum manifold (Millipore).
18. Scintillation liquid.
19. Scintillation counter.
20. Low-speed centrifuge.
21. Microplate spectrophotometer with filter at 595 nm.

2.2. Constitutive Phosphorylation of GPCR CAMs

1. 60-mm Tissue culture dishes.
2. Complete growth medium: minimal essential medium (MEM)-α with 1% antibiotics (penicillin, streptomycin, amphotericin) and 10% fetal bovine serum (FBS).
3. Phosphate-free Dulbecco's modified Eagle's medium (DMEM, GIBCO-Invitrogen).
4. [^{32}P] orthophosphate (high specific activity, 8500–9120 Ci/mmol; NEN-Perkin-Elmer Life Sciences).
5. Protease inhibitors cocktail (Sigma).
6. Phosphatase inhibitor cocktail (Sigma).
7. Lysis buffer: 50 mM Tris-HCl, pH 8.0, 150 mM NaCl, 10 mM EDTA, 1% Triton X-100; 0.1% sodium dodecyl sulfate (SDS), 1% protease inhibitor cocktail, 1% phosphatase inhibitor cocktails.
8. Anti-cMyc mouse monoclonal antibody 9E10.
9. 1.5-mL conical microtubes with sealed cap.

10. Proteine G-sepharose (Pharmacia).
11. SDS polyacrylamide gel electrophoresis sample buffer: Tris-HCl, glycine, SDS, glycerol, bromophenol blue, b-mercaptoethanol.
12. SDS polyacrylamide gel electrophoresis equipment.
13. Gel dryer.
14. Films for autoradiography or phosphor-imager.
15. High-speed refrigerated microcentrifuge.

2.3. Localization of CXCR4 by Immunofluorescence and Confocal Microscopy

1. Tissue culture treated glass cover slips (no. 1, 0.17-mm thick).
2. Glass slides.
3. Methanol.
4. Goat anti-mouse immunoglobulin G labeled with Alexa Fluor 488 (Molecular Probes).
5. TO-PRO-3, solution in dimethylsulfoxide (Molecular Probes).
6. Antifade mounting medium (ProLong, Molecular Probes, or equivalent).
7. Confocal microscope equipped with lasers emitting at 488 nm and 633 nm and at least 40X oil immersion objective.

3. Methods

3.1. γ-S-GTP-Binding Assay to Determine Constitutive Activity of CXCR4

1. Transfectants of CHO cells are grown in 100-mm tissue culture dishes and are subcultured the day before the experiment. The culture should be approx 70 to 80% confluent on the day of the experiment (*see* **Note 1**).
2. Cells are detached from the plate with citrate buffer as follows: wash the cells once with the citrate buffer, aspirate the buffer and add 5 to 6 mL of the same buffer. Put the plates back in the incubator at 37°C for approx 5 min until the cells start detaching. Resuspend the cells with a pipette and transfer the cell suspension into a 15-mL conical tube. Pellet the cells by centrifugation (4 min at 200g), resuspend the cells in 1 mL of PBS, transfer into a 1.5-mL microcentrifuge tube, and pellet the cells again by centrifugation.
3. Disrupt the cells by adding 600 µL of the hypotonic lysis buffer to the pellet. The cell suspension is passed five to six times through an insulin syringe to complete the process.
4. Quantify the total protein concentration of the samples with the Bio-Rad protein assay. Dilute the dye five times in H_2O before use. Prepare standard protein solutions of BSA at 1 mg/mL, 500 µg/mL, 250 µg/mL, 125 µg/mL, 62.5 µg/mL, and 31 µg/mL. In a 96-well plate, add 5 µL of H_2O (blank) or 5 µL of solution of BSA (standard curve) or 5 µL of cell lysate diluted two or four times in H_2O using duplicates for each sample. Add 200 µL of the diluted dye into each well and incubate for 15 to 30 min at room temperature. Measure the optical density at 595 nm with a plate reader. Calculate the protein concentration of each cell lysate.
5. For each variant of the receptor, set up three types of reaction tubes in 5-mL polypropylene tubes (each condition should be performed at least in duplicates):

the nonspecific binding to determine the amount of radioactive γ-[^{35}S]-GTP bound to the cell membranes in the presence of a large excess of unlabeled γ-S-GTP, the basal to determine the level of activation of G proteins in the absence of ligand, and the SDF-1 stimulated to determine the activation of G proteins in the presence of ligand. The reaction mixtures should have a constant total volume of 100 μL (to be completed with the reaction buffer) and should be prepared according to the following table (volumes indicated are based on stock concentrations of 250 μ*M* for γ-S-GTP and 10 m*M* for GDP, each prepared in H$_2$O):

	Lysate	γ-[^{35}S]-GTP	γ-S-GTP	GDP	SDF-1	Reaction buffer
Nonspecific binding	7 μg	0.25 m*M*	10 μ*M* (4 μL)	40 μ*M* (4 μL)	—	to 100 μL
Basal	7 μg	0.25 n*M*	—	40 μ*M* (4 μL)	—	to 100 μL
SDF-1	7 μg	0.25 n*M*	—	40 μ*M* (4 μL)	0.1–100 n*M*	to 100 μL

Incubate for 1 h at 30°C in a water bath.
6. Stop the reaction by placing the tubes on ice.
7. Separate the γ-[^{35}S]-GTP bound to the membrane from the free form by adding 4 mL of ice-cold PBS into each tube and transferring the content of the tube on a GF/C filter placed into a vacuum manifold. Wash the membrane twice with 5 mL of cold PBS and dry the membrane as much as possible.
8. Transfer each filter into a scintillation vial and immerse it in 1.2 mL of scintillation solution. Wait at least 4 h to allow the elution of the radioactivity from the filter, then measure the radioactivity with a scintillation counter. Subtract the count obtained from the non specific binding from every other data point. Typical results are shown in **Fig. 1**.

3.2. Constitutive Phosphorylation of GPCR CAMs

1. Seed CHO transfectants stably expressing CXCR4-wild-type or CAM in a 60-mm tissue culture dish (1×10^6 cells in 4 mL of growth medium) for 24 h. The culture should be approx 70 to 80% confluent at the time of the experiment.
2. The following day, wash each dish twice with 5 mL of phosphate-free DMEM. Aspirate the medium and replace it with 1.5 mL of the same medium containing 150 μCi of [^{32}P] orthophosphate. Grow the cells in an incubator at 37°C for 90 min to allow the cells to metabolically label ATP with the [^{32}P] orthophosphate.
3. For each cell line (wild-type or CAM), keep one dish unstimulated and stimulate the other one for 5 min at 37°C by adding 100 n*M* of SDF-1 (final concentration) into the medium.
4. Stop the reaction by placing the dishes immediately on ice, aspirate the medium, and wash the cells once with 5 mL of ice-cold PBS. From this point on, all the steps are performed on ice or at 4°C.
5. Lyse the cells by adding 1 mL of lysis buffer into each dish. Be certain to recover all the membrane proteins from the plate by thoroughly pipetting the buffer on the surface of the dish or by using a cell scraper. Transfer each lysate into a 1.5-mL microcentrifuge tube. Put the tubes on a rotating wheel for 1 h.

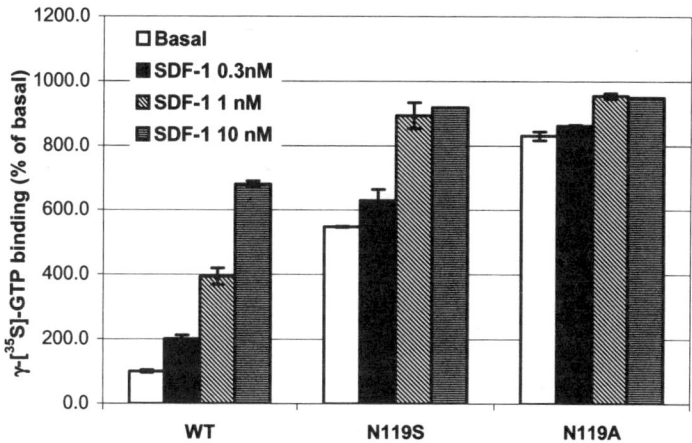

Fig. 1. Point mutants of N119 of CXCR4 exhibit different levels of constitutive and SDF-1-induced activity as assessed by γ-^{35}S-guanosine triphosphate (GTP) binding assay. The constitutively active mutants (N119S and N119A) show a high level of γ-^{35}S-GTP binding in the absence of stromal cell-derived factor (SDF)-1 (none = basal), which can be further increased by SDF-1 in N119S but almost not in 119A, which has the strongest constitutive activity, emphasizing the fact that increased constitutive activity results in an impaired ability of the receptor to respond to its ligand.

6. Centifuge the tubes for 10 min at 12,000g to remove insoluble materials. Transfer the supernatant into a new microcentrifuge tube.
7. Preclear the supernatant by adding 30 μL of a 50% (v/v) suspension of protein G-sepharose beads washed and resuspended in lysis buffer. Incubate for 1 h at 4°C on a rotating wheel. Centrifuge (1 min at 500g) and transfer the supernatant into a new tube.
8. Add 2 μg of anti-cMyc antibody 9E10 into each tube and incubate 2 h at 4°C.
9. Add 30 μL of a 50% (v/v) suspension of protein G-sepharose beads washed and resuspended in lysis buffer. Incubate at for 1 h on a rotating wheel.
10. Centrifuge the tubes (1 min at 500g), discard the supernatant, and wash the beads five times with 1 mL of lysis buffer. Discard the wash buffer and elute the proteins by adding 50 μL of SDS sample buffer to each pellet. Heat to 90°C for 5 min, centrifuge, and collect the supernatant.
11. Separate the proteins on a 10% polyacrylamide gel. Dry the gel and expose it to autoradiography using X-ray film or a screen for phosphorimaging. Typical results are shown in **Fig. 2**.

3.3. Subcellular Localization of CXCR4 in CHO Cell Transfectants by Immunofluorescence and Confocal Microscopy

GPCR typically are phosphorylated on activation resulting from the binding to a specific agonist. This phosphorylation mediated by GRKs results in the

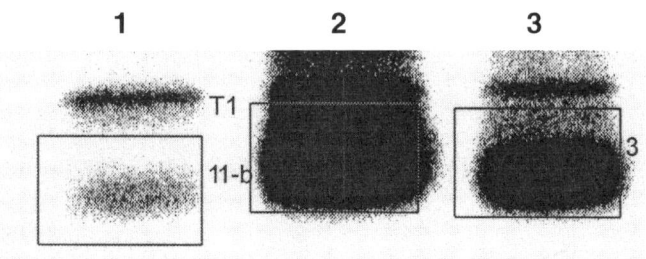

Fig. 2. Phosphorylation of CXCR4 wild-type (WT) and constitutively active mutants (CAMs) in Chinese Hamster Ovary cells. **Lane 1**: WT, unstimulated. **Lane 2**: WT, stimulated with 100 nM stromal cell-derived factor 1 5 min. **Lane 3**: N119S CAM, unstimulated.

interaction of the receptor with β-arrestins and subsequent internalization. CAMs of GPCR are expected to be constitutively internalized in the absence of ligand. The method described here will characterize the distribution pattern of CXCR4 wild-type and CAM before and after stimulation by SDF-1. A monoclonal antibody specific for a Myc-tag expressed at the N-terminal of CXCR4 will be used to localize the receptor at the subcellular level by immunofluorescence and confocal microscopy. In this experiment, the binding of the antibody to the tag is not altered by the prior binding of a natural ligand or antagonist to the receptor.

1. On the day before the experiment, detach the cells with trypsin from a 100-mm dish, count them and adjust the cell suspension to 1 × 10^5 cells/mL in MEM-α 10% FBS. Place an ultraviolet-sterilized tissue culture treated cover slip into each well of a six-well plate. Add 2 mL of the cell suspension into each well and place the plate back into a tissue culture incubator. Allow the cells to attach, spread and grow on the cover slips for at least 24 h.
2. On the next day, check the cell density and morphology. The cells should be less than 50% confluent and their morphology should be comparable to the same cells growing on the plastic at the bottom of each well. Carefully aspirate the medium and replace it with 1 mL of MEM-α containing 0.5% BSA. For each cell type, leave one well unstimulated and stimulate the other one with the appropriate ligand (SDF-1 at 100 nM, final concentration in MEM-α-BSA) for 30 min at 37°C to allow ligand-induced receptor internalization. Put the plate on ice and carefully aspirate the medium and wash the cells twice with cold PBS. Aspirate the PBS and slowly add 2 mL of methanol at –20°C into each well and keep the plate in a freezer at –20°C for 5 min. This step will simultaneously fix the cells and permeabilize the plasma membrane to allow detection of the intracellular proteins with the antibodies. After fixation, wash the cover slips three times with cold PBS containing 0.1% BSA.

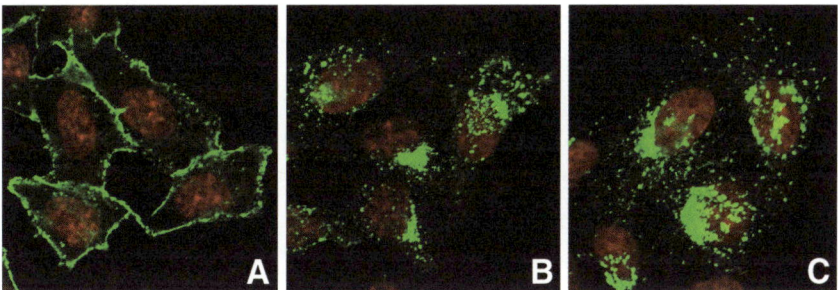

Fig. 3. Cellular localization of CXCR4 wild-type (WT) and N119A constitutively active mutant (CAM) by immunofluorescence and confocal microscopy. (**A**) WT, unstimulated control. (**B**) WT, stimulated by stromal cell-derived factor 1. (**C**) N119A, unstimulated control.

3. For immunostaining, incubate each cover slip with 50 µL of anti-Myc antibody (9E10) diluted at 10 µg/mL in PBS–BSA for 30 min at room temperature. Wash the cover slips twice in cold PBS–BSA. Incubate each coverslip with 50 µL of fluorescent secondary antibody (anti-mouse immunoglobulin G labeled with Alexa Fluor 488 at 10 µg/mL in PBS–BSA) for 30 min at room temperature. Wash the cover slips twice with cold PBS–BSA.

4. To facilitate the localization of the cells, especially when the receptor is predominantly in the endosomal compartment of the cytoplasm and the cell shape can hardly be detected by fluorescence, counterstain the nuclei by incubating the coverslips for 5 min at room temperature with a solution of TO-PRO-3 diluted to 1 µM in PBS. Wash three times in PBS, then mount the cover slips on glass slides thoroughly cleaned with ethanol. Mount each slide using 25 µL of antifade mounting medium. After overnight drying, the cover slips can be sealed, for instance, with nail polish.

5. Analysis: detailed method of analysis depends on the confocal microscope system used. The pictures showed here in **Fig. 3** were generated using a Zeiss LSM510 Meta laser scanning system hooked to a Zeiss Axiovert 200 inverted microscope equipped with a 40X oil immersion objective. Similar results can be obtained with comparable equipments. In any case, the image acquisition should consist of a series of z-scans spanning the thickness of the preparation to allow precise localization of the cell compartments (*see* **Note 2**).

4. Notes

1. The methods described here use CHO cells, but alternative cell lines can be used. However, special attention must be paid not to select cells that would express an autologous GPCR-CAM at a significant level. In that case, the quality of the data would be compromised, especially for the γ-S-GTP binding assay, in which the background binding would be high no matter what receptor is being expressed by transfection.

2. If a confocal microscope is not available, a regular fluorescence microscope equipped with a digital camera can be used instead, even though the discrimination between localization of the fluorescence on the membrane and in the cytoplasm would not be as accurate. In this case, a different combination of fluorochromes would be recommended since the TO-PRO-3 would not be excited very efficiently by a mercury lamp. DAPI or Hoechst would be more appropriate choices for staining of the nuclei.

Acknowledgment

Research described in this chapter was supported in part by Philip Morris USA Inc. and by Philip Morris International.

References

1. Palczewski, K., Kumasaka, T., Hori, T., et al. (2000) Crystal structure of rhodopsin: a G protein-coupled receptor. *Science* **289,** 739–745.
2. Gether, U., Asmar, F., Meinild, A. K., and Rasmussen, S. G. (2002) Structural basis for activation of G-protein-coupled receptors. *Pharmacol. Toxicol.* **91,** 304–312.
3. Ghanouni, P., Steenhuis, J. J., Farrens, D. L., and Kobilka, B. K. (2001) Agonist-induced conformational changes in the G-protein-coupling domain of the beta 2 adrenergic receptor. *Proc. Natl. Acad. Sci. USA* **98,** 5997–6002.
4. Seifert, R. and Wenzel-Seifert, K. (2002) Constitutive activity of G-protein-coupled receptors: cause of disease and common property of wild-type receptors. *Naunyn. Schmiedebergs. Arch. Pharmacol.* **366,** 381–416.
5. Bais, C., Santomasso, B., Coso, O., et al. (1998) G-protein-coupled receptor of Kaposi's sarcoma-associated herpesvirus is a viral oncogene and angiogenesis activator. *Nature* **391,** 86–89.
6. Michelson, S., Dal Monte, P., Zipeto, D., et al. (1997) Modulation of RANTES production by human cytomegalovirus infection of fibroblasts. *J. Virol.* **71,** 6495–6500.
7. Waldhoer, M., Kledal, T. N., Farrell, H., and Schwartz, T. W. (2002) Murine cytomegalovirus (CMV) M33 and human CMV US28 receptors exhibit similar constitutive signaling activities. *J. Virol.* **76,** 8161–8168.
8. Casarosa, P., Bakker, R. A., Verzijl, D., et al. (2001) Constitutive signaling of the human cytomegalovirus-encoded chemokine receptor US28. *J. Biol. Chem.* **276,** 1133–1137.
9. Montaner, S., Sodhi, A., Pece, S., Mesri, E. A., and Gutkind, J. S. (2001) The Kaposi's sarcoma-associated herpesvirus G protein-coupled receptor promotes endothelial cell survival through the activation of Akt/protein kinase B. *Cancer Res.* **61,** 2641–2648.
10. Sodhi, A., Montaner, S., Patel, V., et al. (2000) The Kaposi's sarcoma-associated herpes virus G protein-coupled receptor up-regulates vascular endothelial growth factor expression and secretion through mitogen-activated protein kinase and p38 pathways acting on hypoxia-inducible factor 1alpha. *Cancer Res.* **60,** 4873–4880.

11. Yang, T. Y., Chen, S. C., Leach, M. W., et al. (2000) Transgenic expression of the chemokine receptor encoded by human herpesvirus 8 induces an angioproliferative disease resembling Kaposi's sarcoma. *J. Exp. Med.* **191,** 445–454.

12. Guo, H. G., Sadowska, M., Reid, W., Tschachler, E., Hayward, G., and Reitz, M. (2003) Kaposi's sarcoma-like tumors in a human herpesvirus 8 ORF74 transgenic mouse. *J. Virol.* **77,** 2631–2639.

13. Gershengorn, M. C., Geras-Raaka, E., Varma, A., and Clark-Lewis, I. (1998) Chemokines activate Kaposi's sarcoma-associated herpesvirus G protein-coupled receptor in mammalian cells in culture. *J. Clin. Invest.* **102,** 1469–1472.

14. Geras-Raaka, E., Varma, A., Clark-Lewis, I., and Gershengorn, M. C. (1998) Kaposi's sarcoma-associated herpesvirus (KSHV) chemokine vMIP-II and human SDF-1alpha inhibit signaling by KSHV G protein-coupled receptor. *Biochem. Biophys. Res. Commun.* **253,** 725–727.

15. Geras-Raaka, E., Varma, A., Ho, H., Clark-Lewis, I., and Gershengorn, M. C. (1998) Human interferon-gamma-inducible protein 10 (IP-10) inhibits constitutive signaling of Kaposi's sarcoma-associated herpesvirus G protein-coupled receptor. *J. Exp. Med.* **188,** 405–408.

16. Zhang, W. B., Navenot, J. M., Haribabu, B., et al. (2002) A point mutation that confers constitutive activity to CXCR4 reveals that T140 is an inverse agonist and that AMD3100 and ALX40–4C are weak partial agonists. *J. Biol. Chem.* **277,** 24,515–24,521.

17. Arias, D. A., Navenot, J. M., Zhang, W. B., Broach, J., and Peiper, S. C. (2003) Constitutive activation of CCR5 and CCR2 induced by conformational changes in the conserved TXP motif in transmembrane helix 2. *J. Biol. Chem.* **278,** 36,513–36,521.

18. Lu, M., Grove, E. A., and Miller, R. J. (2002) Abnormal development of the hippocampal dentate gyrus in mice lacking the CXCR4 chemokine receptor. *Proc. Natl. Acad. Sci. USA* **99,** 7090–7095.

19. Zhu, Y., Yu, T., Zhang, X. C., Nagasawa, T., Wu, J. Y., and Rao, Y. (2002) Role of the chemokine SDF-1 as the meningeal attractant for embryonic cerebellar neurons. *Nat. Neurosci.* **5,** 719–720.

20. Tachibana, K., Hirota, S., Iizasa, H., et al. (1998) The chemokine receptor CXCR4 is essential for vascularization of the gastrointestinal tract. *Nature* **393,** 591–594.

21. Nagasawa, T., Hirota, S., Tachibana, K., et al. (1996) Defects of B-cell lymphopoiesis and bone-marrow myelopoiesis in mice lacking the CXC chemokine PBSF/SDF-1. *Nature* **382,** 635–638.

22. Zou, Y. R., Kottmann, A. H., Kuroda, M., Taniuchi, I., and Littman, D. R. (1998) Function of the chemokine receptor CXCR4 in haematopoiesis and in cerebellar development. *Nature* **393,** 595–599.

7

G Protein-Coupled Receptor Dimerization and Signaling

Mario Mellado, Antonio Serrano, Carlos Martínez-A., and José Miguel Rodríguez-Frade

Summary

G protein-coupled receptors are involved in the regulation of many aspects of normal physiology and pathology. Recent research has broadened our view of how the cell transduces ligand binding to cellular responses. It is becoming clear that phenomena that take place both at the cell surface, such as receptor oligomerization, as well as intracellularly, such as interaction between different signaling pathways, have important roles in the response elicited by a ligand. The study of these events requires the combined use of classical biochemical techniques with novel methods that allow analysis of these mechanisms. This chapter gives an overview of both types of techniques, with an emphasis on discussing their main applications and the conclusions that can be drawn in each case.

Key Words: GPCR; dimerization; G protein; kinases; immunoprecipitation; Western blot; resonance energy transfer; fusion proteins.

1. Introduction

G protein-coupled receptors (GPCRs) form the largest single family of cell surface receptors *(1)*. This superfamily of seven transmembrane-spanning proteins responds to a diverse array of stimuli and transduces these stimuli into intracellular second messengers through their ability to recruit and activate heterotrimeric G proteins *(2)*. The GPCRs are involved in regulating many of the body's functions, and their dysfunction has been linked to numerous diseases and disorders. This marks the GPCRs as targets of a growing number of therapeutic treatments, including almost 50% of modern drugs and nearly a quarter of the top 200 best-selling pharmaceutical compounds *(3,4)*. To date, drug screening has been based on testing compounds that block ligand binding or different steps in the signaling cascade. Recent research nonetheless points to novel therapeutic targets based on the early events after ligand stimulation, such as changes in receptor conformation or new molecules involved in GPCR signaling *(5)*.

From: *Methods in Molecular Biology, vol. 332: Transmembrane Signaling Protocols, Second Edition*
Edited by: H. Ali and B. Haribabu © Humana Press Inc., Totowa, NJ

GPCRs generally were assumed to act as monomeric entities. By interacting with its ligand, a monomeric receptor is responsible for coupling to G proteins, whose activation in turn allows the receptor to bind to one or more of an extensive number of effectors. Several reports nonetheless showed oligomerization of nearly all GPCRs examined, and activation of non-G protein-related effectors also has been described *(6,7)*. This led to reconsideration of the conventional models of receptor structure and function, with the generation of novel models better adapted to these observations.

Classical models of GPCR signaling assume that the monomeric receptor acquires an agonist-induced conformational change that enables formation of a receptor–G protein complex *(2,8,9)*. Through the release of guanosine diphosphate (GDP), the heterotrimeric G protein dissociates into α subunits and $\beta\gamma$ dimers, both of which activate several effectors responsible for receptor function *(10,11)*. Some of these second messengers are involved in receptor desensitization, a process controlled by receptor phosphorylation mediated by protein kinases A and C and by the serine/threonine GPCR kinases *(12,13)*.

Receptor activity is regulated not only by receptor desensitization, but also through receptor internalization, degradation, and recycling. The best-characterized mechanism for GPCR internalization acts via clathrin-coated vesicles in a process mediated by β-arrestin interaction with phosphorylated receptors; in addition, clathrin, phosphoinositides and the adaptor protein AP-2 collaborate to target the receptor to clathrin-coated pits at the cell membrane *(13,14)*. The internalized receptors are pinched off the cell surface by the guanosine triphosphatase (GTPase) dynamin and degraded in lysosomes or recycled through endosomes *(15,16)*.

At the postreceptor level, GPCR signaling can be regulated through GTPase-activating proteins (GAPs), such as the regulators of G protein signaling (RGS), although their complete physiological role remains to be elucidated *(17,18)*. Other signaling molecules, such as phosphoinositide-3 kinase (PI3K), cyclic adenosine monophosphate (cAMP), extracellular signal-regulated kinase (ERK), mitogen-activated protein kinase (MAPK), Janus kinase, signal transducer and activator of transcription (STAT) factors, or small GTPases, also link the GPCR family to cell functions, such as proliferation, differentiation, or migration.

Most mechanisms involved in classical GPCR signaling were elucidated using biochemical techniques that are proving to be insufficient for exploring some of the newer hypotheses of GPCR function. For example, although early studies used radioligand binding, crosslinking, or radiation inactivation to predict GPCR oligomerization, it is only now becoming accepted as a general mechanism for receptor function *(19)*. Several questions remain unanswered, including the functional relevance of these oligomers, the role of the ligand in their formation, and the composition of these complexes *(6,20,21)*. Oligomerization has important consequences for our current understanding of the GPCR because it helps to explain both GPCR binding to its numerous effectors, and the vast array of functions triggered through this family of receptors *(6,22)*. Oligomerization has a role in GPCR exit from the endoplasmic reticulum *(23)*, in agonist-promoted receptor internalization *(24)*, and in efficient coupling to signaling cascades *(25)*. Oligomerization also may participate in generating pharmacological diversity and novel functional properties *(19,22)*.

Although the concept that GPCRs exist in an equilibrium of multiple conformations has begun to overcome initial skepticism, further studies are needed to clarify the contribution of these conformations to GPCR biology. New technologies are helping to elucidate aspects, such as the physiological relevance of dimerization, the identification of the residues crucial in forming the dimerization interface, and the establishment of the role of the ligand in modulating oligomerization.

2. Materials

As can be predicted by the controversial results in the literature regarding GPCR oligomerization and its functional consequences, exquisite care must be taken to select the most appropriate materials and methods. The main criticisms of the GPCR oligomerization model arise from discrepancies among the results, which can depend on the experimental conditions used.

1. Cell lines (*see* **Note 1**).
2. Antibodies (Abs) (*see* **Note 2**).
3. Sodium dodecyl sulfate (SDS)-polyacrylamide gel electrophoresis (PAGE) equipment.
4. Buffers for lysis and immunoprecipitation (*see* **Note 3**).
   a. Lysis buffer (10 mM triethanolamine, pH 8.0, 150 mM NaCl, 1 mM ethylenediaminetetraacetic acid (EDTA), 10% glycerol, 2% digitonin.
   b. Washing buffer (50 mM Tris-HCl, pH 7.6).
   c. Secondary Abs coupled to agarose beads.
5. Crosslinking agents.
   a. Bifunctional agents: disuccinimidyl suberate.
6. Reagents for in vitro PI3K activity.
   a. PI3K buffer (50 mM Tris-HCl, pH 7.6).
   b. Adenosine triphosphate (ATP) mix (^{32}P, 1 mM ATP, 100 mM MgCl$_2$. 1:1:8 [w:w:w]).
   c. Thin-layer chromatography (TLC) plates.
   d. Running buffer: a mixture of 45 mL of chloroform, 35 mL of methanol, 3 mL of NH$_3$, 3 mL of H$_2$O.
7. Pull-down assays reagents.

    a. Fusion proteins with glutathione-*S*-transferase (GST); glutathione-Sepharose-GST beads.

    b. Lysis buffer (20% sucrose, 10% glycerol, 50 mM Tris, pH 8.0; 0.2 mM $Na_2S_2O_5$, 2 mM $MgCl_2$, 2 mM 1,4-dithiothreitol [DTT; Cleland's reagent], DTT).

    c. Fish buffer (10% glycerol, 50 mM Tris-HCl, pH 7.4; 100 mM NaCl; 1% NP40; 2 mM $MgCl_2$).

8. Electrophoretic mobility shift assay (EMSA) reagents.

    a. Agarose and DNA equipment.

    b. Buffer A (50 mM NaCl, 0.5 M saccharose).

    c. Buffer B (50 mM NaCl, 50% glycerol).

    d. Buffer C (350 mM NaCl, 25% glycerol).

All buffers (with the exceptions of EMSA buffer and DNA electrophoresis buffer) contain 0.5 mM spermidine, 0.15 mM spermine, 0.1 mM EDTA, 10 mM HEPES, pH 8.0, 0.5 mM phenylmethylsulfonyl fluoride, 2 µg/mL leupeptin, 3 µg/mL pepstatin, 0.2 IU/mL aprotinin, 1.75 mM β-mercaptoethanol, 1 mM pervanadate, and 10 mM NaF.

- EMSA buffer (10 mM Tris-HCl, pH 7.5, 50 mM NaCl, 1 mM DTT, 1 nM EDTA, 5% glycerol).
- DNA electrophoresis buffer (4.5% PAGE, using 0.5X Tris-borate-EDTA).

9. Fluorescence labeling of Abs: dissolving buffer (100 mM NaCl and 35 mM H_3BO_3, pH 8.3); and equilibrating buffer (100 mM NaCl, 50 mM NaH_2PO_4, 1 mM EDTA, pH 7.5).

10. Expression vectors for fluorescent proteins.
11. Tagged receptors.
12. Confocal microscopes.
13. Calcium probes and ionophores.
14. Flow cytometry equipment.
15. Migration chambers: transwell and flow chambers.
16. Radiolabeled ligands.

3. Methods

The methods described allow identification of most GPCR signaling properties. As any single method suffers from intrinsic limitations, most should be used in concert with others to obtain meaningful results. Results using one technique should be verified by an alternative that uses different materials and detection systems. Some methods mentioned are not explained in detail, as they are well described in other publications. Here, we outline the most relevant methods, and those that are not commonly employed for GPCR studies.

Before attempting analysis of GPCR signaling properties, the cell system to be used should be characterized in detail. This should include routine testing, such as analysis of cell cycle status, cell surface receptor expression, and determination of receptor number and affinity constants, especially when cells are transfected with mutant or fluorescently labeled receptors.

3.1. Cell Cycle Analysis

Cell cycle status affects many parameters of a cell's response, and it must be ascertained that this state remains similar following stimulation.

3.1.1. Propidium Iodide Incorporation

1. Incubate 0.5×10^6 cells with 0.5 mL of ice-cold 70% ethanol (5 min, on ice).
2. Add 3 mL of phosphate-buffered saline (PBS) and centrifuge (5 min, 200g).
3. Resuspend the pellet with 0.6 mL of PBS containing 33 µg/mL propidium iodide and 2 µg/mL RNase.
4. Incubate (30 min, 37°C, in the dark) and analyze in the flow cytometer.

3.2. Flow Cytometry Analysis

Definition of cell surface receptor expression should be a prerequisite for signaling studies. In some cases, large intracellular GPCR pools can be observed, as the result of internalization, recycling, or storage; these pools should be differentiated from GPCR that are membrane-expressed and available to ligand. For analysis of heterodimerization levels, expression of both participating receptors should be determined in the same cell.

1. Incubate 2.5×10^5 to 5×10^5 cells in 100 µL of staining PBS in V-bottom plates, using an appropriate dilution of primary Ab (in staining PBS; 30 min, 4°C).
2. Wash cells twice by centrifugation (5 min, 200g).
3. Resuspend pellet with containing secondary, fluorochrome-labeled Abs.
4. Incubate (30 min, 4°C).
5. Wash cell twice by centrifugation (5 min, 200g).
6. Resuspend cells in 100 µL of staining PBS and analyze in a flow cytometer.
7. For intracellular staining, cells must be permeabilized using different agents to facilitate entrance of Abs. These include digitonin (50 µL of 0.005% digitonin in staining PBS, 4 min, room temperature [RT]), saponin (100 mL of 1% paraformaldehyde (PFA), 15 min, 4°C, followed by incubation of the first Ab in staining PBS + 0.2% saponin), or ice-cold ethanol (70%, 15 min, 4°C).

Cells labeled using similar protocols can be sorted using a flow cytometer or magnetic-based separation devices.

3.3. Receptor Number

Several methods are available in which to analyze the number of receptors on the cell surface, based on the use of radiolabeled or fluorescently labeled ligands. Many labeled ligands are now available that use radioactivity or fluorescent compounds. Detailed information of radiolabeling procedures can be found elsewhere *(26)*.

Ligand-binding analyses are performed by competitive inhibition of ^{125}I-ligand binding to GPCR with several concentrations of unlabeled ligand *(27)*.

For these assays, cell membranes or whole cells can be used (*see* **Note 4**). Several computer programs are available to evaluate data and determine binding site number and affinity constants.

For flow cytometry analysis, we use the Dako QIFIKit, designed for quantitative determination of cell surface antigens using indirect immunofluorescence assays (*see* **Note 5**). The system is based on analysis of the capacity of a given Ab to bind its target cell, compared with beads coated with defined amounts of mouse Ab. Assay procedures should follow manufacturer's protocols.

3.4. Western Blot and Immunoprecipitation

Although lysis buffer compositions can vary (*see* **Note 3**), the general protocol used for these assays is similar. Cell number should be adjusted, depending on the amount of receptor and/or signaling molecule to be analyzed.

1. Stimulate cells with the appropriate ligand and dilute them immediately with cold PBS (to terminate stimulation).
2. Centrifuge (20,000g, 2 min, 4°C), wash with cold PBS, resuspend in 200 μL of lysis buffer (10 mM triethanolamine, pH 8.0, 150 mM NaCl; 1 nM EDTA; 10% glycerol; 2% digitonin) and incubate (20 min, 4°C).
3. Preclearing step. Centrifuge (20,000g, 15 min, 4°C) and discard the pellet. Incubate the supernatant containing solubilized proteins with anti-immunoglobulin (Ig) (of the same species as the immunoprecipitating Ab) coupled to agarose (15 min, 4°C).
4. Centrifuge (20,000g, 1 min, 4°C).
5. Incubate the supernatant (90 min, 4°C) with the immunoprecipitating Ab, followed by anti-Ig coupled to agarose (without washing), and incubate (60 min, 4°C).
6. Wash extensively (at least five times) with 200 μL of washing buffer.
7. Resuspend the pellet in sample buffer (50 μL/mL).

Procedures for SDS-PAGE, transfer to nitrocellulose membranes and Western blot are described elsewhere *(26)*, and require no major modifications for GPCR studies (*see* **Notes 6** and **7**).

3.5. Crosslinking

1. Stimulate cells and wash cells by centrifugation (200g, 10 min).
2. Resuspend the pellet in 1 mL of cold PBS and add 10 μL of 100 mM disuccinimidyl suberate.
3. Incubate for 10 min at 4°C with continuous rocking.
4. Terminate the reaction by adding 1 mL of cold PBS and centrifuging (200g, 10 min).
5. Lysis step. Wash three times with cold PBS and resuspend the pellet in lysis buffer for immunoprecipitation.
6. Continue with immunoprecipitation and Western blot as in **Subheading 3.4.** (*see* **Note 8**).

3.6. Tagged Receptors

It can be difficult to trace certain receptors because of a lack of appropriate tools, such as monoclonal Abs, which allow their detection. In these cases, a peptide can be used for which monoclonal antibody (MAb) are available; fused to a receptor, the peptide allows detection of the protein of interest. In the case of GPCR, it is important to recognize artifacts caused by the presence of the tag. Both ligand-binding affinity and receptor distribution should be evaluated before beginning signaling studies (*see* **Note 9**). Although several commercial expression vectors are available, homemade epitopes can be designed and included as fusion proteins with the target receptor using standard molecular biology techniques.

Complementary DNA coding for different chemokine receptors can be cloned in the *Kpn*I-*Xba*I site of pcDNA3. For tagged receptors, eliminate the receptor initiation codon and ligate the oligonucleotides coding for the tags selected.

1. Transfect cells and determine the functional integrity of the individual fusion receptor(s) (*see* functional assays in **Subheading 3.12.**).
2. Cotransfect tagged receptors.

The main concern in co-immunoprecipitation is the formation of nonspecific aggregates (*see* **Note 9**).

3.7. PI3K Activity

The PI3K are a family of intracellular signaling proteins that control a variety of functions, such as proliferation, apoptosis, and migration. The analysis of PI3K activity is based on quantification of ^{32}P incorporation into substrates present in cell lysates.

1. Stimulate cells and immunoprecipitate with the appropriate receptor or PI3K isoform (*see* **Subheading 3.4.**).
2. Wash the remaining cell lysate pellet with PI3K buffer.
3. Prepare 2 µL of phosphatidyl inositol (PI) solution (10 mg/mL) for each sample, desiccate, add 15 µL of PI3K buffer and sonicate (10 min, RT).
4. Add 15 µL of PI solution to the immunoprecipitated pellet and mix.
5. Add 10 µL/sample of ATP mix.
6. Incubate (10 min, RT), add 50 µL of 1 *M* HCl, 132 µL of methanol, and 117 µL of chloroform per sample and mix.
7. Centrifuge (20,000*g*, 10 min, RT) and transfer the lower phase to another Eppendorf tube, desiccate with the Speed Vac.
8. Add 20 to 40 µL of chloroform and load a TLC plate, together with 4 µL of phosphatidyl inositol-3,4,5-triphosphate (PIP$_3$) or phosphatidyl inositol-4-phosphate (PI4P) standard. Pre-stabilize the TLC plate and tank for 2 h with running buffer.
9. Carry out the chromatography for 1 h 40 min, dry plate, and develop standards by incubating with I$_2$ (30 min). Analyze results by autoradiography.

3.8. Pull-Down Assays

These assays are based in the fact that small GTPases act as molecular switches in either an inactive GDP-bound form or an active GTP-bound form *(28)*. Active GTPases bind specific domains, such as the rhotekin-binding domain for Rho, or the Pak family of serine/threonine kinases, which bind Cdc42 and Rac. Pull-down assays include the use of a fusion protein between the substrate and GST.

1. Culture bacteria overnight and induce GST-protein production with 0.1 mM iso-propyl-β-D-thiogalactopyranoside (2 h).
2. Centrifuge (3000g, 10 min, 4°C).
3. Resuspend pellet in 10 mL of lysis buffer.
4. Sonicate and centrifuge (12,000g, 20 min, 4°C).
5. Incubate with 1 mL of 50% glutathione-Sepharose 4B (Amersham Pharmacia Biotech; 30 min, 4°C).
6. Wash three times with lysis buffer and resuspend Sepharose beads (1:1) in Fish buffer.
7. Lyse cells as in **Subheading 3.4.**
8. Mix supernatant and glutathione-Sepharose GST-protein beads, incubate (30 min, 4°C).
9. Wash three times with Fish buffer, resuspend beads in sample buffer, and electrophorese samples in 12.5 to 15% SDS polyacrylamide gels (*see* **Note 10**).

3.9. Electrophoretic Mobility Shift Assay

Activation of transcription factors, such as STAT, is analyzed by electrophoretic mobility assays. Transcription factors are cytoplasmic molecules that change their location after ligand activation, translocating to the nucleus, where they bind to specific DNA sequences and promote gene expression. Several assays demonstrate the activation of transcription factors. In some cases, the use of immunohistochemical techniques allows the cellular location of these molecules to be tracked. Their phosphorylation pattern can be evaluated by immunoprecipitation and Western blot techniques, as activation of most transcription factors requires their phosphorylation in tyrosine and serine/threonine residues. To demonstrate their functional activity, it is necessary to determine their capacity to interact with specific DNA sequences.

1. Stimulate cells and obtain nuclear extracts.
2. Wash the pellet twice with cold PBS and resuspend in 1 mL buffer A.
3. Incubate (2 min, 4°C) and centrifuge (4500g, 3 min, 4°C).
4. Resuspend pellet in buffer B.
5. Centrifuge (4500g, 3 min, 4°C) and incubate pellet with 60 μL of buffer C (30 min, 4°C).
6. Centrifuge (20,000g, 20 min, 4°C).
7. Nuclear extracts can now be stored at –80°C.

8. For EMSA analysis of STAT transcription factors, incubate 10 µg of nuclear extract with 0.5 ng of a ^{32}P-labeled oligonucleotide that contains the serum-inducible element (SIE) sequence of the c-fos promoter. A mutant SIE oligonucleotide that lacks STAT-binding capacity is used as control.

9. Incubate in EMSA buffer (30 min, RT) in a final volume of 10 µL with 1.5 µg of poly(dI-dC) and a 20X molar excess of unlabeled SIE or mutant SIE oligonucleotides.

10. Analyze DNA–protein complexes in 4.5% PAGE using electrophoresis buffer.

3.10. Construction of Fluorescently Labeled Receptors

The use of biochemical procedures to differentiate molecular signaling pathways such as those described requires the use of whole-cell lysates, although cell compartments can be separated using gradient ultracentrifugation. The presence of receptors in different activation states in intracellular reservoirs may affect the conclusions drawn from experiments that use biochemical techniques. This is of particular importance when the study evaluates the effect of a ligand, for example, in determining its role in promoting or stabilizing receptor dimerization.

Microscopy techniques that use fluorescent protein-based constructs allow easy, rapid visualization. These types of constructs are used to analyze receptor interactions through co-localization analysis and energy transfer techniques. Most vectors containing fluorescent proteins are available commercially. Fluorescently labeled receptors can be constructed using standard molecular biology techniques. Insertion of the fluorescent probe in the C-terminal region of the receptor involves elimination of the receptor stop codon, whereas insertion in the N-terminal region requires elimination of the initiation codon. Transfect cells and analyze for receptor expression and function (*see* **Note 11**).

3.11. Fluorescence Labeling of Abs

Because of their intense fluorescence and low hydrophobicity, the Cy dyes are efficient tags for fluorescence labeling.

1. In the standard labeling procedure, the contents of a commercial vial ("to label 1 mg of protein") of Cy2, Cy3, or Cy5 are dissolved in 50 µL of dimethylsulphoxide (DMSO). MAb is dissolved to 1 mg/mL in dissolving buffer.

2. Pipet 10 µL of dye/DMSO mixture into 200 µL of Ab solution and mix by gentle vortexing.

3. Incubate at 25°C in the dark for 30 min.

4. Separate unbound dye by adding 300 µL of 100 mM NaH$_2$PO$_4$. Incubate for 30 min at 25°C and load the sample on a PD-10 column pre-equilibrated with equilibrating buffer.

5. Wash the column twice with equilibration buffer and elute the labeled protein with 2 mL of distilled H$_2$O.

3.12. Co-Localization

Co-localization assays detect light from two different fluorophores and evaluate a digital image for the presence of the same pixel in two distinct channels. Signal co-localization indicates adjacency of fluorophores, and thus of the molecules they label. A high-numerical aperture microscope lens permits resolution near 300 nm, sufficient to locate molecules in different cell compartments, but not to demonstrate molecular association. Controls must be included to assure that the fluorescence of one fluorochrome does not overlap that of the second.

3.13. Resonance Energy Transfer (Flow Cytometry)

Biophysical techniques are used to detect GPCR dimerization in living cells. Both fluorescence resonance energy transfer (FRET) and bioluminescence resonance energy transfer (BRET) measure nonradioactive energy transfer between a luminescent or fluorescent donor and an appropriate fluorescent acceptor. The donor emission spectrum must therefore overlap significantly with the acceptor absorption spectrum; in FRET, the acceptor is excited by the donor fluorophore, which is excited in turn by an external light source. Energy can be transmitted over a very limited distance (2–10 nm), with efficiency dependent on the inverse sixth power of the distance between donor and acceptor fluorophores *(29)*. Acceptors include green fluorescent protein (GFP) and variants, fluorescently labeled Abs; donors include renilla luciferase (for BRET), GFP and variants, fluorescently labeled Abs (for FRET).

1. Cells (2×10^5 cell/0.1 mL) expressing the target receptors are cultured in depletion medium (60 min, 37°C) and stimulated with appropriate ligands.
2. Transfer cells to a 96-well V-bottom plate.
3. Centrifuge cells (200g, 5 min, 4°C).
4. Fix cells with 1% PFA (0.1 mL/2×10^5 cells; 10 min, 4°C) and wash twice with staining PBS.
5. Block with human immunoglobulin before the addition of primary Abs. The amount, duration and species of Ig varies depending on cell line, primary Abs used, etc. These parameters should be analyzed previously in conventional flow cytometry.
6. After incubation with primary antireceptor MAb, add fluorescein isothiocyanate (FITC)- or tetramethylrhodamine isothiocyanate (rhodamine, TRITC)-labeled secondary Abs. Labeled primary Abs also can be used, as can Abs from different species to ensure secondary Ab specificity. When analyzing homodimer formation or if appropriate Abs are unavailable, similar approaches are possible using epitope-tagged or GFP-based transfected receptors.
7. Analysis of the TRITC/FITC ratio. Positive FRET is detected by an increase in this ratio owing to increased TRITC and diminished FITC emission (*see* **Note 12**).

Resonance energy transfer can also be traced using confocal or multiphoton fluorescence microscopy.

3.14. Resonance Energy Transfer (Microscopy)

Modern optical microscopy allows analysis of the way a molecule moves, changes location, or associates with other molecules. Chimeric constructs can be generated in which each protein of interest is fused to a different, modified form of GFP; the most commonly used variants are cyan fluorescent protein (CFP) as a donor and yellow fluorescent protein (YFP) as an acceptor. There are several methods to determine and quantify FRET.

1. Sensitized acceptor fluorescence. The donor fluorescent dye is excited and the acceptor signal is quantitated.
2. Acceptor photobleaching, a method based on quenching donor fluorescence. Some donor photons are used to excite the acceptor, decreasing the energy emission detected.
3. Decrease in donor fluorophore lifetime. The fluorescence lifetime of a donor dye decreases under FRET conditions, independently of fluorophore concentration or excitation intensity. Measuring chromophore fluorescence lifetime on a pseudocolor scale allows analysis of molecule localization and the intermolecular associations triggered (*see* **Note 13** *[30]*).

BRET makes use of nonradiative energy transfer between a light donor and a fluorescent acceptor. In BRET, the bioluminescent energy resulting from the catalytic degradation of coelenterazine by luciferase is transferred to an acceptor fluorophore, which in turn emits a fluorescent signal *(31)*.

3.14.1. FRET Measurement

FRET between CFP and YFP can be determined from the whole image on a pixel-by-pixel basis using a three-filter method *(32–34)*. Autofluorescence, as well as CFP and YFP bleed-through, are calculated on the FRET channel. The corrected FRET image is then obtained by subtracting autofluorescence and bleed-through from original FRET images, and background is subtracted prior to other image treatments. For quantitation, normalized FRET is calculated on the colocalization areas of CFP-receptor 1 and YFP-receptor 2 on a pixel-by-pixel basis *(35)*. Images are processed and FRET calculated using a Visual Basic 6.0 language (Microsoft)-based program.

3.14.2. Fluorescence Lifetime Imaging Microscopy Measurement

Fluorescence lifetime imaging microscopy (FLIM) can be determined in live cells cultured in cover slip chambers or in PFA-fixed cells. FLIM measurements are performed using a confocal microscope with a High-Speed Lifetime Module and 60x PlanApo 1.4 objective or equivalent equipment. Fluorescence lifetime is determined after excitation with a 440 nm pulsed laser (picosecond pulses) and a 470/20 bandpass emission filter, and quantitated using LIMO (Nikon) or similar software (**Fig. 1**).

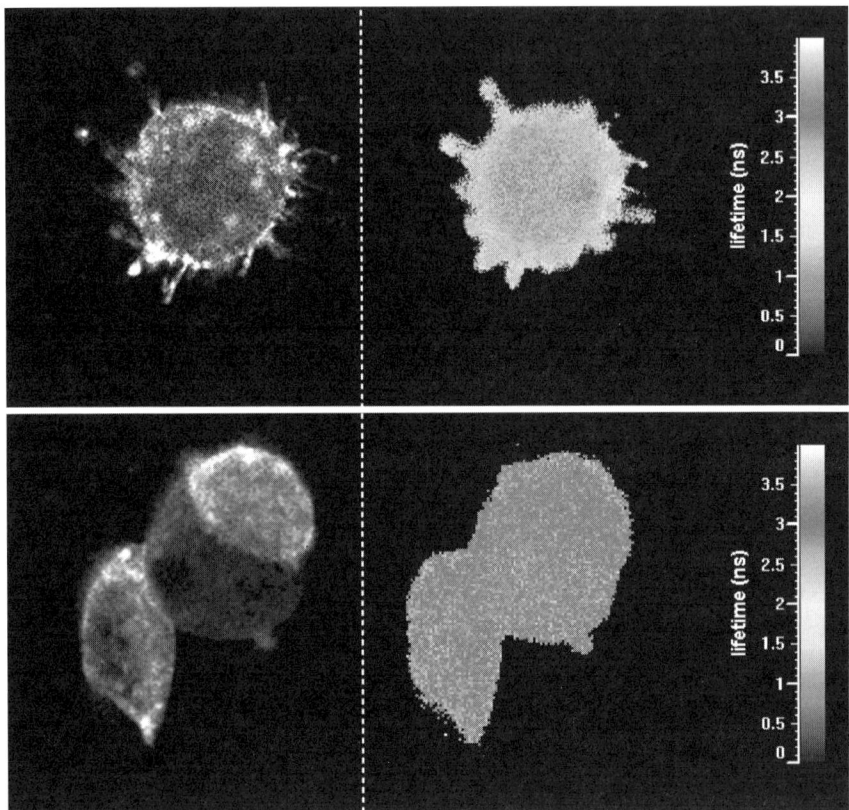

Fig. 1. Cyan fluorescent protein (CFP) fluorescence lifetime images (calculated from the phase shift) of human embryonic kidney-293 cells expressing CXCR4-CFP (upper right panel) or CXCR4-CFP and CXCR4-yellow fluorescent protein (lower right panel). The pseudogray scale ranges from 0 (black) to 4.0 ns (white). As controls, fluorescence images are shown of CXCR4-CFP (left panels). Fluorescence and fluorescence lifetime imagining for all images was determined after excitation with a 440-nm picosecond pulsed laser and a 470/20 emission filter.

3.15. Functional Assays

It is usually necessary to determine not only expression of a given receptor but also its correct function and effector coupling. These attributes are particularly important when using cells transfected with tagged or fluorescent protein-modified receptors, which can be expressed correctly on the cell surface but be nonfunctional. Calcium mobilization is a feature common to most GPCR, and is easily monitored in most cell types. Other assays, such as migration or cell adhesion, are specific to certain GPCR families, such as chemokine receptors.

3.15.1. Calcium Mobilization

1. Resuspend cells in serum-free medium (2.5×10^6 cells/mL).
2. Incubate with 5 to 15 µL/10^6 cells of Fluo-3 AM (0.3 mg/mL in DMSO; 20–30 min, 37°C, with mixing).
3. Wash cells twice with RPMI-1640 medium containing 10% fetal calf serum.
4. Resuspend cells (0.5×10^6 cells/mL) in the same medium supplemented with 2 mM $CaCl_2$.
5. Store the loaded cells at 4°C. Before analysis of Ca^{2+} mobilization in response to the ligand in a flow cytometer, cells should be pre-warmed to 37°C. Always use an ionophore as positive control (*see* **Note 14**).
6. To increase sensitivity, cells can be loaded with two probes, one that increases fluorescence in the presence of Ca^{2+}, such as Fluo-3 AM, and another that decreases fluorescence, such as Fura Red. Analysis of the ratio between the two signals allows amplification of assay sensitivity.

3.15.2. Migration in Transwells

1. Select appropriate transwell pore size (usually 3, 5, or 8 µm) to allow chemotaxis but not random migration.
2. Resuspend cells (2.5×10^5) in 100 µL of serum-free medium and add to the transwell insert.
3. Add the stimulant and controls to the lower transwell chamber (in 0.6 mL of serum-free medium).
4. Incubate (37°C, 5% CO_2, 1–4 h).
5. Cells that migrate to the lower well are counted in a flow cytometer. Controls include the presence of stimulant in upper and lower transwell chambers to differentiate between ligand-induced chemotaxis and nonspecific migration.

4. Notes

1. Heterologous expression systems are the preferred, and in some cases obligate models for study of GPCR oligomerization, although high expression levels can lead to spurious interactions.
2. Most Abs can be obtained commercially. Although MAbs offer unique specificity and reproducibility, polyclonal Abs often have advantages for immunoprecipitation. The use of synthetic peptides as immunogens allows predetermination of the receptor recognition target, for example, intracellular or extracellular loops. When producing your own MAb, be sure that the screening criteria guarantee its use for the techniques desired; an immunoprecipitating Ab will not necessarily work well for Western blot or flow cytometry.
3. Not all buffers preserve receptor associations with signaling molecules or with receptors. Many discrepant observations are easily explained by the use of different buffers.
4. The use of cell membranes requires the presence of ions, such as Mg^{2+} or Ca^{2+} in the binding buffers; for whole cells, complete tissue culture medium should be used. To diminish nonspecific binding and high backgrounds, cells must be filtered and washed extensively. Affinity for the receptor of the radiolabeled ligand may differ from that of the unlabeled ligand, which complicates data analysis.

5. Limitations arise from the availability of appropriate reagents, i.e., Abs.
6. Stripping of nitrocellulose membranes is not recommended because it can alter recognition of the transferred proteins.
7. Immunoprecipitation and Western blot allow the study a number of signaling events, including receptor dimerization or association between signaling molecules and receptors after ligand binding. Discrepancies between results in different studies are commonly attributed to detergent composition of buffers, which can disrupt some binding associations. These techniques have been used successfully to determine receptor dimerization, for which a crosslinking step should be included prior to cell lysis.
8. Care should be taken in cell handling before the lysis step, as the presence of nonintact cells increases the background of nonspecific protein crosslinking. Bifunctional reagent solutions should be freshly prepared.
9. Selection of an appropriate tag and its location in the receptor are both important factors. Amino acids added to the extracellular region can modify ligand binding, whereas modification of intracellular domains can disturb the coupling of signaling molecules. The analysis technique used will also dictate the location of the tag within the receptor. Flow cytometry analysis of intact cells requires tagging extracellular sequences, whereas Western blot, immunoprecipitation, or immunofluorescence are less restrictive as to tag location. It is easy to construct and manipulate double-tagged receptors in a single cell; this method is useful for the study of receptor–receptor interaction; the main disadvantage is its limitation to transfected cells.

    Receptor labeling circumvents the need to raise Abs specific for target receptors, and has been used successfully to establish homodimerization of β2Ars, γ-aminobutyric acid B, mGluR5, δ-opioid, m3-muscarinic, Ca^{2+}, and CCR2 receptors *(36)*. The position of the receptor tag must be chosen carefully so that receptor expression and ligand-binding properties are not affected. FLAG®, c-myc, HA (influenza hemagglutinin), GST (glutathione-*S*-transferase), His (histidine), OMNI (T_7), MBP (maltose-binding protein), S (S_1 probe), Thio (thioredoxin), GFP (green fluorescent protein), CruzTag®, and GAL (galactosidase) are among the tags for which there are commercially-available Abs that immunoprecipitate and recognize the tagged receptor in Western blot. There are many vectors offered containing these tags, which allow expression of tags fused to the protein of interest. With this strategy, we used cells co-transfected with CCR2b receptor cDNA tagged in the N-terminal extracellular domain with Myc or YSKFDT sequences, to show that the ligand induces receptor oligomerization. Finally, when receptor-specific MAb are not available, tagged receptors allow analysis of the signal pathways activated after ligand binding.

    The co-immunoprecipitation strategy is also used for detection of heterodimerization, as receptor-specific Abs allow tracking of heterodimerization not only in transfected cells, but also in cell lines or primary cells. We have used MAb specific for CCR2, CCR5, and CXCR4 in transfectants and primary cells to determine receptor heterodimerization.

10. Temperature in critical to preserve the integrity of GST-protein. Be sure to maintain samples at 4°C during all procedures.

11. The construction of fluorescently labeled receptors uses standard molecular biology techniques. The fluorescent probe must be inserted in the C-terminal region to conserve the original signal peptide of the receptor. The fusion involves elimination of the receptor stop codon and insertion of the receptor into the fluorescent protein vector, separated by several additional amino acids. Transfect cells and analyze for expression and function.

12. Despite the many advantages of these techniques, their use raises several concerns, as in most cases resonance energy transfer requires transfection of donor and acceptor proteins, i.e., receptor overexpression, which may favor oligomer formation. In addition, receptor modification to incorporate donor and acceptor groups can alter receptor signaling. To avoid some of these problems, fluorescently labeled Abs can be used with untransfected cells, although intracellular signaling molecules are not accessible in experiments using living cells.
FRET analysis by flow cytometry is not simple, as the cytometer must be perfectly calibrated to differentiate clearly between donor and acceptor fluorescence signals.

13. For all of these approaches, donor/acceptor choice is critical in determining FRET. Donor emission spectra should ideally have maximum overlap with acceptor absorption spectra, although acceptor and donor emissions should be clearly separable to minimize background interference. Fluorochrome incorporation in the protein to be analyzed must also be considered. The most appropriate donor/acceptor combinations are blue fluorescent protein (BFP)/GFP, CFP/YFP, GFP/dsRed, and YFP/dsRed. In practice, the CFP/YFP combination is the most frequently used, as both molecules are extremely bright and is this combination offers few technical problems. BFP is a poor donor as it is not especially bright, making FRET between BFP and GFP difficult to detect. dsRed is a poor acceptor, as it has a broad absorption spectrum and excites the same wavelength as the donor (GFP or YFP). In addition, constructs that work well with GFP and its variants do not work using dsRed.
Specific Abs conjugated to appropriately selected fluorescent dyes can be used to study fixed cells. For example, although Cy3 as a donor and Cy5 as an acceptor form a suitable fluorochrome pair, the Cy2 donor/Cy3 acceptor pair is often more convenient, as it permits use of the widely available 488-nm argon laser line. Secondary Abs are occasionally necessary, although the increased distance between fluorophores complicates FRET detection; this can be resolved using dye-labeled F(ab') fragments.
The loss of FRET could also be to the result of conformational changes within receptor dimers and, such, does not prove lack of dimerization.

14. Always wash the cytometer with water after analysis of ionophore-treated cells. Cells cannot be stored for long periods of time (>60 min), as they will lose the Fluo-3 AM load. Do not prepare cells for experiments that will require more than 60 min.

References

1. Pierce, K., Premont, R. T., and Lefkowitz, R. J. (2002) Seven-transmembrane receptors. *Nat. Rev. Mol. Cell Biol.* **3**, 639–650.
2. Lefkowitz, R. J. (2000) The superfamily of heptahelical receptors. *Nat. Cell Biol.* **2**, E133–E136.
3. Proudfoot, A. (2002) Chemokine receptors: multifaceted therapeutic targets. *Nat. Rev. Immunol.* **2**, 106–115.
4. Kenakin, T. (2004) Principles: receptor theory in pharmacology. *Trends Pharmacol. Sci.* **25**, 186–192.
5. Rodríguez-Frade, J. M., Martínez-A., C., and Mellado, M. (2005) Chemokine signaling defines novel targets for therapeutic intervention. *Mini. Rev. Med. Chem.* **5**, 781–789.
6. Salahpour, A., Angers, S., and Bouvier, M. (2000) Functional significance of oligomerization of G-protein-coupled receptors. *Trends Endocrinol. Metab.* **11**, 163–168.
7. Lee, S. P., O'Dowd, B. F., and George, S. R. (2003) Homo- and hetero-oligomerization of G protein-coupled receptors. *Life. Sci.* **74**, 173–180.
8. Baldwin, J. M. (1994) Structure and function of receptors coupled to G proteins. *Curr. Opin. Cell Biol.* **6**, 180–290.
9. Karnik, S. S., Gogonea, C., Patil, S., Saad, Y., and Takezako, T. (2003) Activation of G-protein-coupled receptors: a common molecular mechanism. *Trends Endocrinol. Metab.* **14**, 431–417.
10. Hamm, H. E. and Gilchrist, A. (1996) Heterotrimeric G proteins. *Curr. Opin. Cell Biol.* **8**, 189–196.
11. Brady, A. E. and Limbird, LE. (2003) G protein-coupled receptor interacting proteins: emerging roles in localization and signal transduction. *Cell Signal.* **14**, 297–309.
12. Pitcher, J., Freedman, N. J., and Lefkowitz, R. J. (1998) G protein-coupled receptor kinases. *Annu. Rev. Biochem.* **67**, 653–692.
13. Ferguson, S. S. (2001) Evolving concepts in G protein-coupled receptor endocytosis: the role in receptor desensitization and signaling. *Pharmacol. Rev.* **53**, 1–24.
14. Oakley, R., Laporte, S. A., Holt, J. A., Barak, L. S., and Caron, M. G. (1999) Association of b-arrestin with G protein-coupled receptors during clathrin-mediated endocytosis dictates the profile of receptor resensitization. *J. Biol. Chem.* **274**, 32,248–32,257.
15. Zhang, J., Ferguson, S. S. G., Barak, L. S., Menard, L., and Garon, M. G. (1996) Dynamin and b-arrestin reveal distinct mechanism for G protein-coupled receptor internalization. *J. Biol. Chem.* **271**, 18,302–18,305.
16. Bottomley, M., Lo Surdo, P., and Driscoll, P. C. (1999) Endocytosis: how dynamin sets vesicles pHree!!! *Curr. Biol.* **9**, R301–R304.
17. de Vries, L. and Farquhar, M. G. (1999) RGS proteins: more than just GAPs for heterotrimeric G proteins. *Trends Cell Biol.* **9**, 138–144.
18. Ishii, M. and Kurachi, Y. (2003) Physiological actions of regulators of G-protein signaling (RGS) proteins. *Life Sci.* **74**, 163–171.
19. Terrillon, S. and Bouvier, M. (2004) Roles of G-protein-coupled receptor dimerization. *EMBO Rep.* **5**, 30–34.

20. George, S. R., O'Dowd, B. F., and Lee, S. P. (2002) G-protein-coupled receptor oligomerization and its potential for drug discovery. *Nat. Rev. Drug Discov.* **1,** 808–820.
21. Salim, K., Fenton, T., Bacha, J., A., et al. (2002) Oligomerization of G-protein-coupled receptors shown by selective co-immunoprecipitation. *J. Biol. Chem.* **277,** 15,482–15,485.
22. Gomes, I., Jordan, B. A., Gupta, A., Rios, C., Trapaidze, N., and Devi L. A. (2001) G protein-coupled receptor dimerization: implications in modulating receptor function. *J. Mol. Med.* **79,** 226–242.
23. Mueller, A., Kelly, E., and Strange, P. G. (2002) Pathways for internalization and recycling of the chemokine receptor CCR5. *Blood* **99,** 785–791.
24. Cvejic, S. and Devi, L. A. (1997) Dimerization of the d opioid receptor: implication for a role in receptor internalization. *J. Biol. Chem.* **272,** 26,959–26,964.
25. Mellado, M., Rodríguez-Frade, J. M., Mañes, S., and Martínez-A., C. (2001) Chemokine signaling and functional responses: the role of receptor dimerization and TK pathway activation. *Ann. Rev. Immunol.* **19,** 397–421.
26. Harlow, E. and Lane, D. (eds.) (1988) *Antibodies. A Laboratory Manual.* Cold Spring Harbor Laboratories, New York.
27. Chuntharapai, A., Lee, J., Burnier, J., Wood, W. I., Hébert, C., and Kim, K. J. (1994) Neutralizing monoclonal antibodies to human IL-8 receptor A map to the NH_2-terminal region of the receptor. *J. Immunol.* **152,** 1783–1789.
28. Van Aelst, L. and D'Souza-Schorey, C. (1997) Rho GTPases and signaling networks. *Genes Dev.* **15,** 2295–2322.
29. Sekar, R. B. and Periasami, A. (2003) Fluoresecence resonance energy transfer (FRET) microscopy imaging of live cell protein localizations. *J. Cell. Biol.* **160,** 629–633.
30. Periasamy, A., Elangovan, M., Elliott, E., and Brautigan, D. L. (2002) Fluorescence lifetime imaging (FLIM) of green fluorescent fusion proteins in living cells. *Methods Mol. Biol.* **183,** 89–100.
31. McVey, M., Ramsay, D., Kellet, E., et al. (2001) Monitoring receptor oligomerization using time-resolved fluorescence resonance energy transfer and bioluminiscence resonance energy transfer. *J. Biol. Chem.* **276,** 14092–14099.
32. Bastiaens, P. I. and Squire, A. (1999) Fluorescence lifetime imaging microscopy: spatial resolution of biochemical processes in the cell. *Trends Cell. Biol.* **9,** 48–52.
33. Sorkin, A., McClure, M., Huang, F., and Carter, R. (2000) Interaction of EGF receptor and grb2 in living cells visualized by fluorescence visualized by fluorescence resonance energy transfer (FRET) microscopy. *Curr. Biol.* **10,** 1395–1398.
34. Gu, C., Cali, J. J., and Cooper, D. M. (2002) Dimerization of mammalian adenylate cyclases. *Eur. J. Biochem.* **269,** 413–421.
35. Xia, Z. and Liu, Y. (2001) Reliable and global measurement of fluorescence resonance energy transfer using fluorescence resonance energy transfer (FRET) microscopy. *Biophys. J.* **81,** 2395–4028.
36. Angers, S., Salahpour, A., and Bouvier, M. (2002) Dimerization: an emerging concept for G protein-coupled receptor onotgeny and function. *Ann. Rev. Pharmacol. Toxicol.* **42,** 409–435.

8

Real-Time Analysis of G Protein-Coupled Receptor Signaling in Live Cells

Venkatakrishna R. Jala and Bodduluri Haribabu

Summary

Seven transmembrane-spanning receptors, widely referred to as G protein-coupled receptors (GPCRs), mediate a broad spectrum of extracellular signals at the plasma membrane through G proteins, thereby modulating a variety of biological processes. In addition to G proteins, they also interact with a number of other cytoplasmic proteins. Thus, methods to understand GPCR signaling and their interactions with intracellular proteins in real time in live cells are of importance. Recent developments in microscopy methods and the availability of fluorescent proteins facilitated the development of techniques to unravel these interactions more precisely. This chapter describes the methodology for sequential capturing of images of membrane and cytoplasmic proteins fused to different fluorescence probes to understand GPCR interaction with cytosolic proteins and their colocalization.

Key Words: G protein-coupled receptors; live cell video microscopy; green fluorescence protein; red fluorescence protein; internalization; co-localization.

1. Introduction

G protein-coupled receptors (GPCRs) represent a large super family of seven transmembrane-spanning proteins (approx 600 members in human genome) and respond to a diverse array of sensory and chemical stimuli, odor, taste, pheromones, hormones, neurotransmitters, nucleotides, small molecule amines, ions, lipids, amino acids, and peptides. GPCRs transduce the signals from various extracellular elements to the induction of cellular responses via heterotrimeric G proteins (G_α, G_β, G_γ *[1–3]*). Upon activation of GPCR, the G_α subunit of the G protein cmplex exchanges GDP with GTP, followed by dissociation from $G_{\beta\gamma}$ subunits. This exchange allows a variety of G protein-dependent signal transduction pathways, such as activation of adenylate cyclase,

From: *Methods in Molecular Biology, vol. 332: Transmembrane Signaling Protocols, Second Edition*
Edited by: H. Ali and B. Haribabu © Humana Press Inc., Totowa, NJ

phospholipase C, chemotaxis, degranulation, calcium release, and transcriptional regulation of gene expression. In addition, binding of ligand initiates a series of events that attenuate the signals via desensitization, sequestration, and/or internalization of the receptors. The desensitization of GPCRs is regulated by receptor phosophorylation by GPCR kinases and subsequent binding of β-arrestins to phosphorylated receptors *(4,5)*. The receptor–β-arrestin complex associates with clathrin and accessory proteins involved in the formation of clathrin-coated pits, ultimately leading to receptor internalization *(6)*. The list of GPCR-interacting proteins is growing rapidly and includes other GPCRs, GPCR kinases, arrestins, kinases, chaperone proteins, receptor activity-modifying proteins, and PDZ, PH domain-containing proteins *(7,8)*. Previously, crosslinking, coimmunoprecipitation, and yeast two hybrid systems were used to investigate interactions between proteins.

These techniques determine the end result of the interaction and but are not useful to determine the kinetics or transient interactions in vivo. Recently, the availability of various flurochromes (green, cyan, yellow, and red fluorescence proteins, i.e., GFP, CFP, YFP, and RFP, respectively) to tag the proteins in addition to very sensitive, state-of-the-art microscopes revolutionized the research on GPCRs and their interacting proteins. Fluorescence resonance energy transfer and bioluminescence resonance transfer were used efficiently to understand the dimerization of GPCRs' live cells *(7)*. These techniques are now expanding to investigate the other interactions of GPCRs with subcellular proteins. We developed a methodology to monitor the kinetics of receptor internalization and translocation of cytosolic proteins and their interactions with receptors. Using these methods, we have shown that D6, a chemokine decoy receptor, constitutively associates with the cytoplasmic adaptor β-arrestin and that this interaction is essential for D6 internalization *(9)*. We also demonstrated receptor phosphorylation independent β-arrestin translocation and internalization of leukotriene B_4 receptors (hBLT1) using real-time spinning disc confocal analysis of cells expressing GFP-β-arrestin and hBLT1–RFP *(10)*. This methodology has the advantage over the conventional confocal microscopy in that a laser light source is not required and photo bleaching and cell damage during the imaging period is not a concern. The current methodology uses a mercury lamp as light source and allows sequential monitoring of multiple wavelengths in single live cells to delineate the molecular basis of protein–protein interactions.

2. Materials
2.1. Cell Culture

1. Rat basophilic leukemia cell line (RBL-2H3; any adherent cell lines).
2. Delbecco's modified Eagle's medium (DMEM).
3. Phenol red-free RPMI or DMEM.

4. Fetal bovine serum (FBS).
5. L-Glutamine (200 mM).
6. Penicillin-streptomycin (10,000 U/mL).
7. Trypsin-ethylene diamine tetraacetic acid (EDTA): 1X, 0.05% trypsin, 0.53 mM EDTA.
8. Versene (1:5000).
9. T75 tissue culture flasks.
10. 100-mm Tissue culture dishes.
11. 30-mm Sterile glass cover slip-bottomed dishes (0.17-mm thick; WillCo-dish).
12. 5- and 10-mL sterile plastic serological pipets.
13. Sterile 15-mL centrifuge tubes.

2.2. Transfection

1. Sterile Gene Pulser Cuvette (0.4-cm electrode gap; Bio-Rad).
2. Gene Pulser II electroporater (Bio-Rad).
3. HEPES (1 M).
4. 1-mL Sterile plastic serological pipets.
5. Complementary DNA of GPCR (hBLT1 or human D6 in the present study) tagged with red fluorescence protein at C-terminus (hBLT1–RFP or hD6–RFP).
6. Complementary DNA of cytosolic protein tagged with GFP (β-arrestin1–GFP in the present study).

2.3. Microscopy

1. TE-FM Epi-Fluorescence system attached to Nikon Inverted Microscope Eclipse TE300. The microscope equipped with heating stage. A cool snap HQ digital B/W charge-coupled device (Roper Scientific) camera and LAMDA 10–2 optical filter changer (Sutter Instruments) is attached to CARV II™ spinning disk confocal fluorescence system (ATTO Biosciences). The choice of the filters used here are: filter sets S480/20×, S525/40m and S565/25×, S620/60m for GFP and RFP, respectively; enhanced green fluorescent protein (EGFP)/DsRed dual dichroic beam splitter (Chroma Technology). This filter wheel can accommodate up to six filter sets. Microscope is attached with Prior Proscan stage controller with Joystick. All of these hardware attachments can be controlled by Metamorph (Universal Imaging) software.
2. 30-mm Glass cover slip-bottomed dishes.
3. Phenol red-free RPMI or DMEM.

2.4. Data Analysis

1. Metamorph software from Universal Imaging.

3. Methods
3.1. Cell Culture

1. RBL-2H3 cells are maintained in T-75 cell culture flasks at 37°C in a humidified atmosphere of 95% air, 5% CO_2 as monolayer cultures in DMEM supplemented with 15% FBS, 2 mM L-glutamine, 100 U/mL penicillin, and 100 U/mL streptomycin.

2. RBL-2H3 cells grown to 75% confluence are detached from culture dishes with trypsin-EDTA (6 mL per T-75 flask).
3. After 5 min of incubation at 37°C, cells are gently resuspended in an equal volume of growth medium.
4. Cells are reseeded at a density of 6×10^6 cells per T-75 tissue culture flasks in growth medium to obtain fresh cells for transfection next day.

3.2. Transfection (See Notes 1 and 2)

1. Next day, aspirate the medium and wash with 5 mL of versene (1:10,000) and incubate with 5 mL of versene for 10 min at 37°C in a humidified atmosphere of 95% air, 5% CO_2.
2. Count the cells using a hemocytometer and centrifuge 4×10^6 cells in a 15-mL tube at 480g for 3 min and resuspend in 200 µL of transfection medium (DMEM, 20% FBS, 50 mM HEPES).
3. Add 25 µg of DNA of hBLT1–RFP expression vector and/or 15 µg of β-arrestin1–GFP vector to sterile electroporation cuvets (0.4-cm electrode gap).
4. Add the above resuspended cells (200 µL) to cuvets containing either hBLT1–RFP or β-arrestin1–GFP or both.
5. Mix them gently with a 1-mL sterile pipet.
6. Keep at room temperature for 10 min.
7. Electroporate these cells in Gene Pulser II (set the voltage at 250 and the capacitance at 500 µF).
8. Let sit for 10 min undisturbed at room temperature.
9. Add 1 mL of regular growth medium to the cuvet and distribute 300 µL of this mixture into 30-mm tissue culture dishes (*see* **Note 3**) containing 3 mL of regular growth medium and incubate at 37°C in a CO_2 incubator.
10. Change medium after 1 h of incubation and let the cell stay at 37°C for 18–24 h.

3.3. Microscopy (See Notes 4 and 5)

1. Lift the cells from 30-mm culture dishes using versene and plate the cells (3×10^5 cells) onto 30-mm glass cover slip-bottomed culture dishes.
2. Incubate cells with 3 mL of regular growth medium at 37°C for 1–2 h to allow the cells to adhere glass bottom dishes.
3. Wash the cells two or three times with a warm phenol red-free RPMI buffer (phenol red-free RPMI buffer + 10 mM HEPES, pH 7.55).
4. Place 30-mm glass cover slip-bottomed culture dishes containing RBL-2H3 cells, transfected with hBLT1–RFP and β-arrestin1–GFP on the heated microscope stage (37°C).
5. Watch and focus the cells (*see* **Note 6**) using regular bright field transmitted light using Nikon Plan Apo 60x/1.4 numerical aperture oil immersion lens before shifting to fluorescent light through charge-coupled device camera attached to computer.
6. Switch from bright field (transmitted light) to fluorescence filter (*see* **Note 7**; RFP filter, if you expect to observe receptor or GFP, if you expect to observe β-arrestin1) where fluorescence of the cells observed.

7. Choose a bright and healthy cell, which is expressing GPCR–RFP on the cell surface using RFP filter and capture the image.
8. Then shift to GFP filter and make sure that β-arrestin–GFP expressed in the cytoplasm and capture the image. It is desirable to pseudocolor both the images, before collecting images of live cells and make sure that bleed through of fluorescence is not occurring between RFP and GFP (*see* **Note 8**).

3.4. Acquisition of Images

1. After selecting a healthy cell, program the acquisition of images according to your needs. In the case of example shown in **Fig. 1**, 16-bit images are acquired with the camera binning set to 1×1 combined with 60X objective Nikon Plan Apo 60X/1.4 numerical aperture oil immersion lens (*see* **Note 9**). We start collecting the images for RFP and GFP fluorescence simultaneously at 30-s time interval using filter wheels controlled by Metamorph software. Normally add the desired ligand to activate the GPCR after collecting two to three base line images (1–2 min).
2. Camera exposure set to 1000 ms for RFP and GFP. Set the specific exposure times according to the fluorescence intensity of RFP and GFP. Exposure times may have to be increased when switched to spinning disk confocal CARV II system.
3. Images are collected for desired length of time (we usually collect images for up to 60 min after adding the ligand) and stored as TIFF images with increasing order of the file names and can be made as stack files using the Metamorph software.

3.5. Image Processing and Quantification of Fluorescence

1. After collecting the fluorescence images separate RFP and GFP images subtract from back ground image (*see* **Note 10**) and measure the fluorescence intensity of RFP and GFP at different time points from defined regions to measure the translocation of receptors and β-arrestin using Metamorph quantitative analysis.
2. Plot the amount fluorescence as the function of time. This will provide information about the kinetics of translocation of a given molecule.

4. Notes

1. The efficiency of transfection is high when cells were not fully confluent. It is desirable to lift the cells at approx 60 to 80% confluence.
2. RBL-2H3 cells can also be transfected with lipofectamine.
3. RBL-2H3 cells also can be cultured directly in glass-bottom culture dishes for overnight.
4. Make sure that cells are spread out nicely over the dish. It is important that cells should be separated well to determine the response to ligands and obtain clean images. If too many cells were present in 30 m*M* culture dish, it is advisable to dilute the cells and replate them.
5. The set up of the microscope used in this study is described in the **Subheading 2.3.** Make sure that the heating plate has reached 37°C before you place the cells on microscope. One can also use different fluorescence proteins (such as CFP,

hBLT1-RFP	β-arrestin1-GFP	Overlay	Time (min)

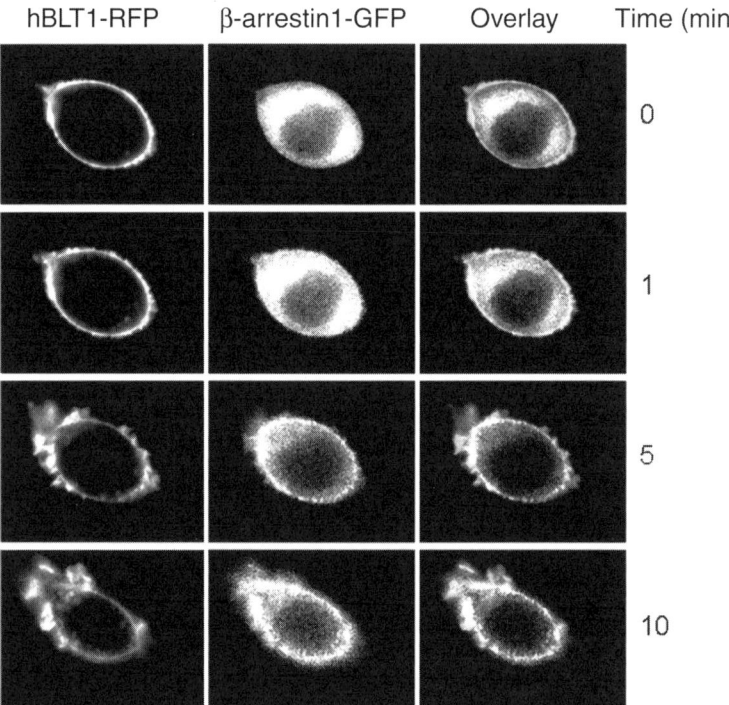

Fig. 1. Real-time analysis of leukotriene B_4 (LTB_4)-induced co-localization of LTB_4 receptors (hBLT1) and β-arrestin. β-Arrestin–green fluorescent protein (GFP) (15 μg) is cotransfected along with hBLT1–wild-type red fluorescent protein (RFP) (25 μg) into rat basophilic leukemia cell line cells as described. After the addition of LTB_4 (0-time), Red and green fluorescence images were collected every 10 s for 70 min using appropriate filters. With time, co-localization of β-arrestin-GFP and receptor RFP is seen. Videos of the real-time analysis of β-arrestin and G protein-coupled receptor interactions are available as an online supplement *(10)*.

YFP, etc.) and their respective filters in the filter wheel according to their purpose. YFP and CFP are used to measure the fluorescence resonance energy transfer.
6. Select healthy cells that are adherent to the glass cover slip-bottomed dish for analysis.
7. Excitation and emission wavelengths are controlled with filter wheels attached to ATTO spinning disk head and controlled by Lamba 10–2 filter wheel controller (Sutter Instruments). Exposure time 500 ms should be enough to view RFP or GFP in live cells. Hardware control and acquisition of images are controlled by Metamorph software.
8. We did not observe any bleed through of fluorescence between RFP and GFP. This is likely a function of GFP and RFP variants, as well as the filters used.

High-specificity filters may be obtained from Chroma Technologies. We used specific filter sets (cat. no. 86007, Chroma Technologies) to measure the fluorescence of DsRed and EGFP.

9. Nikon Plan Apo 100X/1.4 numerical aperture oil immersion lens can be also used to get more insights into the cellular interaction. This microscope is attached to CARV II spinning disk confocal fluorescence system, where we could obtain confocal images as well. A limitation for the usage of CARV II system is it requires higher exposure times compared with regular fluorescence microscope.

10. Acquire the images with plain media (without cells) with RFP and GFP fluorescent filters. Subtract these images from actual data images.

Acknowledgments

Research in our laboratory is supported by National Institutes of Health grants AI-43184 and AI-52381 and Kentucky Lung Cancer Research Board and institutional support from The James Graham Brown Cancer Center to B.H.

References

1. Gether, U. (2000) Uncovering molecular mechanisms involved in activation of G protein-coupled receptors. *Endocr. Rev.* **21,** 90–113.
2. Gether, U., Asmar, F., Meinild, A. K., and Rasmussen, S. G. (2002) Structural basis for activation of G-protein-coupled receptors. *Pharmacol. Toxicol.* **91,** 304–312.
3. Pierce, K.L., Premont, R. T., and Lefkowitz, R. J. (2002) Seven-transmembrane receptors. *Nat. Rev. Mol. Cell Biol.* **3,** 639–650.
4. Lefkowitz, R. J. (1998) G protein-coupled receptors. III. New roles for receptor kinases and beta-arrestins in receptor signaling and desensitization. *J. Biol. Chem.* **273,** 18,677–18,680.
5. Shenoy, S. K. and Lefkowitz, R. J. (2003) Multifaceted roles of beta-arrestins in the regulation of seven-membrane-spanning receptor trafficking and signaling. *Biochem. J.* **375,** 503–515.
6. Goodman, O. B., Jr., Krupnick, J. G., Santini, F., et al. (1996) Beta-arrestin acts as a clathrin adaptor in endocytosis of the beta2-adrenergic receptor. *Nature* **383,** 447–450.
7. Brady, A. E. and Limbird, L. E. (2002) G protein-coupled receptor interacting proteins: emerging roles in localization and signal transduction. *Cell Signal.* **14,** 297–309.
8. Premont, R. T. and Hall, R. A. (2002) Identification of novel G protein-coupled receptor-interacting proteins. *Methods Enzymol.* **343,** 611–621.
9. Galliera, E., Jala, V. R., Trent, J. O., et al. (2004) β-Arrestin-dependent constitutive internalization of the human chemokine decoy receptor D6. *J. Biol. Chem.* **279,** 25,590–25,597.
10. Jala, V. R., Shao, W. H., and Haribabu, B. (2005) Phosphorylation-independent beta-Arrestin Translocation and Internalization of Leukotriene B4 Receptors. *J. Biol. Chem.* **280,** 4880–4887.

II ——————————————————————

SPECIFIC TOPICS

B. Lipid Raft and Caveolae in Transmembrane Signaling

9

Isolation of Membrane Rafts and Signaling Complexes

Kathleen Boesze-Battaglia

Summary

Traditionally, lipid rafts have been defined by their insolubility in ice-cold Triton X-100 and low-buoyant density. These low-density membrane microdomains have been referred to as detergent-resistant membranes, Triton-insoluble membranes, and Triton-insoluble floating fraction. They are enriched in cholesterol, often sphingomyelin and various gangliosides (GM1, GM2, and GM3). The ability of the B-subunit of cholera toxin to bind GM1 has been exploited to visualize membrane rafts by confocal microscopy in patching and capping experiments. Biochemically, membrane rafts are isolated by solubolization in ice-cold Triton X-100 and separation of the low-buoyant density fractions from soluble material on sucrose density gradients. We describe the isolation of Jurkat cell-specific membrane rafts using 2% Triton X-100. This procedure yielded a consistent raft product that was enriched in cholesterol, gangliosides sphingomyelin and membrane raft protein markers including lck and lat 1. Moreover, rafts were visualized using Alexa Fluor 647 cholera toxin capped with anti-cholera toxin antibody. Co-localization of the C subunit of cytolethal distending toxin to rafts was determined using patching techniques.

Key Words: Membrane microdomains; detergent resistance; Jurkat cells; Triton X-100.

1. Introduction

The physical relationship between hydrophilic aqueous medium and hydrophobic fat-like molecules has intrigued scientists since the 1770s, when Benjamin Franklin observed that any oily substance clearly covered half the surface area when compared with an equal volume of aqueous solution, implying a bilayer structure. As early as 1925, Gorter and Grendel *(1)* proposed the now-classic deduction that membrane lipids are arranged in a bilayer configuration in which parallel sheets of phospholipids have polar or charged headgroups oriented toward the aqueous environment and acyl chains interacting within the hydrophobic membrane core. In 1972, Singer and Nicolson provided a model that took into consideration the dynamic nature of lipid–protein interac-

From: *Methods in Molecular Biology, vol. 332: Transmembrane Signaling Protocols, Second Edition*
Edited by: H. Ali and B. Haribabu © Humana Press Inc., Totowa, NJ

tions, providing a matrix in which proteins have a degree of motion that, in turn, can have a dramatic impact on activity. Thus, the fluid mosaic model *(2)* became the framework and benchmark for our current understanding of membrane bilayers and their physiological function. The assumed homogeneous nature of membrane bilayers proposed in this model was called into question in the 1970s, when it was observed that membranes contain a unique composition of lipid and protein components that are specific to cell type and subcellular localization. A heterogeneous distribution of lipids and proteins is even observed within spatially separated regions of the same membrane of Golgi *(3)* or apical and basolateral plasma membranes of polarized cells *(4)*. Within the past decade, a unifying theme describing the organization of lipid and membrane proteins has focused on localized regions within the membrane known collectively as membrane microdomains. The last 5 yr have seen an emergence of interest in a specific type of microdomain, known colloquially as a membrane raft. More precisely, these regions are globally defined as cholesterol-rich domains in the liquid-ordered phase. These microdomains are proposed to be involved in a wide variety of cellular processes including, protein sorting *(5)*, signal transduction *(6)*, calcium homeostasis *(7)*, transcytosis *(8)*, potocytosis *(9)*, alternative routes of endocytosis *(10)*, internalization of toxins, bacteria, and viruses *(11–13)*, HIV-1 assembly and release *(14)*, and cholesterol transport *(15,16)*.

The association of cytolethal distending toxin (cdt) with Jurkat cells will be used to illustrate the methods used to analyze membrane raft functionality using biochemical and microscopic techniques.

2. Materials

2.1. Isolation of Triton X-100 Resistant Membrane Microdomains

1. 0.5 to 2% Triton X-100 in MOPS buffer.
2. 0.5 to 2% octyglucopyranoside (wt/v) in MOPS buffer.
3. MOPS buffer: 10 mM MOPS, pH 7.2, 60 mM KCl, 30 mM NaCl, 5 mM MgCl$_2$, 1 mM dithiothreitol (DTT; *see* **Note 1**), 5 µM aprotinin, and 1 µM leupeptin.
4. 0.5, 0.6, 0.65, 0.7, 0.75, 0.80, and 2.4 M sucrose in MOPS buffer.
5. Beckman Optima LE 80K Ultracentrifuge, including SW-41 rotor, SW-41 titanium buckets, and ultra-clear tubes.
6. Wheaton glass-homogenizer; 7.5-mL or 15-mL volume.
7. T-cell leukemia cell line Jurkat (E6-1; T1B152, lot no. 2113016, ATCC) or other cells of interest (cell count approx 5 × 108).

2.2. Visualization of Membrane Rafts in Jurkat Cells and Localization of Proteins to Raft Microdomains

1. Hank's balanced salt solution (HBSS): Gibco 10X HBSS, dilute with H$_2$O to make 1X HBSS.

2. Cholera toxin B subunit-Alexa Fluor 647. Reconstituted with 100 μL of phosphate-buffered saline (PBS). Stock concentration of 1 mg/mL. Use 1 μL of stock per 1 mL of cell suspension/appropriate tube.

3. Anticholera toxin, subunit B, *Vibrio cholera*, (goat). Reconstituted with 100 μL of dH$_2$O, 10 μL of reconstituted antitoxin + 240 μL of HBSS = 1:25 dilution. Use 100 μL of the antitoxin dilution per appropriate tube.

4. CGM: RPMI-1640 Glutamax, 10% fetal bovine serum, 2% Pen/Strep, 1 m*M* sodium pyruvate.

5. ABChis (*see* **Note 2**): B:021004, 163 μg/mL: 37 μL of stock + 263 μL of CGM = 20 μg/mL.

6. Buffer: PBS/1% bovine serum album (ice-cold).

7. Goat immunoglobulin (Ig; Southern Biotech) diluted to 0.5 mg/mL in PBS (stock). Make 1:50 dilution of the stock in PBS.

8. Anti-ABChis purified Ig monoclonal antibody: anti-ABChis, 17A1.15, use 10 μL of a 100 μg/mL stock to 1 μg/mL.

9. Goat anti-mouse Ig biotin. Dilution: 25 μL of stock solution at 500 μg/mL + 225 μL of buffer = 50 μg/mL. Use 20 μL of 50 μg/mL dilution/appropriate tube = 1 μg/appropriate tube.

10. Alexa Fluor 488-streptavidin reconstitued with 1 mL of PBS, 030904 TLM. Dilute 4 μL of stock + 996 μL of buffer to equal 1:250 dilution. (Centrifuge for 5 min at 10,000*g* before use.)

11. Radiance 2000 laser confocal microscope with argon, green He/Ne, Red diode and Blue diode lasers.

3. Methods

The Methods described below outline the biochemical isolation of rafts and the visualization and localization of proteins to membrane rafts.

3.1. Detergents Used in the Isolation of Membrane Rafts

The isolation of membrane microdomains or rafts relies on the relative insolubility of the less-fluid cholesterol-rich liquid-ordered membrane regions in Triton X-100. Recently, the repertoire of detergents used to isolate low-buoyant density membrane microdomains and signaling complexes has been expanded to include Brij 98, NP-40, CHAPS, and Lubrol. These detergents differ in their critical micelle concentration (CMC) and thus are postulated to solubilize mixed "raft-like domains" and tetraspanin protein complexes *(17)*. Lastly, although not reviewed in this chapter, detergent-independent modes of raft and/ or caveolae isolation have been developed. These include sodium carbonate lysis, sonication and sucrose gradient centrifugation (*see* **Note 3** *[18]*), and the isolation of plasma membrane-specific rafts using Percoll gradient-purified membranes, which are sonicated and rafts isolated by floatation in continuous Opti-prep gradients *(19)*.

3.2. Isolation of Signaling Complexes in Detergent-Resistant Membranes

The isolation of detergent-resistant membranes (DRMs) from bacterial toxin-treated cells, HIV-infected cells, and cells stimulated with a variety of ligands have provided valuable information on the mode of action of these agonists. Stimulation of cells with growth factors and the isolation rafts has allowed investigators to determine in which compartment a specific signaling event has occurred. Using this approach, tyrosine kinases appear to activate mitogen-activated protein kinase (MAPK) and other signal transduction pathways from within rafts; phosphatidylinositol turnover occurs in lipid rafts in response to growth factor stimulation; and cholera toxin's mode of action requires the association of the B unit with gangliosides enriched in membrane rafts. Because membrane rafts are stabilized through the interaction of cholesterol with other lipid components, as a complementary approach to studying raft function is the depletion of cell membrane cholesterol, membrane raft integrity can be disrupted with the addition of β-methylcyclodextrin (Sigma), an agent that sequesters and removes membrane cholesterol *(20)*. Using this cholesterol depletion approach, the role of lipid rafts in specific signaling events can be studied directly in intact cells (for review, *see* **ref. 28**). Although there are countless examples of the role of membrane rafts in a variety of biological processes, the basic techniques used to isolate rafts are largely similar to those described here.

3.3. Isolation of Jurkat Cell Membrane Rafts on Toxin Association

The T-cell leukemia cell line Jurkat (E6-1) was maintained in RPMI-1640 supplemented as described *(21)*. Cells were harvested in mid-log growth phase, and for membrane raft preparations, the cells were grown at 2.0×10^6 cells/mL in T-75 flasks. The cells were exposed to medium or toxin for 2 h, isolated, and washed in MOPS buffer *(22)*. To distinguish between Triton X-100-resistant membranes and simply a partial detergent-dependent nonraft-specific solubilzation, control cells were homogenized in 2% octyl glucopyranoside: parallel to the Triton X-100-treated samples. Membrane rafts were isolated from the Jurkat cells as described in the following steps.

1. Resuspend isolated cells in a final volume of 1.1 mL of MOPS buffer; if the cells appear to aggregate, resulting in a nonhomogenous suspension, add an additional 1 mL of MOPS buffer (*see* **Note 4**).
2. Transfer 1 mL of cells to a Wheaton glass homogenizer (on ice), add 0.77 mL of ice-cold 2% (v/v) Triton X-100 in MOPS buffer (*see* **Note 5**). For control cells, add 0.77 mL of 2% octyl glucopyranoside.
3. Homogenize five strokes on ice, taking care to keep bubbles and foaming to a minimum.
4. Let sit on ice for 15 min.

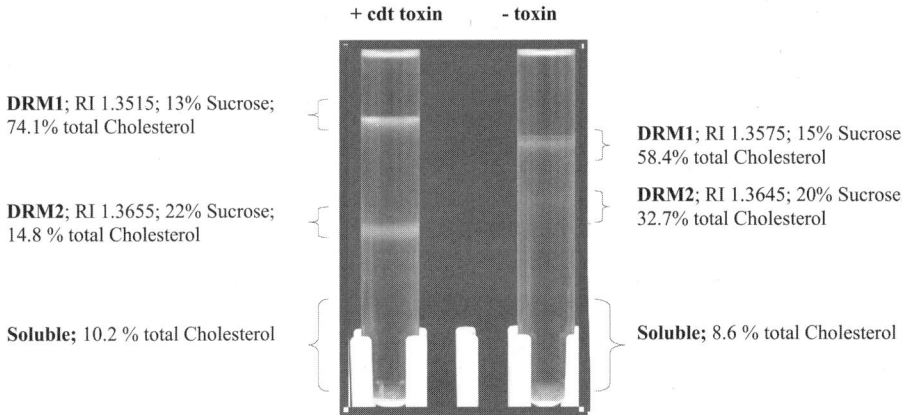

+ cdt toxin   **- toxin**

DRM1; RI 1.3515; 13% Sucrose; 74.1% total Cholesterol

DRM2; RI 1.3655; 22% Sucrose; 14.8 % total Cholesterol

Soluble; 10.2 % total Cholesterol

DRM1; RI 1.3575; 15% Sucrose; 58.4% total Cholesterol

DRM2; RI 1.3645; 20% Sucrose; 32.7% total Cholesterol

Soluble; 8.6 % total Cholesterol

Fig. 1. Cdt toxin treatment alters the buoyant density of Jurkat cell membrane rafts. In order to determine whether the toxin, or any of its subunits, localize to lipid microdomains, detergent-resistant membranes (DRMs) from both untreated Jurkat cells and cells exposed to CdtABC were isolated. After a 2-h incubation, Jurkat cells were disrupted, homogenized in ice-cold Triton X-100, and separated on a sucrose gradient. Two distinct low-buoyant density bands, designated DRM1 and DRM2, were obtained and the position the sucrose density gradient determined as a measure of refractive index as indicated. The cholesterol content composition of these bands was analyzed *(24)* and is presented as percentage of the total membrane cholesterol.

5. Add 1.24 mL of 2.4 *M* sucrose in MOPS buffer (*see* **Note 6**), vortexing immediately.
6. Transfer samples to clear SW-41 centrifuge tubes.
7. Sequentially layer 1 mL of each of the following: 0.8, 0.7, 0.65, 0.6, and 0.5 *M* sucrose solutions onto sample to create a sucrose step gradient (*see* **Note 7**).
8. Top samples off with requisite volume of MOPS buffer so that the tubes are filled approx 0.10 cm from the top.
9. Place samples in SW-41 buckets.
10. Spin at 400,000*g* for 20 h at 4°C.

3.4. Fractionation of Sucrose Density Gradient

Detergent-insoluble membrane fractions are isolated as low-buoyant density fractions, as shown in **Fig. 1**. Jurkat cells both with or without toxin exhibit a characteristic low-buoyant density band. These bands are collected either directly with a Pasteur pipet or the sucrose density gradient fractionated as described *(23)*. The low-buoyant density bands, DRMs, are analyzed for cholesterol *(24)*, phospholipid *(25)*, and total protein (Bio-Rad). To further confirm that these membranes are rafts, the level of GM1 is analyzed by immunoblotting using anticholera toxin antibody. In addition, or as an alternative to analysis of the gangliosides, total lipid extracts may be prepared as

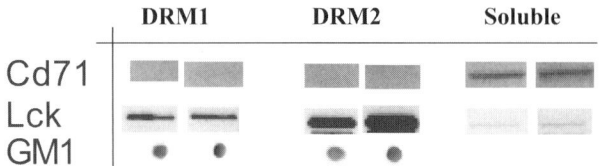

Fig. 2. DRM1 and DRM2 are enriched in membrane raft markers: GM1 and Lck. As shown, both DRM1 and DRM2 were enriched in GM1 (dot blot) and in total cholesterol. Moreover, the raft-associated protein Lck was enriched in these fractions. In contrast, the transferrin receptor (CD71), a nonraft-associated protein, was found in the soluble fraction.

described *(26)* and analyzed by thin-layer chromatography (TLC) for total sphingolipid content *(27)*. The protein markers used to confirm that a DRM is a raft are described immediately below in **Subheading 3.5.**

3.5. Membrane Raft Protein Markers

In addition to higher levels of cholesterol, GM1, phospholipids with saturated fatty acyl side chains, and sphingomyelin, lipid rafts can be characterized based on the presence of specific lipid raft marker proteins. Although these proteins vary from cell to cell, often flotilin-1 and 2, LAT, Thy, and as a rule of thumb, most GPI-anchored proteins and src family kinases are membrane raft-associated. A comprehensive list of such proteins can be found in **refs.** *28*, *28a*, and *29*. In addition, caveolin 1 and 2 (*see* **Note 8**) are often are associated with high cholesterol caveolae or membrane caves, a subset of the membrane microdomain family. Conversely, the transferrin receptor and geranylated proteins are routinely nonraft markers. As shown in **Fig. 2**, the DRMs isolated from Jurkat cells (with or without cdt toxin) were enriched in the raft marker, Lck and deficient in the nonraft marker, CD71. Collectively in these studies, both DRM1 and 2 were identified as membrane rafts based on position in the sucrose density gradient, the increased levels of GM1, high percentage of cholesterol, increased sphingomyelin, and the presence of raft-specific protein markers (shown is Lck).

3.6. Visualization and Localization of Proteins to Membrane Rafts

3.6.1. Membrane Raft Capping and Patching Techniques

The visualization of membrane rafts in cells is limited by the resolution of the techniques used. Rafts have been visualized in model membranes, giant unilammellar vesicles (GUVs) composed of distinct raftophilic lipids using fluorescent membrane probes designed to detect lipid ordering *(30)*. Visualization of rafts in intact cells is somewhat more difficult for a number of reasons.

GPI-anchored proteins and GM1 markers appear uniformly distributed since their concentration may be only three- to fourfold higher in rafts than in the remaining membrane or membrane rafts may be transient complexes formed in response to agonists or antigens *(31)*. Thus, to overcome these problems, membrane rafts are visualized buy exploiting the fact that GPI-anchored proteins and GM1 molecules will cluster in response to antibodies. In the method described below, GM1 is crosslinked by the cholera toxin B subunit (this process is referred to as capping), and the capped cholera toxin is subsequently treated with anti-cholera toxin antibody resulting in the clustering of GM1 in a process referred to as patching *(32)*. Moreover, we describe the co-localization of the C-subunit of cdt toxin to these membrane clusters using biotin–streptavadin labeling techniques.

1. Harvest Jurkat cells, wash one time with HBSS, centrifuge at 800*g* for 8 min, discard supernatant, resuspend cell pellet in HBSS, and count.
2. Prepare a tube with 6 mL of cells in HBSS at 2×10^6 cells/mL.
3. Add 6 µL of Alexa Fluor 647 cholera toxin (stock 1 mg/mL) to the 6 mL of cells (final 1 µg/mL).
4. Incubate the cells on ice for 30 min.
5. Add 1 mL of the cholera toxin-treated cell suspension to each of five tubes and label them 1 through 5.
6. Wash cells with 2 mL of HBSS, centrifuge at 800*g* for 8 min, and discard supernatant.
7. Repeat **step 6**.
8. Add 100 µL of HBSS to tube 1 (no capping control).
9. Add 100 µL of 1:25 dilution of anticholera toxin to all remaining tubes (positive capping).
10. Incubate for 30 min on ice; incubate at 37°C for 40 min.
11. Wash cells with 2 mL of HBSS, centrifuge at 800*g* for 8 min, and discard supernatant.
12. Repeat **step 11**.
13. Resupend each tube with 500 µL of CGM.
14. Add 500 µL of CGM to the "cells only" tubes.
15. Add 400 µL of CGM + 100 µL of ABChis at 20 µg/mL to the "cells + ABChis" tubes. (ABChis, B:021004, stock at 163 µg/mL.)
16. Incubate all tubes at 37°C for 2 h.
17. Wash cells with 2 mL of buffer, centrifuge at 1000*g* for 8 min, and discard supernatant.
18. Repeat **step 17**.
19. Add 10 µL of goat Ig in buffer at 10 µg/mL to tubes 1 through 5.
20. Incubate on ice for 10 min.
21. Add 10 µL of either buffer to tubes 1 and 2, and 10 µL of Anti-ABChis purified Ig 17A1.15 at 100 µg/mL to tubes 3 through 5 (final, 2 µg/tube).
22. Incubate on ice for 30 min.

23. Wash all tubes with 2 mL of buffer, centrifuge at 1000*g*, for 8 min, and discard supernatant.
24. Repeat **step 23**.
25. Add 20 µL of 50 (g/mL dilution of goat anti-mouse Ig biotin to all tubes (1 µg/tube).
26. Incubate on ice for 30 min.
27. Wash all tubes with 2 mL of buffer, centrifuge at 1000*g* for 8 min, and discard supernatant.
28. Repeat **step 27**.
29. Add 50 µL of a 1:250 dilution of Alexa Fluor 488-SA to all tubes.
30. Incubate on ice for 30 min.
31. Wash cells with 2 mL of buffer, centrifuge at 800*g* for 8 min, and discard supernatant.
32. Repeat **step 31**.
33. Resuspend cells in 500 µL of 2% formaldehyde.

As shown in **Fig. 3**, the CdtC subunit localizes to membrane lipid rafts. Utilizing confocal fluorescence microscopy we demonstrate co-localization of the C subunit with the cholera toxin B subunit (CtB) bound to GM1. To control for nonspecific staining, isotype matched control IgG was used instead of the anti-Cdt monoclonal antibody. As shown below, virtually all fluorescence associated with either CdtC co-localizes with GM1 (i.e., CtB fluorescence).

4. Notes

1. Buffers containing DTT should be prepared fresh daily. Routinely, MOPS buffer is prepared in the absence of DTT. DTT is added to the desired volume prior to the start of each experiment.
2. The ABChis is the active holotoxin of the cytolethal distending toxin. In individualized experiments, this may be a ligand for a receptor or any bacterial toxin of interest.
3. The sodium carbonate lysis method relies on a pH of 11.0 and is often is used to remove excess peripheral proteins from the membranes. This method is described in detail in Chapter 10.
4. It is important that the cells be a homogenous suspension. Thus, the volume of buffer used in the resuspension may be increased; however, with this increase there must be an increase in the amount of 2% ice-cold Triton X-100 added. For example, for 1 mL of cell suspension, we add 0.77 mL of Triton; for 2 mL of a cell suspension, 1.44 mL of Triton X-100 is required, etc. (*see* **Note 6** for sucrose amounts).
5. Different concentrations of Triton X-100 have been used by a variety of investigators to isolate DRMs; the relative solubility of the components of interest determines the protein composition of the membrane raft. The final concentration of Triton X-100 in the sample is expected to be less than 1% and limited solubility is seen with decreasing amounts of Triton X-100 (*see* **ref. 21**).

| | CdtC
Alexa Fluor 488 | Cholera Toxin β
Alexa Fluor 647 | Merged
Image |

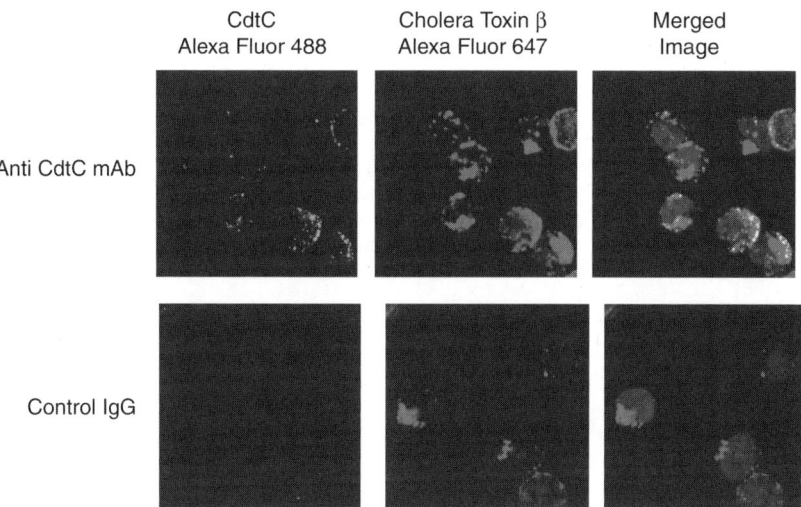

Fig. 3. Visualization and co-localization of cdt to membrane rafts. To demonstrate that the Cdt subunits localize to membrane lipid rafts, confocal fluorescence microscopy to demonstrate co-localization of the C subunit with the cholera toxin B subunit (CtB) bound to GM1 was used. Jurkat cells were first exposed to CtB-Alexa Fluor 647 for 20 min; the cells were then treated with anti-CtB antisera to induce patch formation. Cells were then exposed to CdtABC for 2 h, washed, and sequentially stained with monoclonal antibody (MAb) to CdtC, goat anti-mouse Ig conjugated to biotin, and streptavidin-Alexa Fluor 488. To control for nonspecific staining, isotype-matched control IgG was used instead of the anti-Cdt MAb. As shown, virtually all fluorescence associated with either CdtC colocalized with GM1 (i.e., CtB fluorescence).

6. If the volume of cells was increased (as per **Note 4**) to maintain an appropriate sucrose concentration (i.e., >0.8 M), then the amount of 2.4 M sucrose must be increased. For example, to a preparation containing 1 mL of cells and 0.77 mL of Triton X-100, add 1.25 mL of 2.4 M sucrose; to a preparation containing 2 mL of cells and 1.44 mL of Triton X-100, add 2.5 mL of 2.4 M sucrose.
7. As an alternative to sucrose step gradients, some investigators prefer a continuous gradient from 5 to 30% sucrose.
8. A number of cells, Jurkats included, do not contain caveolin. In addition, membrane caves, i.e., membrane microdomains enriched in caveolae, are isolated using a nondetergent-based Opti-prep gradient *(19)* or sodium carbonate lysis procedures *(18)*. The isolation of caveolin-enriched microdomains is described in Chapter 10.

Acknowledgments

The author would like to thank Cheryl Gretzula and Lisa Pankoski for their critical reading of the manuscript and Terry McKay for her expert technical assistance. Supported by NIH: DE06014 and EY10420.

References

1. Gorter, E. and Grendel, F. (1925) On biomolecular layers of lipids on the chromatocytes of blood. *J. Exp. Med.* **41,** 439–443.
2. Singer, S. J. and Nicolson, G. L. (1972) The fluid mosaic model of the structure of cell membranes. *Science* **175,** 720–731.
3. Simons, K. and van Meer, G. (1988) Lipid sorting in epithelial cells. *Biochemistry* **27,** 6197–6202.
4. Rodriguez-Boulan, E. and Nelson, W. J. (1989) Morphogenesis of the polarized epithelial cell phenotype. *Science* **245,** 718–725.
5. Simons, K. and Ikonen, E. (1997) Functional rafts in cell membranes. *Nature* **387,** 569–572.
6. Zajchowski, L. D. and Robbins, M. (2002) Lipid rafts and little caves: compartmentalized signalling in membrane microdomains. *Eur. J. Biochem.* **269,** 737–752.
7. Isshiki, M. and Anderson, R. G. W. (1999) Calcium signal transduction from caveolae. *Cell Calcium* **26,** 201–208.
8. Simionescu, N. (1983) Cellular aspects of transcapillary exchange. *Physiol. Rev.* **63,** 1536–1560.
9. Anderson, R. G. W., Kamen, B. A., Rothberg, K. G., and Lacey, S. W. (1992) Potocytosis: sequestration and transport of small molecules by caveolae. *Science* **255,** 410–411.
10. Smart, E. J., Graf, G. A., McNiven, M. A., et al. (1999) Caveolins, liquid-ordered domains, and signal transduction. *Mol. Cell. Biol.* **19,** 7289–7304.
11. Parton, R. G., Joggerst, B., and Simons., K. (1994) Regulated internalization of caveolae. *J. Cell Biol.* **127,** 1199–1215.
12. Fivaz, M., Abrami, L., and van der Goot, F. G. (1999) Landing on lipid rafts. *Trends Cell Biol.* **9,** 212–213.
13. Shin, J.-S., Gao, Z., and Abraham, S. N. (2000) Involvement of cellular caveolae in bacterial entry into mast cells. *Science* **289,** 785–788.
14. Ono, A. and Freed, E. O. (2001) Plasma membrane rafts play a critical role in HIV-1 assembly and release. *Proc. Natl. Acad. Sci. USA* **98,** 13,925–13,930.
15. Oram, J. F. and Yokoyama, S. (1996) Apolipoprotein-mediated removal of cellular cholesterol and phospholipids. *J. Lipid Res.* **37,** 2473–2491.
16. Smart, E. J., Ying, Y.-S., Donzell, W. C., and Anderson, R. G. W. (1996) A role for caveolin in transport of cholesterol from endoplasmic reticulum to plasma membrane membrane. *J. Biol. Chem.* **271,** 29,427–29,435.
17. Claas, C., Stipp, C. S., and Hemler, M. E. (2001) Evaluation of prototype transmembrane 4 superfamily protein complexes and their relation to lipid rafts. *J. Biol. Chem.* **276,** 7974–7984.
18. Song, K. S., Li, S., Okamoto, T., Quiliam, L. A., Sargiacomo, M., and Lisanti, M. P. (1996) Co-purification and direct interaction of Ras with caveolin, an integral membrane protein of caveolae microdomains. *J. Biol. Chem.* **271,** 9690–9697.
19. Smart, E. J., Ying, Y.-S., Mineo, C., and Anderson, R. G. W. (1995) A detergent-free method for purifying caveolae membrane from tissue culture cells. *Proc. Natl. Acad. Sci. USA* **92,** 10,104–10,108.

20. Christian, A. E., Haynes, M. P., Phillips, M. C., and Rothblat, G. H. (1997) Use of cyclodextrins for manipulating cellular cholesterol content. *J. Lipid Res.* **38**, 2264–2274.
21. Pike, L. J. (2003) Lipid rafts: bringing order to chaos. *J. Lipid Res.* **44**, 655–667.
22. Shenker, B. J., Besack, D., Mc Kay, T., Pankoski, L., Zekavat, A., and Demuth, D. (2004) actinobacillus actinomycetemcomitans cytolethal Distending Toxin (Cdt): Evidence evidence that the holotoxin is composed of three subunits: CdtA, CdtB, and CdtC. *J. Immunol.* **172**, 410–417.
23. Boesze-Battaglia, K., Besack, D., Pankoski, L., Mc Kay, T., Jordan-Sciutto, K., and Shenker, B. J. (2004) Association of actinobacillus actinomycetemcomitans cytolethal distending toxin (cdt) with membrane rafts, submitted.
24. Boesze-Battaglia, K., Hennessey, T., and Albert, A. D. (1989) Cholesterol heterogeneity in bovine retinal rod outer segment disk membranes. *J. Biol. Chem.* **264**, 8151–8155.
25. Allain, C. C., Poon, L. S., Chan, C. S., Richmond, W., and Fu, P. C. (1974) Enzymatic determination of total serum cholesterol. *Clin. Chem.* **20**, 470–475.
26. Bartlett, G. R. (1959) Phosphorous assay in column chromatography. *J. Biol. Chem.* **234**, 466–473.
27. Folch, J., Lees, M., and Sloane-Stanley, G. H. (1957) A simple method for the isolation and purification of total lipides from animal tissue. *J. Biol. Chem.* **226**, 497–509.
28. Boesze-Battaglia, K., Dispoto, J., and Kahoe, M. A. (2002) Association of a photoreceptor-specific tetraspanin protein, ROM-1, with triton X-100-resistant membrane rafts from rod outer segment disk membranes. *J. Biol. Chem.* **277**, 41,843–41,849.
28a. Pike, L. J. (2003) Lipid rafts: bringing order to chaos. *J. Lipid Res.* **44**, 655–667.
29. van der Goot, F. G. and Harder, T. (2001) Raft membrane domains: from a liquid-ordered membrane phase to a site of pathogen attack. *Semin. Immunol.* **13**, 89–97.
30. Dietrich, C., Volovyk, Z. N., Levi, M., Thompson, N. L., and Jacobson, K. (2001) Partitioning of Thy-1, GM1, and cross-linked phospholipid analogs into lipid rafts reconstituted in supported model membrane monolayers. *Proc. Natl. Acad. Sci. USA* **98**, 10,642–10,647.
31. Brown, D. A. and London, E. (2000) Structure and function of sphingolipid- and cholesterol-rich membrane rafts. *J. Biol. Chem.* **275**, 17,221–17,224.
32. Mitchell, J. S., Kanca, O., and Mc Intyre, B. W. (2002) Lipid microdomain clustering induces a resdistribution of antigen recognition and adhesion molecules on human T lymphocytes. *J. Immunol.* **168**, 2738–2744.

10

Methods for the Study of Signaling Molecules in Membrane Lipid Rafts and Caveolae

Rennolds S. Ostrom and Paul A. Insel

Summary

Lipid rafts and caveolae are cholesterol- and sphingolipid-rich microdomains of the plasma membrane that concentrate components of certain signal transduction pathways. Interest in and exploration of these microdomains has grown in recent years, especially after the discovery of the biochemical marker of caveolae, caveolin, and the recognition that caveolin interacts with many different signaling molecules via its scaffolding domain. There are three major types of caveolins (1, 2, and 3), with some selectivity in their expression in different tissues. Results assessing lipid raft/caveolae co-localization of molecules in signal transduction pathways have provided support for the idea that signaling components are compartmentalized or preassembled together. This chapter describes nondetergent- and detergent-based methods for isolating lipid rafts and caveolae for biochemical studies. We also describe a method for immunoisolation (using antibodies to caveolins) of detergent-insoluble membranes that selectively isolates caveolae vs lipid rafts. Together, these methods are useful for assessment of the role of lipid rafts and caveolae in transmembrane signaling.

Key Words: Lipid rafts; caveolae; membrane microdomains; plasma membrane vesicles; density gradient centrifugation; immunoprecipitation; immunoisolation.

1. Introduction

Recent investigations have identified caveolae and lipid rafts as important microdomains of the plasma membrane (PM) that concentrate and perhaps "preassemble" the components of signal transduction pathways *(1,2)*. The enrichment of a large array of signal transduction components in those domains *(3)* suggests that caveolae and lipid rafts are signaling centers, whose role in physiological and pathophysiological processes is the subject of intense investigation. Caveolae ("little caves") were morphologically identified in the 1950s

From: *Methods in Molecular Biology, vol. 332: Transmembrane Signaling Protocols, Second Edition*
Edited by: H. Ali and B. Haribabu © Humana Press Inc., Totowa, NJ

as flask-like, 50- to 100-nm invaginations of the PM of endothelial cells and later shown to be involved in potocytosis, endocytosis, and transcellular movement of molecules *(4,5)*. Endocytosis by caveolae differs from that mediated by another specialized region of the PM, clathrin-coated pits. These two vesicular structures differ biochemically and transport different types of molecules, thus representing parallel, but distinct, pathways *(5)*.

Lipid rafts form via the coalescence of particular lipids, most prominently sphingolipids and cholesterol *(12)*. Caveolae have a similar lipid composition to rafts but also contain a "coat" of caveolin proteins on the inner leaflet of the membrane bilayer *(6)*. Although all mammalian cells appear to contain PM lipid rafts *(7)*, only cells expressing caveolins express caveolae (for example, leukocytes contain lipid rafts but no caveolae). There are three isoforms of caveolin, caveolin-1, caveolin-2, and caveolin-3; caveolae can form if cells express either caveolin-1, the predominant isoform, or caveolin-3, the striated muscle-specific isoform *(8,9)*. Caveolin-2 does not appear to induce caveolae formation but is found in hetero-oligomers with caveolin-1 and caveolin-3 *(10–14)*. Although caveolae generally are considered subsets of lipid rafts because of their similar lipid composition, the two entities likely differ in a variety of ways *(15–17)*. By their original definition, caveolae are discrete morphological structures identifiable at the electron microscopic level; however, lipid rafts cannot readily be identified by microscopic techniques (except by atomic force microscopy *[18]*).

Lipid rafts and caveolae are most readily studied with biochemical approaches. Cells are disrupted and the lipid rafts and caveolae are then extracted from other cellular material, generally based on their relative insolubility in certain detergent or nondetergent conditions. Most commonly, lipid rafts and caveolae are then isolated by virtue of their high buoyancy when centrifuged on a density gradient consisting of either a discontinuous gradient of sucrose *(5,19–21)*, as we describe here, or a continuous gradient of Optiprep, as described elsewhere *(22,23)*. These approaches, which rely on properties common to both lipid rafts and caveolae, do not distinguish between these domains. However, caveolae can be preferentially isolated from lipid rafts by using immunological approaches to trap caveolin-rich membrane domains *(24,25)*. We describe one general method for isolating caveolae from cells and tissues in this chapter. Another method developed by Oh and Schnitzer isolates caveolae using immunoisolation of caveolin; this method is particularly suitable for cells or tissues with high levels of expression of caveolins and caveolae, such as pulmonary vascular endothelia *(24)*.

Caveolins also can act as scaffolding proteins, whereby other proteins, including proteins involved in signal transduction, bind; such binding generally is associated with an inhibition in signaling activity *(2)*. Immunoprecipitation of caveolin proteins or expression of peptides that interfere with the caveolin-binding motif, the domain on caveolin-1 and caveolin-3 that binds those other proteins *(2)*, can be used to assess the role of the putative caveolin scaffold. Caveolin protein overexpression and knockout also has been used to examine the role of these proteins in physiology and signal transduction *(26)*. Because caveolins act as regulators of several signal transduction pathways *(6)*, alteration in their expression by such approaches cannot be considered a pure "probe" of the compartmentation of a particular signaling pathway in caveolae. The function of lipid rafts and caveolae in signaling or other cellular processes also can be inferred from studies in which the microdomains are disrupted. Methyl-β-cyclodextrin, a chemical that does not enter cells but can bind cholesterol and remove it from the PM, disrupts lipid rafts and caveolae *(27)*. Filipin, a polyene antibiotic and sterol-binding agent, also disrupts lipid rafts and caveolae *(28,29)*. This chapter does not describe these methodologies but interested readers may wish to consult articles describing their use *(19,28,30)*.

Microscopic studies are useful and sometimes important for corroboration of results from biochemical studies of lipid rafts and caveolae. However, such approaches, not described here, are limited by their inability to detect lipid rafts (as noted previously), the poor resolution of light and fluorescent microscopy for the identification of caveolae, and the variable ability and availability of suitable antibodies to detect the proteins of interest. With suitable antibodies, one can assess protein localization using double immunostaining and electron microscopy. Cholera toxin, which binds to GM1 ganglioside enriched in lipid rafts *(28)*, can be used as a marker of lipid rafts and caveolae whereas caveolin can be used as a marker of caveolae. Atomic force microscopy, which can detect the surface topology of cell membranes, is a technique with future promise for visualizing lipid rafts *(18)*. Other technologies, such as fluorescence resonance energy transfer and bioluminescence energy transfer, are powerful tools for examination of interaction of components, but their use is limited to the examination of exogenously expressed proteins with fluorescent/luminescent tags *(31)*.

This chapter describes nondetergent- and detergent-based methods for isolating lipid rafts/caveolae and caveolae. We describe some variations of this general approach that can be useful in specific circumstances and we present complementary methods to address the question of signaling in lipid rafts or caveolae. We also describe a method for immunoisolation of detergent-insoluble membranes that selectively isolates caveolae vs lipid rafts. Each method has advantages and disadvantages that should be considered when choosing an experimental approach to answer a particular biological question. Nondetergent fractionation of cells retains certain proteins in lipid raft/caveolar fractions that can be lost upon detergent solubilization *(22,32)*. However, detergent-based approaches can allow the measurement of protein function, such as enzyme activity, whereas the nondetergent methods (because of high pH and high energy sonication) do not. We believe that it generally is desirable to use a combination of different, complementary approaches, along with immunoprecipitation (i.e., "pull-down" of proteins in detergent conditions that solubilize all membrane proteins [not described herein]) to assess protein–protein interactions to study signal transduction in lipid rafts and caveolae.

2. Materials

1. Phosphate-buffered saline (PBS).
2. 500 mM Na$_2$CO$_3$ (should be approx pH 11.0, but do not adjust).
3. 2-(N-morpholino)ethanesulfonic acid (MES)-buffered saline (MBS): 25 mM MES, 150 mM NaCl, pH 6.0.
4. MBS/Na$_2$CO$_3$: 250 mM Na$_2$CO$_3$ in MBS.
5. 90% Sucrose/MBS: dissolve 45 g of sucrose with MBS until volume equals 50 mL. Heat in a microwave oven (in 10-s intervals) to dissolve/melt.
6. 35% Sucrose in MBS/Na$_2$CO$_3$: 5.83 mL of 90% sucrose/MBS plus 9.17 mL of MBS/Na$_2$CO$_3$.
7. 5% Sucrose in MBS/Na$_2$CO$_3$: 0.83 mL of 90% sucrose/MBS plus 14.17 mL of MBS/Na$_2$CO$_3$.
8. Triton X-100 buffer: MBS, 1% Triton X-100, protease inhibitor mix (Sigma P-8340, diluted 1:100).
9. 35% Sucrose in MBS/Triton X-100: 5.83 mL of 90% sucrose/MBS plus 9.17 mL of Triton X-100 buffer.
10. 5% Sucrose in MBS/Na$_2$CO$_3$: 0.83 mL of 90% sucrose/MBS plus 14.17 mL of Triton X-100 buffer.
11. Membrane buffer: 0.25 M of sucrose, 1 mM of ethylene diamine tetraacetic acid, 20 mM tricine, pH 7.8
12. 30% Percoll: 3 mL of Percoll stock solution diluted in 9 mL of PBS.
13. Modified lysis buffer: 50 mM of Tris-HCl, pH 7.5; 150 mM NaCl; 1 mM EGTA; 10 mM MgCl$_2$; 0.5% Triton X-100; protease inhibitor mix (Sigma P-8340, diluted 1:100).
14. Protein A-agarose and protein G-agarose (50 µL per sample).
15. Immunoprecipitation (IP) wash buffer 1: 50 mM Tris-HCl, pH 7.5, 500 mM NaCl, 0.2% Triton X-100.
16. IP wash buffer 2: 10 mM Tris-HCl, pH 7.5, 0.2% Triton X-100.

3. Methods

3.1. Isolation of Lipid Rafts and Caveolae by Sucrose Density Centrifugation

This method relies on the unique lipid composition (with enrichment in sphingolipid and cholesterol) of lipid rafts and caveolae, which makes these membrane domains resistant to solubilization in certain conditions, as well as more buoyant than other cellular lipids. We describe the method for adherent cells in tissue culture but it can be readily adapted for cells in suspension or for tissue samples. We have found that two 150-mm plates of cells are adequate for one preparation.

3.1.1. Sodium Carbonate Isolation of Lipid Rafts and Caveolae

1. Check that cells are at least 70% confluent. One preferably grows cells on 150-mm tissue culture plates (dishes) to optimize the amount of starting material. Aspirate medium and wash three times with ice-cold PBS. On the last wash, be sure to remove all PBS by tilting the plate at a steep angle for 30 s then aspirating all liquid.
2. Apply 1 mL of 500 mM Na$_2$CO$_3$ to each 150-mm plate and make sure it covers the entire monolayer. Scrape cells from the plate with a cell scraper, making sure to retain as much cellular material as possible.
3. Transfer the cells from two plates (2 mL total) to a prechilled Dounce (glass–glass) homogenizer and homogenize the cells with 20 strokes on ice.
4. Transfer the homogenate to a cold 50-mL Falcon tube and homogenize with a polytron three times for 10 s with intervals of 10–15 s. Rinse the polytron blade with 0.5 mL of 500 mM Na$_2$CO$_3$ into the sample to recover all possible material.
5. Homogenize the sample using an ultrasonic cell disruptor equipped with a stainless-steel probe at high power three times for 20 s each with a full 60 s of rest between each homogenization.
6. Proceed to **Subheading 3.1.4.**

3.1.2. Detergent Isolation of Lipid Rafts and Caveolae

1. Check that cells are at least 70% confluent. Aspirate medium and wash three times with ice-cold PBS. On the last wash, be sure to remove all PBS by tilting the plate at a steep angle for 30 s and then aspirating all liquid.
2. Apply 1 mL of 1% Triton-X 100 buffer to each 150-mm plate so that it covers the entire monolayer (*see* **Note 1**). Scrape cells from the plate with a cell scraper making sure to retain as much cellular material as possible.
3. Transfer the cells from two plates (2 mL total) to a prechilled Dounce (glass–glass) homogenizer and incubate on ice for 20 min. Homogenize the cells with 20 strokes on ice.
4. Proceed to **Subheading 3.1.4.**

3.1.3. Variation: Isolation of Lipid Rafts and Caveolae From PMs

1. Check that cells are at least 70% confluent. Aspirate medium and wash three times with ice-cold PBS. On the last wash, be sure to remove all PBS by tilting the plate at a steep angle for 30 s and then aspirating all liquid.
2. Apply 1 mL of membrane buffer to each 150-mm plate so that it covers the entire monolayer. Scrape cells from the plate with a cell scraper making sure to retain as much cellular material as possible.
3. Collect the cells from two plates (2 mL total) and homogenize cells with 20 strokes in a Dounce (glass–glass) or Teflon-glass homogenizer on ice then centrifuge at 300g for 5 min and collect the supernatant.
4. Layer the supernatant on top of 30% Percoll and centrifuge at 64,000g (19,000 rpm on a SW41 rotor for a Beckman ultracentrifuge) for 30 min.

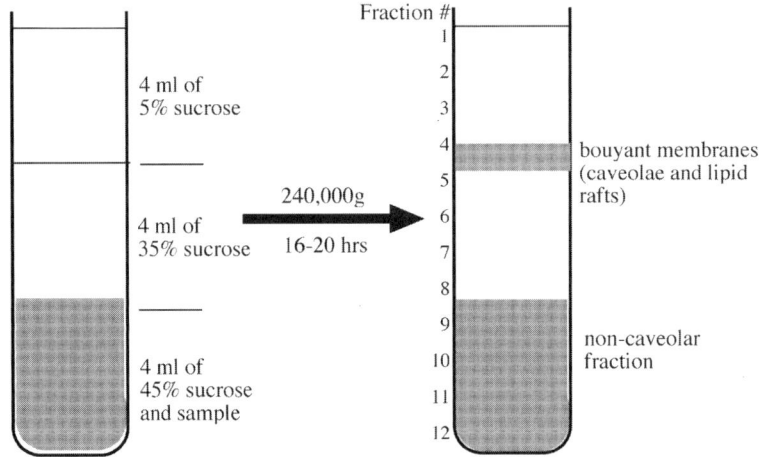

Fig. 1. Schematic diagram of sucrose density centrifugation to isolate buoyant fractions consisting of lipid rafts and caveolae (*see* **Subheading 3.1.4.**). Homogenized cell or tissues lysates in a final concentration of 45% sucrose are bottom loaded in an ultracentrifuge tube. A discontinuous sucrose gradient consisting of 35 and 5% sucrose is formed on top of the sample and the gradient is centrifuged for 16 to 20 h at approx 240,000g. Buoyant lipid raft and caveolar material will float up to the 35 to 5% sucrose interface whereas the rest of the cellular material will remain in the 45% sucrose layer. The gradient is typically collected in 1-mL fractions starting at the top.

5. Collect the opaque band near the top of the Percoll layer as PM.
6. Adjust PM to 500 mM Na$_2$CO$_3$ by adding an equal volume of 1 M Na$_2$CO$_3$ and sonicate three times for 20 s with full 60-s rests between intervals.

3.1.4. Sucrose Density Centrifugation to Fractionate Cell Homogenates

Once cells or tissues are homogenized using one of the above approaches (**Subheadings 3.1.1–3.1.3.**), the lipid raft/caveolar fraction can be isolated using sucrose density centrifugation. As explained previously, this method relies on the unique lipid composition of the lipid raft/caveolae (rich in sphingolipid and cholesterol), which makes these domains of the membrane more buoyant than other cellular components. A schematic diagram of this fractionation, as shown in **Fig. 1** and as described subsequently, involves use of a discontinuous gradient, although a continuous gradient also can be used.

1. Mix 2 mL of homogenized sample (leaving any foam behind) with 2 mL of 90% sucrose/MBS in an ultracentrifuge tube. Save any remaining sample as whole cell lysate.
2. Carefully layer 4 mL of either 35% sucrose in MBS/Na$_2$CO$_3$ (if sample was homogenized by nondetergent method, **Subheadings 3.1.1.** or **3.1.3.**) or 35%

sucrose in MBS/Triton X-100 buffer (if sample was homogenized by detergent method, **Subheading 3.1.2.**) on top of the sample/90% sucrose/MBS layer. A visible interface should exist between the two density layers.

3. Carefully layer 4 mL of either 5% sucrose in MBS/Na$_2$CO$_3$ (if sample was homogenized by nondetergent method, **Subheadings 3.1.1.** or **3.1.3.**) or 5% sucrose in MBS/Triton X-100 buffer (if sample was homogenized by detergent method, **Subheading 3.1.2.**) on top of the 35% sucrose layer. A second interface should be visible between the 35% sucrose and the 5% sucrose layers and the ultracentrifuge tube should be nearly full.

4. Centrifuge for 16–20 h at 39,000 rpm (4°C) in SW41Ti rotor (Beckman), equivalent to a maximum force (bottom of tube) of approx 260,000g and an average force (middle of the tube) of approx 188,000g (*see* **Note 2**).

5. At the completion of the centrifugation, carefully remove the ultracentrifuge tube from the bucket. A faint light-scattering band, which consists of the buoyant lipid raft/caveolar material, is often visible at the 35% sucrose–5% sucrose interface.

6. Collect samples from the gradient from the top down as 1-mL fractions, being careful to keep the pipet at the top of the liquid in order to draw each fraction appropriately.

7. Fractions can then be analyzed by sodium dodecyl sulfate-polyacrylamide gel electrophoresis (SDS-PAGE) and immunoblotting (*see* **Note 3**).

3.2. Immunoisolation of Caveolae

With this method, one takes advantage of the reduced solubility of lipid rafts and caveolae to detergent in order to isolate these domains from the rest of the cellular material (as in **Subheading 3.1.2.**) and one then selectively "traps" caveolae (vs lipid rafts) by using an antibody to immunoprecipitate caveolin (and associated lipids and proteins). Antibodies to any of the caveolin isoforms can be used, but they should be chosen carefully based on expression of caveolins in the cell or tissue of interest. With this method, one has the potential to maintain enzyme and receptor binding activity, thus allowing one to assess protein function. We have used this approach to assay receptor-activated adenylyl cyclase activity in caveolae from cardiac myocytes *(25)* and other cells; thus, the function of other proteins are likely also maintained.

1. Check that cells are at least 70% confluent. Aspirate medium and wash three times with ice-cold PBS. On the last wash, be sure to remove all PBS by tilting plate at a steep angle for 30 s and then aspirating all liquid.

2. Add 2 mL of modified lysis buffer to each 15-cm plate. Homogenize cells with 20 strokes in a Dounce (glass–glass) homogenizer.

3. Transfer to a 1.5-mL microtube and add 50 µL of either protein G- or Protein A-agarose suspension (*see* Note 4). Incubate at 4°C on a rocking platform for 1 h.

4. Centrifuge in a microcentrifuge at maximum speed (12,000–14,000g) for 30 s to pellet agarose and then transfer the supernatant to a new tube.

5. Add primary antibody (1–3 µL, depending on the antibody concentration) and gently mix (preferably by rocking) at 4°C for 1 h.

6. Add 50 µL of protein A- or protein G-agarose to tube and incubate at 4°C on a rocking platform for 1 h.
7. Centrifuge in a microcentrifuge at maximum speed (12,000–14,000*g*) for 30 s to pellet agarose. Supernatant should be saved as the IP supernatant and used as a control.
8. Add 1 mL of modified lysis buffer to pellet, mix, and rock at 4°C for 5 min.
9. Centrifuge in a microcentrifuge at maximum speed (12,000–14,000*g*) for 30 s to pellet agarose, remove supernatant, and add 1 mL of wash buffer 1 to pellet, mix, and rock at 4°C for 5 min.
10. Centrifuge in a microcentrifuge at maximum speed (12,000–14,000*g*) for 30 s to pellet agarose, remove supernatant, and add 1 mL of wash buffer 2 to pellet, mix, and rock at 4°C for 5 min.
11. The final pellet should then be suspended in a suitable assay buffer (if enzyme activity is to be measured) and/or in sample buffer for analysis by SDS-PAGE (for immunoblotting). Immunoblot analysis should be performed on a portion of the immunoprecipitated pellet and supernatants to confirm appropriate IP and to assess which proteins have been coprecipitated with the target protein.

4. Notes

1. Nonionic detergents other than Triton X-100, including NP-40, octylglucoside, CHAPS, Lubrol, and Brij 98, can be used to solubilize cells and isolate lipid raft and caveolar domains (*19*). In addition, some investigators have used concentrations of Triton X-100 less than 1% in protocols similar to that described in this chapter.
2. Other rotors can be used for the sucrose density centrifugation, including a Beckman SW55Ti rotor with 5-mL buckets. In this case, 1 mL of cell homogenate is mixed with 1 mL of 90% sucrose and 2 mL of of 35% sucrose with 1 mL of 5% sucrose layered on top. The rotation speed is adjusted to maintain equivalent g forces. Fractions are collected in 0.5-mL aliquots to yield 10 fractions.
3. It is critical to determine the appropriateness of each fractionation by performing immunoblot analysis for markers of cellular organelles. Fractions from the 5% sucrose/35% sucrose interface (numbers 4 and 5) should contain the bulk of any caveolin immunoreactivity but exclude markers of clathrin-coated pits (such as adaptin-β), Golgi (such as mannosidase II), and other cellular organelles for the markers discussed previously. The total protein in each fraction also can be used as an indicator of appropriate fractionation. The buoyant fractions from most cells contain approx 5% or less of the total cellular protein. Individual fractions from the gradient can be pooled into a buoyant fraction (collected from the 5 to 35% sucrose interface) and a nonbuoyant fraction (consisting of the entire 45% sucrose layer), if desired, to facilitate rapid screening of antibodies. This approach simplifies immunoblot analysis when the localization of many different proteins needs to be assessed. However, it is wise to collect individual fractions as described previously and then to combine a portion of these fractions to form pooled fractions, saving some portion of the individual fractions at –70°C for

future use (**Fig. 1**). When immunoblot analysis of fractions is planned, it is best to add SDS-PAGE sample buffer to each fraction and to denature at –70°C for 10 min immediately after collecting the gradient, which will ensure more reproducible results when storing frozen samples for extended periods. For detection of low abundance proteins, samples also can be concentrated in a Speed-vac (or similar type) concentrator before addition of sample buffer. However, the fractions from the bottom of the gradient will not concentrate as well because of the presence of higher concentrations of sucrose. Dialysis also can be used to remove sucrose and to concentrate the samples.

4. When performing any type of immunoprecipitation, one should carefully choose between protein A-agarose and protein G-agarose for precipitation of the primary antibody. Protein A has high affinity for human, rabbit, guinea pig, and pig immunoglobulin (Ig)Gs and moderate affinity for mouse, horse, and cow IgGs. Protein G has high affinity for human, horse, cow, pig, and rabbit IgGs and moderate affinity for sheep, goat, chicken, hamster, guinea pig, rat, and mouse IgGs. Protein A and protein G also differ in their affinities for different subclasses of IgGs. More information on the different affinities of protein A and protein G can be found in the product information sheet from Roche Applied Science.

Acknowledgments

Work in the authors' laboratories related to this topic is supported by grants from the National Institutes of Health and the American Heart Association.

References

1. Ostrom, R. S. and Insel, P. A. (2004) The evolving role of lipid rafts and caveolae in G protein-coupled receptor signaling: Implications for molecular pharmacology. *Br. J. Pharmacol.* **143**, 235–245.
2. Okamoto, T., Schlegel, A., Scherer, P. E., and Lisanti, M. P. (1998) Caveolins, a family of scaffolding proteins for organizing "preassembled signaling complexes" at the plasma membrane. *J. Biol. Chem.* **273**, 5419–5422.
3. Shaul, P. W. and Anderson, R. G. (1998) Role of plasmalemmal caveolae in signal transduction. *Am. J. Physiol.* **275**, L843–L851.
4. Palade, G. (1953) Fine structure of blood capilaries. *J. Appl. Physiol.* **24**, 1424.
5. Anderson, R. G. (1998) The caveolae membrane system. *Annu. Rev. Biochem.* **67**, 199–225.
6. Razani, B., Woodman, S. E., and Lisanti, M. P. (2002) Caveolae: from cell biology to animal physiology. *Pharmacol. Rev.* **54**, 431–467.
7. Hooper, N. M. (1999) Detergent-insoluble glycosphingolipid/cholesterol-rich membrane domains, lipid rafts and caveolae (review). *Mol. Membr. Biol.* **16**, 145–156.
8. Song, K. S., Scherer, P. E., Tang, Z., et al. (1996) Expression of caveolin-3 in skeletal, cardiac, and smooth muscle cells. Caveolin-3 is a component of the sarcolemma and co-fractionates with dystrophin and dystrophin-associated glycoproteins. *J. Biol. Chem.* **271**, 15,160–15,165.

9. Tang, Z., Scherer, P. E., Okamoto, T., et al. (1996) Molecular cloning of caveolin-3, a novel member of the caveolin gene family expressed predominantly in muscle. *J. Biol. Chem.* **271,** 2255–2261.

10. Scherer, P. E., Okamoto, T., Chun, M., et al. (1996) Identification, sequence, and expression of caveolin-2 defines a caveolin gene family. *Proc. Natl. Acad. Sci. USA* **93,** 131–135.

11. Scherer, P. E., Lewis, R. Y., Volonté, D., et al. (1997) Cell-type and tissue-specific expression of caveolin-2. Caveolins 1 and 2 co-localize and form a stable hetero-oligomeric complex in vivo. *J. Biol. Chem.* **272,** 29,337–29,346.

12. Razani, B., Wang, X. B., Engelman, J. A., et al. (2002) Caveolin-2-deficient mice show evidence of severe pulmonary dysfunction without disruption of caveolae. *Mol. Cell Biol.* **22,** 2329–2344.

13. Rybin, V. O., Grabham, P. W., Elouardighi, H., and Steinberg, S. F. (2003) Caveolae-associated proteins in cardiomyocytes: caveolin-2 expression and interactions with caveolin-3. *Am. J. Physiol. Heart Circ. Physiol.* **285,** H325–H332.

14. Lahtinen, U., Honsho, M., Parton, R. G., Simons, K., and Verkade, P. (2003) Involvement of caveolin-2 in caveolar biogenesis in MDCK cells. *FEBS Lett.* **538,** 85–88.

15. Sowa, G., Pypaert, M., and Sessa, W. C. (2001) Distinction between signaling mechanisms in lipid rafts vs. caveolae. *Proc. Natl. Acad. Sci. USA* **98,** 14,072–14,077.

16. Williams, T. M. and Lisanti, M. P. (2004) The caveolin proteins. *Genome Biol.* **5,** 214.

17. Oh, P. and Schnitzer, J. E. (2001) Segregation of heterotrimeric G proteins in cell surface microdomains. G(q) binds caveolin to concentrate in caveolae, whereas g(i) and g(s) target lipid rafts by default. *Mol. Biol. Cell.* **12,** 685–698.

18. Henderson, R. M., Edwardson, J. M., Geisse, N. A., and Saslowsky, D. E. (2004) Lipid rafts: feeling is believing. *News Physiol. Sci.* **19,** 39–43.

19. Pike, L. J. (2003) Lipid rafts: bringing order to chaos. *J. Lipid Res.* **44,** 655–667.

20. Ostrom, R. S., Post, S. R., and Insel, P. A. (2000) Stoichiometry and compartmentation in G protein-coupled receptor signaling: implications for therapeutic interventions involving Gs. *J. Pharmacol. Exp. Ther.* **294,** 407–412.

21. Song, S. K., Li, S., Okamoto, T., et al. (1996) Co-purification and direct interaction of Ras with caveolin, an integral membrane protein of caveolae microdomains. Detergent-free purification of caveolae microdomains. *J. Biol. Chem.* **271,** 9690–9697.

22. Smart, E. J., Ying, Y. S., Mineo, C., and Anderson, R. G. (1995) A detergent-free method for purifying caveolae membrane from tissue culture cells. *Proc. Natl. Acad. Sci. USA* **92,** 10,104–10,108.

23. Rybin, V. O., Xu, X., Lisanti, M. P., and Steinberg, S. F. (2000) Differential targeting of beta-adrenergic receptor subtypes and adenylyl cyclase to cardiomyocyte caveolae. A mechanism to functionally regulate the cAMP signaling pathway. *J. Biol. Chem.* **275,** 41,447–41,457.

24. Oh, P. and Schnitzer, J. E. (1999) Immunoisolation of caveolae with high affinity antibody binding to the oligomeric caveolin cage. Toward understanding the basis of purification. *J. Biol. Chem.* **274,** 23,144–23,154.

25. Ostrom, R. S., Gregorian, C., Drenan, R. M., et al. (2001) Receptor number and caveolar co-localization determine receptor coupling efficiency to adenylyl cyclase. *J. Biol. Chem.* **276,** 42,063–42,069.

26. Razani, B. and Lisanti, M. P. (2001) Caveolin-deficient mice: insights into caveolar function in human disease. *J. Clin. Invest.* **108,** 1553–1561.

27. Smart, E. J. and Anderson, R. G. (2002) Alterations in membrane cholesterol that affect structure and function of caveolae. *Methods Enzymol.* **353,** 131–139.

28. Orlandi, P. A. and Fishman, P. H. (1998) Filipin-dependent inhibition of cholera toxin: evidence for toxin internalization and activation through caveolae-like domains. *J. Cell Biol.* **141,** 905–915.

29. Schnitzer, J. E., Oh, P., Pinney, E., and Allard, J. (1994) Filipin-sensitive caveolae-mediated transport in endothelium: reduced transcytosis, scavenger endocytosis, and capillary permeability of select macromolecules. *J. Cell Biol.* **127,** 1217–1232.

30. Ostrom, R. S., Bundey, R. A., and Insel, P. A. (2004) Nitric oxide inhibition of adenylyl cyclase type 6 activity is dependent upon lipid rafts and caveolin signaling complexes. *J. Biol. Chem.* **279,** 19,846–19,853.

31. Zacharias, D. A., Violin, J. D., Newton, A. C., and Tsien, R. Y. (2002) Partitioning of lipid-modified monomeric GFPs into membrane microdomains of live cells. *Science* **296,** 913–916.

32. Rybin, V. O., Xu, X., and Steinberg, S. F. (1999) Activated protein kinase C isoforms target to cardiomyocyte caveolae: stimulation of local protein phosphorylation. *Circ. Res.* **84,** 980–988.

II

SPECIFIC TOPICS

C. Protein–Protein Interaction in Transmembrane Signaling

11

Bioluminescence Resonance Energy Transfer to Monitor Protein–Protein Interactions

Tarik Issad and Ralf Jockers

Summary

The bioluminescence resonance energy transfer (BRET) methodology allows for the study of protein–protein interactions as well as conformational changes within proteins or molecular complexes. BRET is a highly versatile technique that can be applied to in vitro studies using purified proteins, crude cell membranes, cell fractions obtained by centrifugation on a density gradient, as well as permeabilized cells. Importantly, BRET also allows for monitoring of protein–protein interactions, in real time, in intact living cells that can be submitted to various stimuli. Moreover, quantitative BRET analysis also permits a pharmacological approach of protein–protein interactions, allowing one to determine whether a given stimulus induces a conformational change within preassociated partners or increases the association (recruitment) between two separated partners. Determination of the proportion of the dimeric vs monomeric form of a protein in the cell also is possible. Therefore, the BRET technology can be considered as a new and powerful tool in the field of protein–protein interactions.

Key Words: Resonance energy transfer; cell signaling; receptor activation; oligomerization; protein–protein interactions.

1. Introduction

Unraveling signaling pathways activated by membrane receptors is of fundamental importance for a detailed understanding of physiological and pathological processes. Transmembrane signaling generally involves the activation of a receptor by a ligand. Binding of the ligand induces conformational changes within receptors that often are organized as homo- or hetero-oligomers. Receptor activation may be associated with the stimulation of an enzymatic activity associated with the receptor, recruitment of signaling proteins, and/or relocalization of the receptor to a different cell compartment.

From: *Methods in Molecular Biology, vol. 332: Transmembrane Signaling Protocols, Second Edition*
Edited by: H. Ali and B. Haribabu © Humana Press Inc., Totowa, NJ

Regardless of the specific activation mechanism, changes in protein–protein interactions play a key role in the stimulation of intracellular signaling pathways that will eventually result in activation of biological processes. Thus, a detailed description of the dynamics of protein–protein interactions appears to be crucial for a better understanding of signaling pathways. Ideally, these interactions should be studied in the natural environment of the partners, in intact cells, that can be stimulated by different hormones, metabolites, and pharmacological drugs. Indeed, under these conditions, interactions that depend on the subcellular localization of the proteins and/or on their posttranslational modifications (e.g., phosphorylations/dephosphorylations) will be preserved. These studies can now be performed using a very powerful technique, the bioluminescence resonance energy transfer (BRET) methodology.

1.1. Principle of the BRET Methodology

BRET is a natural phenomenon, observed in marine organisms, in which an energy transfer occurs between luminescent donor and fluorescent acceptor proteins. Oxidation of coelenterazine by *Renilla* luciferase (Rluc) produces light with a wavelength of 480 nm. However, in the sea pansy *Renilla*, the close proximity of a green fluorescent protein allows a nonradiative energy transfer that results in light emission at 509 nm by the green fluorescent protein *(1,2)*. Resonance energy transfer occurs when part of the energy of an excited donor is transferred to an acceptor fluorophore, which re-emits light at another wavelength. Resonance energy transfer only takes place if the emission spectrum of the donor molecule and the absorption spectrum of the acceptor molecule overlap sufficiently. It also depends on the distance between the donor and the acceptor (which should range between 10 and 100 Å) and on their relative orientation *(3,4)*.

To study protein–protein interactions by BRET, one of the partners of interest is fused to Rluc, whereas the other partner is fused to a fluorescent protein (e.g., yellow fluorescent protein [YFP]; **Fig. 1**). If the two partners do not interact, only one signal, emitted by the luciferase, can be detected after addition of its substrate, coelenterazine. If the two partners interact, resonance energy transfer occurs between the luciferase and the YFP, and an additional signal, emitted by the YFP, can be detected. In addition, because resonance energy transfer not only depends on the distance between the luminescent and the fluorescent protein, but also on their relative orientation, conformational changes that occur between two interacting proteins, or within a single protein, also can be monitored *(5)*. This methodology has been first described in a study on the dimerization of the bacterial Kai B clock protein *(6)*. BRET was subsequently used to study the dimerization and/or ligand induced conformational changes of G protein-coupled receptors (GPCRs) *(7–11)*, a tyrosine kinase receptor *(12)*, and a cytokine receptor *(13)*. BRET also has been used to study the in-

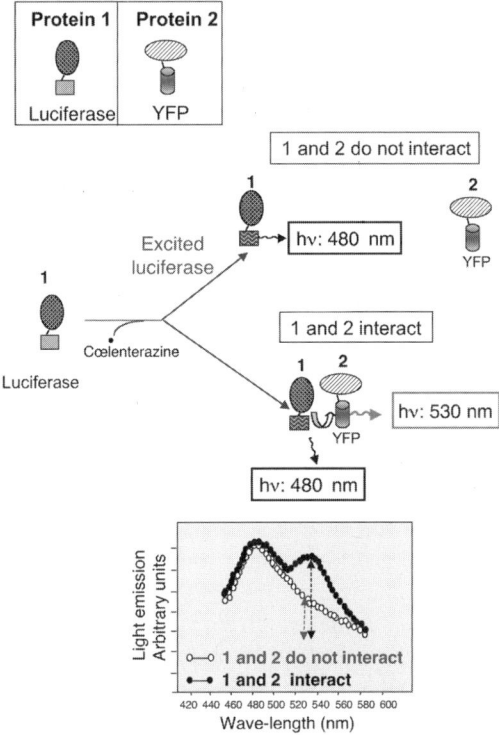

Fig. 1. Principle of the BRET technology. To study the interaction between two proteins, protein 1 is fused to *Renilla* luciferase (Rluc) and protein 2 is fused to a yellow fluorescent protein (YFP). The reaction is initiated by addition of the substrate of luciferase, coelenterazine. If the distance between 1 and 2 is greater than 100 Å, light is emitted with an emission spectrum characteristic of the Rluc. If the distance between 1 and 2 is 10 to 100 Å, part of the energy of the excited Rluc is transfered to the YFP, resulting in an additional signal emitted by the YFP.

teraction of membrane receptors with intracellular partners, such as the association of β-arrestin *(7)* and GPCR kinase 2 *(14)* with GPCRs, and the interaction of a tyrosine kinase receptor with protein tyrosine phosphatases *(15,16)*.

BRET is a very versatile technique that can be used in vitro, in acellular systems, as well as in vivo, in intact living cells. Two BRET systems (BRET1 and BRET2) can be used. The respective advantages and disadvantages of BRET1 vs BRET2 have been discussed elsewhere *(17)*. We will describe here in vitro and in vivo protocols that have been used in our laboratory to monitor conformational changes or protein–protein interactions using the BRET1 technology and BRET1 tools (YFP as the fluorescent protein, coelenterazine H as the luciferin, and emission filters of 485 and 530 nm). These protocols can easily be applied to BRET2 by using BRET2-specific tools.

2. Materials

1. Dual wavelength microplate reader. For BRET measurements, we generally use one of the first BRET-designed apparatus (Fusion from Perkin Elmer), but other adequate equipment have since been developed (Mithras, Berthold; Envision, Perkin Elmer). The following optimized filter settings are used to measure Rluc light emission (485/20 nm) and YFP light emission (530/25 nm). For fluorescence measurements, the reader should be equipped with a fluorescence module to quantify the amount of the YFP fusion protein (filter settings: excitation 480/20 nm, emission 530/25nm).
2. White 96- or 384-well microplates (reusable optiplates, Perkin Elmer) for both in vitro experiments and in vivo experiments using cells in suspension.
3. White culturPlate-96 (Perkin Elmer) for in vivo experiments with adherent cells.
4. Black 96-well plates (Homogeneous time-resolved fluorescence [HTRF] plates, Perkin Elmer) for fluorescence measurements.
5. Coelenterazine H (Interchim, Montlucon, France). Store stock solution (1 mM) in ethanol at $-20°C$ in the dark.
6. Buffer A: 1% (w/v) Triton X-100, 20 mM 3-(N-Morpholino) propanesulfonic acid (MOPS), 2.5 mM benzamidine, 1 mM ethylene diamine tetraacetic acid (EDTA), 1 mM-4-(2-aminoethyl) benzenesulfonyl fluoride hydrochloride (AEBSF), and 1 μg/mL each aprotinin, pepstatin, antipain, and leupeptin.
7. Buffer B: 0.1% (w/v) Triton X-100, 20 mM MOPS, 2.5 mM benzamidine, 1 mM EDTA, and 1 μg/mL each of pepstatin, antipain, and leupeptin.
8. Buffer C: 20 mM MOPS, pH, 7.4.
9. Buffer D: 5 mM Tris-HCl, pH 7.4, 2 mM EDTA, 5 μg/mL soybean trypsin inhibitor, 5 μg/mL leupeptin, and 10 μg/mL benzamidine.
10. Buffer E: 75 mM Tris-HCl, pH 7.4, 12.5 mM MgCl$_2$, 5 mM EDTA.
11. Buffer F: 20 mM HEPES, pH 7.4, 2 mM EDTA, 2 mM ethylenebis (oxyethylenenitrilo)tetraacetic acid, 6 mM magnesium chloride, 1 mM AEBSF; and 1 μg/mL each aprotinin, pepstatin, antipain, and leupeptin.

3. Methods

3.1. cDNA Constructs for BRET Experiments

The partners of interest are typically fused at either N-terminus or C-terminus with Rluc and the fluorescent protein. The choice of the fusion protein (C-terminal vs N-terminal) may sometimes be dictated by the nature of the protein (for instance, transmembrane protein) or by pre-existing information concerning specific regions of the protein of interest (domains of interaction with other proteins, targeting sequences, regulatory domains, etc.).

Because the BRET signal not only depends on the distance between the two partners but also on the relative orientation of the energy donor and acceptor, it is always informative to have, when possible (e.g., for soluble cytosolic proteins), fusions in both orientations (N-terminal and C-terminal). This allows one to determine which conformation gives the best results and can bring information concerning the domains of the partners involved in the interaction.

3.2. Cell Transfections

For BRET experiments, we have mostly used human embryonic kidney (HEK) 293 and COS-7 cells, but any cell type that is easily transfected can be used (*see* **Note 1**). In each experiment, we generally perform two sets of transfections using the transfection reagent FuGene 6 (Roche, Basel, Switzerland) according to the supplier instructions:

1. Transfection of one set of cells only with the complementary DNA (cDNA) coding for the partner fused to Rluc.
2. Cotransfection of another set of cells with both the cDNA coding for the partner fused to Rluc and the cDNA coding for the partner fused to YFP.

These sets allow one to correct for the luciferase signal in the 535-nm detection window.

3.3. BRET Measurements

BRET measurements should be performed under temperature-controlled conditions to obtain reproducible results. We usually perform all our experiments in a room where the temperature is controlled by air conditioning (+20°C). Other temperatures can be used because BRET designed-apparatus generally include temperature-controlled units.

1. Use 96-well culture plates for adherent cells or distribute resuspended cells, membranes, or extracts containing Rluc and YFP fusion proteins into reusable microplates.
2. Initiate BRET reaction by adding the luciferase substrate coelenterazine H (5 μM final concentration). Coelenterazine can be added before ligand stimulation, at the same time, or after preincubation with different ligands. Depending on the assay (*see* **Note 2**), data acquisition is started immediately after coelenterazine addition or later on.
3. Sequentially integrate the luminescence signals at 485 and 530 nm for 1 s. This can be repeated every minute for the next 10 to 45 min (*see* **Note 2**).
4. Calculate BRET ratio (*see* **Note 3**).

3.4. BRET on Partially Purified Receptors

We will exemplify in vitro BRET experiments with our work on partially purified insulin receptors *(12)*. To monitor the effect of insulin on the conformation of the insulin receptor, we use the cDNA coding for the insulin receptor fused in frame with either Rluc or YFP.

3.4.1. Partial Purification of Insulin Receptors

1. HEK 293 cells are transfected in 10-cm diameter culture dishes with either insulin receptor-Rluc (IR-Rluc) or IR-Rluc and IR-YFP.
2. Two days after transfection, cells are washed in ice-cold phosphate-buffered saline (PBS) and extracted on ice in buffer A.

3. The insoluble material is removed by centrifugation at 4°C for 10 min at 10,000*g*.

4. The soluble extract is gently mixed with 3 mL of wheat germ lectin sepharose for 2 h at 4°C.

5. The mixture is poured into a column and washed at 4°C with 60 mL of buffer B.

6. Partially purified insulin receptors are eluted with 15 mL of buffer B containing 0.3 *M* *N*-acetyl-glucosamine *(18)*. Fifteen fractions (1 mL each) are collected and rapidly tested for their content in insulin receptors by measuring the luciferase activity on a 10 μL aliquot in the presence of coelenterazine in a final volume of 50 μL of buffer C.

7. The fractions enriched in luciferase activity are pooled, concentrated using centrifugal filter devices (Amicon Ultra, Millipore), aliquoted, and stored at –80°C for subsequent use.

3.4.2. In Vitro BRET Measurement

1. Preincubate 5 μL of partially purified receptors (approx 2 μg of proteins) in 96-well microplates at 20°C in a total volume of 60 μL of buffer C with ligands for the appropriate time (*see* **Note 4**).

2. Add coelenterazine and start BRET measurements immediately.

3.5. BRET on Cell Fractions

3.5.1. BRET on Crude Membrane Preparations

1. Transfect cells with the cDNAs coding for Rluc or Rluc and YFP fusion proteins in 10-cm diameter culture dishes.

2. Two days after transfection, cells are placed on ice, washed twice with ice-cold PBS, and then detached mechanically in buffer D.

3. Homogenize cells with a polytron (Janke and Kunkel Ultra-Turrax T25) three times for 5 s.

4. Centrifuge the lysate at 450*g* for 5 min at 4°C.

5. Centrifuge the supernatant at 48,000*g* for 30 min at 4°C.

6. Resuspend the pellet in 0.5 mL of buffer E.

7. Perform BRET experiments in white microplates in the same buffer with 25 μL of membrane/well in the presence of coelenterazine in a final volume of 30 to 50 μL. Start readings immediately and repeat measurements for the next 10 min.

3.5.2. BRET on Subcellular Fractions Obtained by Sucrose Density Gradient

BRET measurements can also be performed on different subcellular fractions after fractionation of the cells on a sucrose gradient (*see* **Note 5**), which may be very useful in determining in which cell compartment an interaction takes place *(15)*.

1. Transfect cells with the cDNAs coding for Rluc or Rluc and YFP fusion proteins in 10-cm diameter culture dishes.

2. Prepare 8 mL of sucrose gradient (0.25–2.0 *M*) in a Beckman ultraclear tube (344057). Let the tube rest overnight at 4°C for stabilization of the gradient.

3. Place the culture dish on ice and lyse the cells with 1 mL of ice-cold buffer F.
4. Homogenize by eight passes through a ball-bearing homogenizer (Cell cracker EMBL).
5. Lay carefully 0.5 mL of the lysate on the sucrose gradient. Centrifuge at 75,000g in a Beckman SW55 rotor for 6 h at 4°C. With a 1-mL pipet, collect carefully fractions of 300 µL from the top of the tube. Perform BRET measurements to assess the interaction between the two partners in each fraction on 20-µL aliquots in a final volume of 50 µL containing coelenterazine. Because BRET measurements always imply recording of luciferase activity in each well, these measurements also permit to evaluate the distribution of the Rluc fused protein in each fraction (reading at 485 nm; *see* **Note 6**). Determine the distribution of the YFP fused partner in each fraction by measuring the fluorescence obtained upon exogenous YFP excitation at the appropriate wavelength using 20-µL aliquots distributed in black 96-well HTRF plates.
6. Use the remaining of each fraction for analysis of markers of different cell compartments (plasma membrane, endoplasmic reticulum) by classical biochemical procedures (Western blotting, enzymatic determination, etc. *[15]*).

3.6. BRET on Permeabilized Cells

For some membrane receptors (e.g, cytokine receptors), considerable amounts of protein may be retained in intracellular compartments where they are not accessible to stimulation with hydrophilic ligands. To get access to these receptors without disturbing the overall architecture of the cell, permeabilization of the membrane may be adequate. We have employed this assay to monitor the leptin-induced conformational change within the short isoform of the leptin receptor, which is predominantly localized in intracellular compartments *(13)*.

1. Equimolar amounts of C-terminal Luc- and YFP–leptin receptor fusion proteins are coexpressed in COS-7 cells.
2. Forty-eight hours after transfection, cells are washed once with PBS, and incubated for 5 min with trypsin/EDTA.
3. Cells are collected in serum-containing medium and washed twice with PBS.
4. Cells are resuspended in PBS, distributed in white microplates (1–2×10^5 cells/ well) and preincubated with ligand for 10 min at room temperature in the presence of 0.015% saponin (optimized concentration to obtain cell permeabilization without modification of luciferase activity and YFP fluorescence).
5. BRET measurements are initiated by coelenterazine addition to a final volume of 30 to 50 µL and reading is started immediately.

3.7. Using BRET on Intact Cells

3.7.1. Cells in Suspension

Cells are transfected and prepared exactly as described in **Subheading 3.6.** BRET measurements are performed in PBS in the absence of saponin to keep the cells intact.

3.7.2. Adherent Cultured Cells

The use of adherent cells instead of cells in suspension may sometimes appear to be necessary. Indeed, the architecture of the cell may be altered when removing cell attachment to the culture dish, and this can markedly affect the dynamics of protein–protein interactions. For instance, we definitively observed that insulin-induced interaction of the insulin receptor with intracellular partners was markedly affected when BRET experiments were performed in nonadherent cells.

1. Transfect cells in a six-well culture plate exactly as described previously for BRET measurement on cells in suspension.
2. One day after transfection, transfer the cells into microplates (white culturPlate-96) at a density of 30,000 cells per well (*see* **Note 7**).
3. On the following day (48 h after tranfection), remove the cell culture medium, and wash the cells with 40 to 50 µL of PBS before BRET measurements.
4. Perform BRET measurements directly in these 96-well microplates in PBS (final volume, 50 µL) containing coelenterazine (*see* **Note 8**).

3.8. BRET Donor Saturation Experiments

The BRET donor saturation assay has been developed by us and others to extend the information obtained from basic BRET experiments toward a more quantitative and detailed interpretation of the BRET signal *(13,16,19)*. The assay has been particularly used: to determine whether ligand-induced BRET signals correspond to conformational changes between preassociated BRET donor and acceptor fusion proteins, or to a change in the number of BRET competent complexes (association or dissociation of interacting partners *[13,16]*); to compare the relative affinities of different couples of interaction partners *(19)*; and to estimate the proportion of the dimeric vs monomeric fraction of the two BRET partners *(13,19)*.

1. Perform several independent transfections using a constant amount of cDNA coding for the BRET donor and increasing quantities of cDNA coding for the BRET acceptor (i.e., 0, 10, 20, 50, 100, 200, 300, 500, 1000 ng/well in a 6-well plate; *see* **Note 9**).
2. Preincubate cells in the absence or presence of a saturating agonist concentration if ligand-induced BRET signals will be analyzed. Otherwise, proceed directly to **step 3**.
3. Perform BRET measurements in white microplates or cell culture plates as described in **Subheading 3.7.** and determine the amount of BRET donor by measuring the maximal luciferase activity (reading at 485 nm).
4. Determine the amount of BRET acceptor for each transfection by measuring the fluorescence obtained on exogenous YFP excitation in black 96-well HTRF plates and subtract background fluorescence determined in wells containing untransfected cells (*see* **Note 10**).

5. Calculate YFP/Rluc fusion protein ratio for each data point. Depending on the application, it may be necessary to convert luminescence and fluorescence values into actual amounts of interacting partners using standard curves correlating luminescence and fluorescence signals with protein amounts (*see* **Note 11**).

6. Plot BRET values as a function of the YFP/Rluc fusion protein ratio.

7. Fit data using a nonlinear regression equation assuming a single binding site (GraphPad Prism) and determine $BRET_{max}$ and $BRET_{50}$ values (*see* **Note 12**) or $BRET_{1/1}$ values (*see* **Note 13**).

4. Notes

1. Overexpression of proteins, especially those that are localized in the same cellular compartment (membrane, cytosol), may give rise to nonspecific protein interactions and BRET signals. To avoid these nonphysiological conditions, proteins should be expressed at low levels. Furthermore, the relative amount of YFP to luciferase fusion proteins has to be considered because the BRET signal depends on this parameter (*see* **Subheading 3.8.**). We typically use a 1:1 ratio between the YFP and the Rluc fusion proteins whenever determination of the amount of BRET partners is possible using a quantitative technique (i.e., radioligand binding assay). Preliminary experiments should be designed to adjust these parameters for the proteins to be studied. A noninteracting fluorescent or luminescent protein can also be used as a control for nonspecific interactions. Rluc or YFP alone can be used as noninteracting cytosolic control proteins. An irrelevant protein, fused to Rluc or YFP, targeted to the same compartment as the protein of interest, can be used as a non-interacting control for this subcellular compartment.

2. Luciferase signal (light emission by luciferase after addition of coelenterazine) increases rapidly, reaches a maximum and decreases slowly with time. Because BRET is a ratiometric measurement (530 nm/485 nm), this does not affect the BRET ratio (provided the signal is still sufficient for a reliable measurement of light emission above background). Depending on the assay (in vitro, cell in suspension, adherent cells), the kinetics of increase and decrease in luciferase signal can be quite different. Therefore, depending on the assay, a preincubation period with coelenterazine may be necessary in order to start BRET measurements when the luciferase signal is maximal. The rate of decrease of the luciferase signal also depends on the specific assay. Thus, depending on the experiments, light emission acquisition will be possible for relatively short periods of time (10 min for partially purified preparations) or much longer periods of time (40–50 min for intact cells).

3. The BRET ratio has been defined previously *(7)* as: [(emission at 530 nm) − (emission at 485 nm) × Cf]/(emission at 485 nm), were Cf corresponded to (emission at 530 nm) / (emission at 485 nm) for the Rluc fusion protein expressed alone in the same experimental conditions. Developing this simple equation shows that the BRET ratio corresponds to the ratio 530 nm/485 nm obtained when the two partners are present minus the ratio 530 nm/485 nm obtained under the same experimental conditions when only the partner fused to Rluc is present in the assay.

4. Different reaction buffers can be used, but pilot experiments should be conducted because some chemicals may inhibit the luciferase activity.

5. Sucrose gradients rather than commercially available Nycodenz® (Nycomed AS) solutions should be used, as we find that the latter markedly reduces the luciferase signal.

6. Reliable quantification of Rluc by measuring light emission at 485 nm is possible under conditions of energy transfer between Rluc and YFP fusion proteins, because the amount of energy transfered from the Rluc to the YFP is generally negligible compared with the energy released by the Rluc as light at 485 nm.

7. We do not recommend direct transfection of the cells in 96-well culture plates. Indeed, in our experience, this has proven to introduce important variability in the level of expression of the different partners (because of inaccuracy of transfecting very small amounts of DNA associated with the variability in cell number in each well). Some cells adhere poorly in 96-well culture dishes when washed and incubated in PBS. Therefore, polylysine coating of the wells prior to transferring the transfected cells is highly recommended for some cell types (e.g., HEK 293 cells).

8. Note that for adherent cells, penetration of coelenterazine into the cells is probably slower than for cells in suspension, since the luciferase signal reaches its maximal value only 10 to 20 min after addition of coelenterazine (vs 2–3 min in nonadherent cells). We usually perform a preincubation of the cells for 15 min with coelenterazine before starting BRET measurements on adherent cells.

9. If BRET donor saturation curves of different BRET partners are compared to determine their relative affinities, keep the amount of expressed BRET donor constant for all conditions to obtain comparable results.

10. For adherent cells, fluorescence can be determined directly in white culture plates at the end of the BRET experiment. Although background fluorescence is higher under these conditions, the signal above background is generally sufficient to allow for accurate determination of the relative amount of the fluorescent partner.

11. If the purpose of the experiment is to discriminate between a ligand-induced conformational change between two interacting partners and a ligand-induced change in the number of interacting partners (for instance, an increased recruitment of one partner to the other, owing to an increased affinity between the two partners), luminescence and fluorescence values may be directly used to calculate the YFP:Rluc ratio. In contrast, if the purpose of the experiment is to compare the relative affinities between two different pairs of partners, or to determine the proportion of the dimeric vs the monomeric fraction of two BRET partners, it is necessary to convert luminescence and fluorescence values to actual protein amounts using independently established standard curves correlating both parameters. Correlation curves have to be established for each construct because this correlation is an intrinsic characteristic of each fusion protein. For all fusion proteins tested so far, a linear correlation has been observed *(11,13,19–22)*. Correlation curves may be established by transfecting cells separately with increasing DNA concentrations of Rluc or YFP fusion protein constructs. The amount of protein should be determined with a quantitative technique, such as the

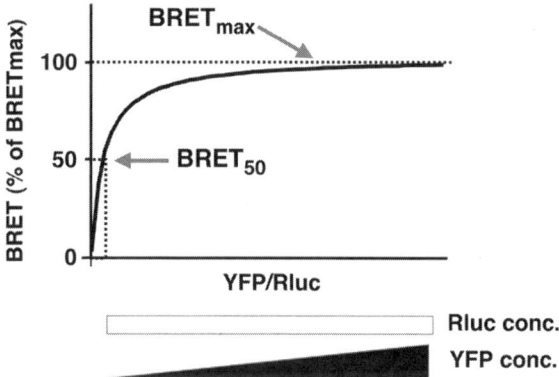

Fig. 2. BRET donor saturation curve. Cells are transfected with a constant amount of Rluc-fusion protein and increasing amounts of yellow fluorescent protein-fusion protein. BRET values increase as a hyperbolic function reaching an asymptote ($BRET_{max}$). The half-maximal BRET ($BRET_{50}$) for a given amount of donor reflects the relative affinity of the two partners.

radioligand binding assay in the case of membrane receptors, in Rluc-expressing cells on the one hand and YFP-expressing cells on the other hand. On the same batches of Rluc-transfected or YFP-transfected cells, maximal luminescence and YFP fluorescence can be determined, respectively. Luminescence and fluorescence values can then be plotted against the number of binding sites to generate linear regression curves.

12. As the amount of BRET acceptor increases, the BRET signal increases as a hyperbolic function for a given amount of the BRET donor and reaches an asymptote ($BRET_{max}$), which corresponds to the saturation of all BRET donor molecules by acceptor molecules. Assuming that the association of interacting proteins, fused to the BRET donor and the BRET acceptor respectively, is random, the amount of acceptor required to obtain the half-maximal BRET ($BRET_{50}$) for a given amount of donor reflects the relative affinity of the two partners (*see* **Fig. 2** *[19]*). Saturation of the curve is important because the accuracy of the determination of the $BRET_{max}$ value governs the precision of the $BRET_{50}$ value. If the BRET curve does not saturate, express lower BRET donor amounts and/or higher BRET acceptor amounts in order to reach saturation. Comparison of $BRET_{50}$ values obtained from cells incubated in the absence or presence of agonist may shed light on the activation mechanism of the corresponding proteins. Identical $BRET_{50}$ values indicate that conformational changes within BRET partners rather than the recruitment or dissociation of BRET partners are involved (*see* **Fig. 3**). A shift of the $BRET_{50}$ toward lower values in the presence of agonist indicates the recruitment of BRET partners because of an increase in affinity between Rluc and YFP fusion proteins. A shift toward higher values indicates a decrease in the affinity of the BRET partners and their dissociation. Provided that

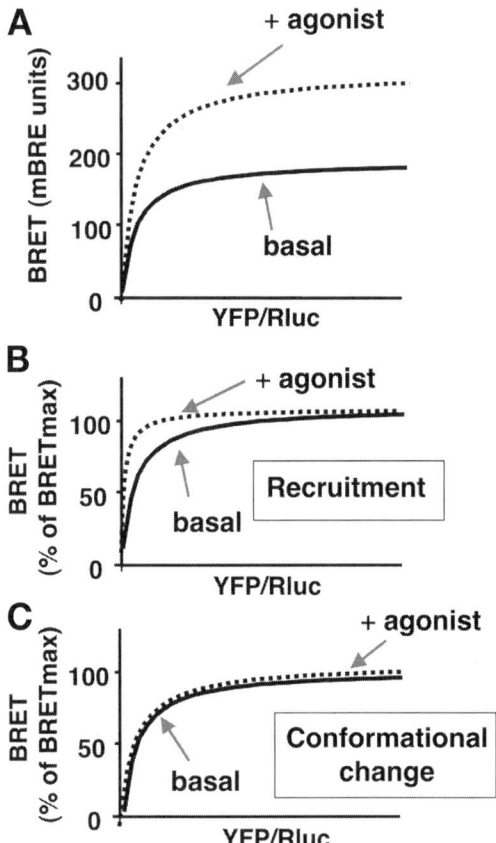

Fig. 3. Interpretation of ligand induced BRET signals. Basal- and ligand-stimulated BRET values (**A**) are expressed as the percent of $BRET_{max}$ (**B,C**), to determine whether the effect of a ligand corresponds to a conformational change or a recruitment between partners. If the curve obtained in the presence of agonist is shifted to the left compared with the basal curve, ligand-induced BRET signal can be interpreted as the recruitment of BRET partners on agonist stimulation (**B**). If both curves are superimposible, the signal induced by the ligand corresponds to a conformational change (**C**).

BRET donor saturation curves have been carried out at the same BRET donor concentration and that effectively expressed protein amounts have been determined, $BRET_{50}$ values give a direct measure for the relative affinity of different BRET couples (*see* **Fig. 4**).

13. If a free equilibrium governs the association of BRET donor and acceptor monomers, one can predict that, in the case of a 1:1 molecular ratio of the two BRET partners, only 50% of the dimers (donor/acceptor) would produce BRET, whereas dimers that contain only BRET donors or acceptors, would represent 25% each of total dimers (donor/donor and acceptor/acceptor). Accordingly, the BRET value

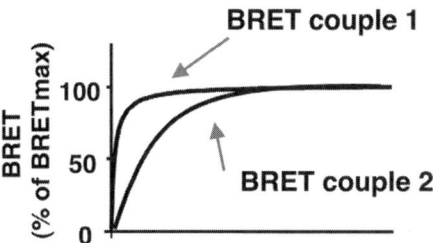

Fig. 4. Comparison of the relative affinity of two couples of interacting partners. Saturations curves are obtained by plotting BRET values against the ratio of donor and accepor fused partners, after conversion of the *Renilla* luciferase and yellow fluorescent protein signals into actual amounts of each partner. In this example, the partners in couple 1 interact with higher affinity than the partners in couple 2.

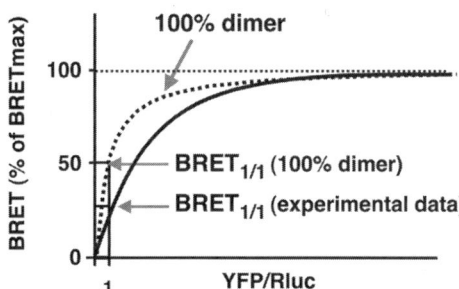

Fig. 5. Determination of the monomeric vs dimeric fraction of interaction partners. Saturation curve obtained from experimental data (after conversion of the *Renilla* luciferase and yellow fluorescent protein signals into actual amounts of each partner) is plotted on the same graph as the theoretically expected curve for a homogenous population of BRET competent dimers.

observed under these conditions ($BRET_{1/1}$) should reach 50% of maximal BRET, the value corresponding to the complete saturation of BRET donor by BRET acceptor. Depending on the proportion of monomers and dimers, the $BRET_{1/1}$ will theoretically vary between 50% of $BRET_{max}$ (100% dimer) to 0% of $BRET_{max}$ (100% monomer; *see* **Fig. 5**).

References

1. Morin, J. G., and Hastings, J. W. (1971) Energy transfer in a bioluminescent system. *J. Cell Physiol.* **77**, 313–318.
2. Lorenz, W. W., McCann, R. O., Longiaru, M., and Cormier, M. J. (1991) Isolation and expression of a cDNA encoding Renilla reniformis luciferase. *Proc. Natl. Acad. Sci. USA* **88**, 4438–4442.
3. Tsien, R. Y., Bacskai, B. J., and Adams, S. R. (1993) FRET for studying intracellular signalling. *Trends Cell Biol.* **3**, 243–245.

4. Wu, P. and Brand, L. (1994) Resonance energy transfer: methods and applications. *Anal. Biochem.* **218,** 1–13.

5. Boute, N., Jockers, R., and Issad, T. (2002) The use of resonance energy transfer in high-throughput screening: BRET versus FRET. *Trends Pharmacol. Sci.* **23,** 351.

6. Xu, Y., Piston, D. W., and Johnson, C. H. (1999) A bioluminescence resonance energy transfer (BRET) system: application to interacting circadian clock proteins. *Proc. Natl. Acad. Sci. USA* **96,** 151–156.

7. Angers, S., Salahpour, A., Joly, E., Hilairet, S., Chelsky, D., Dennis, M., and Bouvier, M. (2000) Detection of beta 2-adrenergic receptor dimerization in living cells using bioluminescence resonance energy transfer (BRET). *Proc. Natl. Acad. Sci. USA* **97,** 3684–3689.

8. McVey, M., Ramsay, D., Kellett, E., et al. (2001) Monitoring receptor oligomerization using time-resolved fluorescence resonance energy transfer and bioluminescence resonance energy transfer. The human delta-opioid receptor displays constitutive oligomerization at the cell surface, which is not regulated by receptor occupancy. *J. Biol. Chem.* **276,** 14092–14099.

9. Kroeger, K. M., Hanyaloglu, A. C., Seeber, R. M., Miles, L. E., and Eidne, K. A. (2001) Constitutive and agonist-dependent homo-oligomerization of the thyrotropin-releasing hormone receptor. Detection in living cells using bioluminescence resonance energy transfer. *J. Biol. Chem* .**276,** 12,736–12,743.

10. Cheng, Z. J. and Miller, L. J. (2001) Agonist-dependent dissociation of oligomeric complexes of G protein-coupled cholecystokinin receptors demonstrated in living cells using bioluminescence resonance energy transfer. *J. Biol. Chem.* **276,** 48,040–48,047.

11. Ayoub, M. A., Couturier, C., Lucas-Meunier, E., et al. (2002) Monitoring of ligand-independent dimerization and ligand-induced conformational changes of melatonin receptors in living cells by bioluminescence resonance energy transfer. *J. Biol. Chem.* **277,** 21,522–21,528.

12. Boute, N., Pernet, K., and Issad, T. (2001) Monitoring the activation state of the insulin receptor using bioluminescence resonance energy transfer. *Mol. Pharmacol.* **60,** 640–645.

13. Couturier, C. and Jockers, R. (2003) Activation of the leptin receptor by a ligand-induced conformational change of constitutive receptor dimers. *J. Biol. Chem.* **278,** 26,604–26,611.

14. Hasbi, A., Devost, D., Laporte, S. A., and Zingg, H. H. (2004) Real-time detection of interactions between the human oxytocin receptor and G protein-coupled receptor kinase-2. *Mol. Endocrinol.* **18,** 1277–1286.

15. Boute, N., Boubekeur, S., Lacasa, D., and Issad, T. (2003) Dynamics of the interaction between the insulin receptor and protein tyrosine-phosphatase 1B in living cells. *EMBO Rep.* **4,** 313–319.

16. Lacasa, D., Boute, N., and Issad, T. (2005) Interaction of the insulin receptor with the receptor-like protein tyrosine-phosphatases PTPα and PTPε in living cells. *Mol. Pharmacol.* **67,** 1206–1213.

17. Milligan, G. (2004) Applications of bioluminescence- and fluorescence resonance energy transfer to drug discovery at G protein-coupled receptors. *Eur. J. Pharm. Sci.* **21,** 397–405.

18. Issad, T., Tavaré, J., and Denton, R. M. (1991) Analysis of insulin receptor phosphorylation sites in intact rat liver cells by two-dimensional phosphopeptides mapping. Predominance of the trisphosphorylated form of the kinase domain after stimulation by insulin. *Biochem. J.* **275,** 15–21.
19. Mercier, J. F., Salahpour, A., Angers, S., Breit, A., and Bouvier, M. (2002) Quantitative assessment of beta 1- and beta 2-adrenergic receptor homo- and heterodimerization by bioluminescence resonance energy transfer. *J. Biol. Chem.* **277,** 44,925–44,931.
20. Ramsay, D., Carr, I. C., Pediani, J., et al. (2004) High-affinity interactions between human alpha1A-adrenoceptor C-terminal splice variants produce homo- and heterodimers but do not generate the alpha1L-adrenoceptor. *Mol. Pharmacol.* **66,** 228–239.
21. Ramsay, D., Kellett, E., McVey, M., Rees, S., and Milligan, G. (2002) Homo- and hetero-oligomeric interactions between G-protein-coupled receptors in living cells monitored by two variants of bioluminescence resonance energy transfer (BRET): hetero-oligomers between receptor subtypes form more efficiently than between less closely related sequences. *Biochem. J.* **365,** 429–440.
22. Ayoub, M. A., Levoye, A., Delagrange, P., and Jockers, R. (2004) Preferential formation of MT1/MT2 melatonin receptor heterodimers with distinct ligand interaction properties compared with MT2 homodimers. *Mol. Pharmacol.* **66,** 312–321.

12

Identification of Interacting Proteins Using the Yeast Two-Hybrid Screen

Kelly L. Jordan-Sciutto and Marshall B. Montgomery

Summary

Transmembrane-signaling events are mediated and regulated by protein–protein interactions. The yeast two-hybrid screen has proven to be an effective approach for studying interaction between signaling molecules, such as ras and raf. This approach can be used to identify new binding partners for a protein of interest or define the interaction domains and relative affinity between two proteins known to interact. To determine interaction, one protein is produced as a fusion protein with a known DNA-binding domain and a second protein is produced as a fusion protein with an acidic activation in yeast. If there is interaction between the two proteins of interest, the DNA-binding domain is brought into the vicinity of the acidic-activation domain, which recreates a functional transcriptional activator, which drives transcription of reporter genes allowing for selection and/or quantification of interaction between the two proteins. Here we describe a two-hybrid yeast system that has been used to successfully characterize protein interactions among signaling molecules.

Key Words: Yeast two-hybrid; interaction trap; protein–protein interaction; yeast transformation; auxotrophy; β-galactosidase activity; matchmaker.

1. Introduction

The two-hybrid screen was devised as a method of identifying novel binding partners for a protein of interest *(1)*. The screen itself is based on the separable nature of the DNA-binding and acidic-activation domains in transcription factors *(2,3)*. By generating a fusion protein between a protein of interest (called the bait) and the DNA-binding domain (DBD) of a known transcription factor, a protein "hybrid" is produced that will bind to a specific promoter sequence, but will have no transcriptional activity (**Fig. 1A**). The consensus DNA-binding sites for the DBD are present in the promoters of two reporter

From: *Methods in Molecular Biology, vol. 332: Transmembrane Signaling Protocols, Second Edition*
Edited by: H. Ali and B. Haribabu © Humana Press Inc., Totowa, NJ

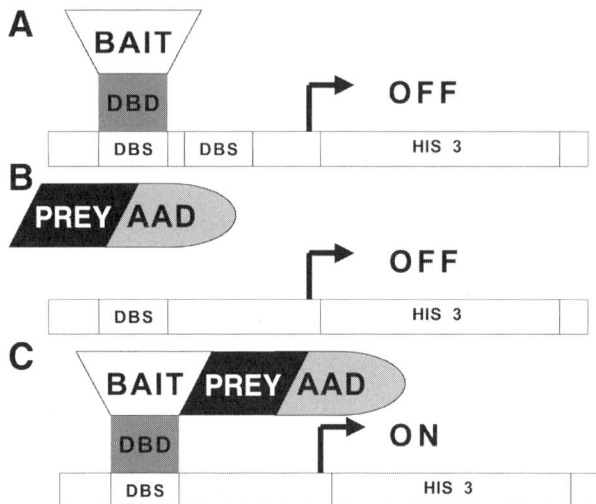

Fig. 1. Scheme of the two-hybrid yeast screen. (**A**) Bait: The protein for which you would like to identify or characterize binding partners is produced as a fusion protein with a known DNA-binding domain (DBD). The DBD binds the DNA-binding domain consensus sequence (DBS), which is located in the promoter upstream of the HIS3 enzyme. Transcription of this enzyme is "OFF" when only the bait is present. (**B**) Prey: A cDNA library is cloned in frame with an acidic-activation domain (AAD). The fusion protein produced is termed the prey. In the absence of an interacting bait, transcription is "OFF." (**C**) If the bait and prey interact, the AAD is brought into the vicinity of the DBD bound to the DBS in the promoter of the *HIS3* gene, leading to production of the *HIS3* transcript and ultimately the protein that allows for selection of yeast with the ability to grow in the absence of supplemental histidine.

genes, usually one for autotrophy for a specific amino acid (ability to grow in media without that amino acid, such as histidine [his]) and Lac Z, an enzyme with a colorimetric read out. A second fusion protein is introduced containing a hybrid of a known acidic-activation domain (AAD) and a library protein (**Fig. 1B**). If a library protein interacts with the bait, a functional transcriptional activator is recreated at the promoter (**Fig. 1C**). Transcriptional activation will produce functional enzymes that allow for selection of interacting proteins (*see* **Fig. 1C**). For use as a screen for novel proteins that interact with the "bait," a complementary (c)DNA library is cloned in frame with the coding region for the AAD. Only cDNAs encoding interacting proteins will activate expression of the reporter genes allowing for selection of proteins that bind to the bait.

Since its inception in 1989, several versions of the two-hybrid screen have been generated and used to identify interacting proteins *(4–7)*. Presently, there are commercially available kits for the two-hybrid screen that contain further modifications that ease isolation and analysis of the identified proteins (*see* www.clontech.com, www.invitrogen.com, and www.origene.com). The goal of this chapter is not to replace the manual for such kits (nor can it provide specific directions for all noncommercial variants because there are multiple versions, each with unique features). We provide methods for a yeast two-hybrid system that has been used to identify interacting partners for the G protein ras *(7)*. However, given the protocol herein, it should be relatively easy to extrapolate this protocol to other systems because they are based on the same principles, although the vector names and selectable markers may change. A final goal of this chapter is to provide some insight into the problems that can arise in using this screen and how they can be addressed.

2. Materials

2.1. Growth of Yeast for the Two-Hybrid Screen

1. Yeast two-hybrid vectors: BTM116, BTM116-VP16, BTM116-daughterless, pVP16, pVP16-MyoD, and cDNA library in pVP16 or compatible vector (has LEU2 selectable yeast marker).
2. Yeast strain: L40 (MATa, his3Δ200, tryptophane [trp]1-901, leucine [leu]2-3, 112 adenine [ade]2, LYS2::[lexAop]$_4$-HIS3, URA::[lexAop]$_8$-lacZ, GAL4, gal80) constructed by Stan Hollenberg.
3. Media and agar plate recipes: a key to success in using the yeast two-hybrid screen is proper preparation of selective drop-out media (SD). Maintenance of the plasmids and reporter genes in the yeast cells is dependent on converting auxotrophic yeast strains (yeast that require specific amino acids or nucleic acids to grow) to an autotrophic state (the ability to grow in the absence of an amino acid or nucleic acid). The gene allowing the conversion to autotrophy is encoded on the plasmid. To select for presence of the plasmid, make SD lacking the appropriate amino acid. In the recipe in this section, we list the amount of each amino acid needed for 1 L. However, if you make this up each time you need it, it can be quite time consuming. We recommend making "drop-out" amino acid and nucleic acid stocks. To do this, we recommend multiplying the amount of each amino acid needed by a common factor (e.g., 10, which will be enough for 10 L) and adding them into a common stock omitting the unneeded metabolite. Mix well by rotating in a sterile Erlenmeyer flask for 1 h. When making the media, add the appropriate amount of the stock. For example, if making media lacking his, trp, and leu, the amount of drop-out stock would be 1.0 g. (All the amino acids add up to 1.25 g. Without the 0.05 g of his, 0.10 g of leu, and 0.10 g of trp, only 1.0 g is needed.) Make up drop-out stocks for the bait vector only (i.e., trp⁻), the prey vector only (i.e., leu⁻), the bait and prey vector (i.e., uracil (ura)⁻, trp⁻, and leu⁻), and the bait, prey, and selectable reporter (i.e., trp⁻, leu⁻, ura⁻, lysine (lys)⁻, and his⁻). In the system described here, ura and lys are left out as they select for integration of the *HIS3* and *Lac Z* reporter genes in the L40 yeast strain, respectively.

a. SD media: for 1 L: add 1.2 g yeast nitrogen base (without amino acids and [NH$_4$]$_2$SO$_4$), 5 g of ammonium sulfate, 10 g of succinic acid, 6 g of NaOH, and SD in appropriate quantity. (SD per liter [or add the premade stock]: 0.1 g each of ade, arginine, cysteine, leu, lys, threonine, trp, and ura; and 0.05 g each of valine, tyrosine, serine, praline, phenylalanine, methionine, isoleucine, his, and aspartic acid.)

Bring to 900 mL with water. (If making agar plates also add 20 g of bactoagar/L.) Sterilize by autoclaving. For agar plates, cool to 55°C, add 100 mL of sterile filtered 20% glucose, and pour into Petri dishes (approx 40 mL/plate). For liquid media, the autoclaved solution can be stored until needed in the autoclaved bottles. Before use, add one-tenth of the volume of 20% glucose to nine-tenths volume of the media.

To generate your specific selective media, omit the amino acid of which the plasmid encodes the gene for synthesis (i.e., if the plasmid contains the Leu2 selectable marker, then omit leu from the selective media and only yeast containing the plasmid will grow.)

b. YPAD media: add 10 g of yeast extract, 20 g of peptone, and 0.1 g of adenine to 900 mL of water. Autoclave, cool to 55°C, and add 100 mL of sterile filtered 20% glucose. YPA is the same as YPAD but with water substituted for the glucose (or dextrose, as the D indicates).

2.2. Yeast Transformation Via Lithium Acetate

1. 10X TE (autoclaved): 100 mM Tris-HCl, pH 7.5, 10 mM ethylene diamine tetraacetic acid (EDTA).
2. 10X LiOAc (autoclaved): 1 M LiOAc.
3. 50% Polyethylene glycol (PEG): 50% (w/v) PEG 3350, filter-sterilized (slow, but necessary).
4. 10 mg/mL Salmon sperm DNA.
5. Dimethyl sulfoxide.

2.3. Yeast Transformation Via Electroporation

1. 1 M Sorbitol (ice-cold, filter-sterilized).
2. Sterile H$_2$O (ice-cold).

2.4. β-Galactosidase Filter Assay

1. Nitrocellulose filters (circular, same size as Petri dishes).
2. Whatman filter paper no. 1.
3. Liquid nitrogen.
4. Z-buffer: 60 mM Na$_2$HPO$_4$, 40 mM NaH$_2$PO$_4$, 10 mM KCl, 1 mM MgSO$_4$, pH 7.0 in H$_2$O.
5. 50 mg/mL X-gal (5-bromo-4-chloro-3indolyl-β-D-galactopyranoside) in H$_2$O.

2.5. β-Galactosidase Quantitative Liquid Assay

1. Breaking buffer: 100 mM Tris-HCl, pH 8.0, 1 mM dithiothreitol, 20% glycerol in H$_2$O.

2. 125 µ*M* Phenylmethylsulfonylfluoride in isopropanol.
3. Glass beads (0.45–0.5 mm)
4. Z-buffer: (for quantitative β-galactosidase assay) 60 m*M* Na$_2$HPO$_4$, 40 m*M* NaH$_2$PO$_4$, 10 m*M* KCl, 1 m*M* MgSO$_4$, and 30 m*M* β-mercaptoethanol (added fresh).
5. 4 mg/mL *O*-nitrophenyl β-D-galactopyranoside (ONPG).
6. 1 *M* Na$_2$CO$_3$.
7. Bradford Reagent (Bio-Rad).
8. Plastic cuvets (for visible range spectrophotometry).

2.6. Plasmid Isolation From Yeast (Smash and Grab)

1. Phenol:chloroform.
2. Smash and Grab lysis buffer: 2.5 *M* LiCl, 50 m*M* Tris-HCl (pH 8.0), 4% Triton X-100, 62.5 m*M* Na$_2$ EDTA.
3. Ethanol (amount will vary based on number of samples tested).
4. 3 *M* sodium acetate.
5. Luria Broth agar plates with ampicillin (100 µg/mL).

2.7. Other Reagents

1. Y-Per® yeast protein extraction kit (Pierce) for protein extraction.

2.8. Major Equipment

1. Incubators for growing liquid cultures and plates at 30°C.
2. Agitator or roller drum for aerating liquid cultures.
3. Water bath at 42°C.
4. Centrifuge for spinning large volumes (up to 500-mL bottles).
5. Microcentrifuge.
6. Plating wheel.
7. Electroporator equipped for bacterial and yeast transformation with 0.2-cm gap cuvets for yeast and 0.1-cm gap cuvets for bacteria.
8. Spectrophotometer.

3. Methods

3.1. Identifying Interacting Proteins Using the Yeast Two-Hybrid Screen

Because this volume provides methods for studying signaling pathways, we have chosen to describe the protocol for a yeast two-hybrid screen used to identify interacting proteins for a protein involved in these processes. In addition to identifying raf as a ras interacting protein *(7)*, this screen also has been used successfully to identify interacting proteins for the daughterless transcription factor, the E2F1 transcription factor *(8)* and a novel developmental protein that shuttles between the nucleus and cytoplasm, FAC1 *(9,10)*. The protocols provided here are applicable to any yeast two-hybrid system with minor modification. Whether using a commercial kit or another system described in the literature, we hope these protocols augment instruction materials in the literature and provide alternative approaches.

The basic outline of the protocol is to (1) generate a bait vector and choose a library (prey); (2) transform the bait and prey vectors into a reporter yeast strain; (3) screen the resulting colonies for interaction by assessing activation of both reporters; (4) remove false-positives; and (5) identify real-positives and confirm interaction.

3.2. Choosing the Bait Plasmid

Although screening your clones for true-positives is critical, the most important aspect for a successful yeast two-hybrid screen is choosing the appropriate bait. Before initiating the screen, it is necessary to first determine whether the protein of interest will work as bait in the system. First, you must choose a bait vector. For the screen described here, we will use the BTM116 vector (**Fig. 2**), designed by Paul Bartel and Stan Fields *(11)*. The features of this vector include: origins of replication for both yeast and bacteria; ampicillin-resistance gene for selection in bacteria; the *TRP1* gene, which will allow growth of auxotrophic yeast strains on plates lacking the amino acid trp; and the Lex A DBD followed by a polyinker for inserting your bait cDNA (**Fig. 2**, *see* **Note 1** for choosing DBDs).

At this point, we suggest analyzing the bait with regard to the following parameters:

1. Endogenous transcriptional activation.
2. Interaction with the AAD.
3. Production and stability of fusion protein.
4. Nuclear entry (*see* **Note 2**).
5. Hydrophobic domains.

Other concerns are discussed in the under **Subheading 4.** (*see* **Notes 3–5**). To address these concerns, we suggest creating a panel of "bait" vectors that can be assessed for each of the above criteria. In addition to the full-length protein, we suggest making overlapping deletion mutants based on known functional motifs and predicted secondary structure (an example is shown in **Fig. 3**). Once you have chosen the "baits" you plan to make, clone them via standard cloning methods into BTM116. Be sure that your cDNA sequence is in the same translational reading frame as the Lex A DBD. The panel of baits will not only help you determine what will work best in the screen, but will provide you with an immediate tool to evaluate the interaction domain for any identified binding partners at the end of the screen.

To determine whether your protein has endogenous transactivation abilities, transform BTM116-bait into yeast cells of the L40 yeast strain using the small-scale lithium acetate transformation (*see* **Subheading 3.4.1.**). Plate half the

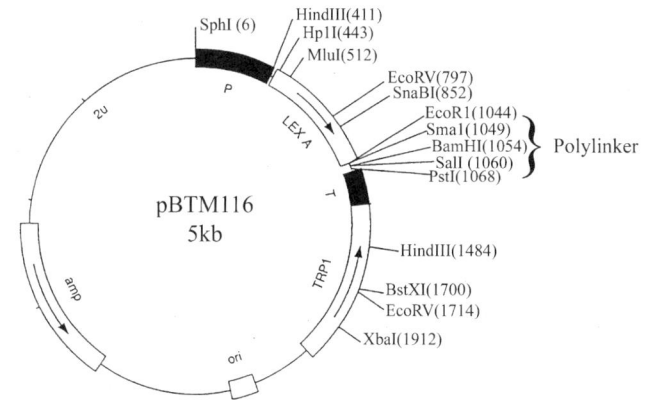

Polylinker reading frame:

GAA TTC CCG GGG ATC CGT CGA CCT GCA G

| EcoRI | SmaI | BamHI | SalI | PstI |

Adapted from Stan Hollenberg

Fig. 2. Example of bait vector features. The map for the bait vector described in this protocol is shown here. The cDNA for the Lex A DNA-binding domain (Lex A) is upstream of a polylinker for inserting the cDNA of interest. The reading frame of the polylinker is shown below the map with unique restriction sites for cloning. This vector also encodes the yeast *TRP1* gene for autotrophy on tryptophane lacking plates, yeast, and bacterial origins of replication (ori and 2μ), and an ampicillin-resistance gene for maintenance in bacteria.

LAC Z activity

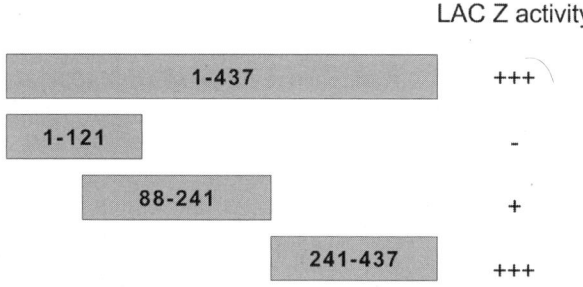

Fig. 3. Example of bait analysis with regard to transactivation of the two-hybrid yeast reporter, Lac Z. Bait plasmids encoding amino acids 1–437, 1–121, 88–241, or 241–437 of a hypothetical transcription factor were introduced into yeast and assayed for β-galactosidase activity (Lac Z activity). Shown is the relative activation of Lac Z using various deletion constructs. Based on the basis of these results, the bait that will work best in the screen will encode amino acids 1–121, whereas 241–437 likely contains an acidic activation domain.

transformation on agar plates lacking trp and half on a plate that lacks trp and his (the reporter in L40 for activation from the Lex A binding sites). If the bait alone leads to growth on both the bait selected media (SD/trp⁻) and the bait/reporter selected media (SD/trp⁻, his⁻), it is activating transcription of the reporter gene in the absence of the library. To determine the strength of activation, we recommend using the quantitative β-galactosidase activity assay (*see* **Subheading 3.6.2.2.**). For the screen, choose a bait that allows minimal or, ideally, no activation of Lac Z or the autotrophic reporter, his. As a control, include yeast transformed with BTM116 alone. This should be completely silent in the screen. In **Fig. 3**, the ideal construct to be used as a bait would include amino acids 1-121 of the hypothetical proteins shown. It is also valuable to have a positive control, such as a bait vector with the AAD fused to the DBD. We use BTM116-VP16, which has the VP16 AAD cloned in frame with the Lex A DBD. This will give positive growth on both plates and high activation in the β-galactosidase assay.

To determine whether your bait interacts with the transcriptional activation domain, transform yeast (again using the small-scale lithium acetate transformation protocol; *see* **Subheading 3.4.1.**) with BTM116-bait and the plasmid used for the library, but with no inserted cDNA (pVP16). Plate half the transformation on agar plates selecting for these two plasmids (SD/trp⁻ and leu⁻). Plate the other half on a plate that lacks the two amino acids that allow selection of the bait and library vector, as well as the reporter amino acid (SD/trp⁻, leu⁻, and his⁻). If a bait allows growth on the plate lacking his, as well as the control plate (lacking only the plasmid markers), it is likely that the bait is interacting with the AAD. As a control, include the bait vector without your protein of interest (BTM116) in combination with the library vector (pVP16). If you have a control that contains two known interacting proteins (one in the bait and one in the prey vector) this would serve as a good positive control and assist in identifying false-positives. We use BTM116-daughterless and pVP16-myo D.

To determine whether your bait fusion protein is produced and stable, make protein extracts from yeast transformed with the bait plasmid, BTM116-bait using the Y-Per yeast protein extraction kit (Pierce) following the manufacturer's instructions. Immunoblot the protein extracts for the DBD or the bait if an antibody is available for the protein domain you are investigating. As a control, assess expression of the DBD from the bait vector (pBTM116) without your gene of interest.

Methods to determine whether your protein is getting into the nucleus have been devised but are not available for the screen described here *(12)*. If such an assay is available for your system, please use it to determine nuclear localization of your protein of interest. In the assay described here, a nuclear localization signal (NLS) is present in the library vector (pVP16), which should transport the bait protein to the nucleus if it is interacting with the protein produced from the library cDNA contained in the vector. To determine whether the NLS is sufficient for delivery of the bait to the nucleus, we recommend cloning the bait in frame with the AAD of the library vector (pVP16-bait), which will produce a bait-VP16 fusion protein. Again, transform yeast with this vector and plate half on SD/leu⁻ and half on SD/leu⁻– his⁻ agar plates. If your protein is still excluded from the yeast nucleus, then it will not activate transcription of the reporters leading to inability to grow on the plates lacking his and reducing Lac Z activity via the quantitative β-galactosidase activity (*see* **Subheading 3.6.2.2.**). This activity should be compared with yeast transformed using a vector containing the AAD alone (pVP16 alone) as a negative control and the DBD fused to the AAD (pVP16-LexA) as a positive control.

Hydrophobic domains are notoriously promiscuous in the yeast two-hybrid screen. We recommend two ways of assessing them. The first is to use known algorithms to predict hydrophobicity of the bait protein (Kite Doolittle, etc). The second is to do a pilot transformation with the library and bait. Using the small-scale lithium acetate transformation (yielding <10,000 transformants), transform L40 yeast cells with BTM116-bait and the pVP16-library. Plate a small aliquot to determine the number of transformants on SD/trp⁻, leu⁻, and ura⁻ agar plates (ura⁻ selects for cells to maintain the His3 reporter contruct). Plate the remaining transformation on SD/trp⁻, leu⁻, his⁻, ura⁻, and lys⁻) agar plates. Few interactors should be identified. If more than 0.1% of the transformants are positive on the plates lacking his, then the background of your screen is too high for that bait, which may be attributable to hydrophobic or other "sticky" domains. By doing a pilot first, you will not waste time sifting through the 6000 or more clones that would come up in a full genome-wide screen. Once you have identified a bait that is usable in the screen and minimizes false-positives, you are now ready to consider what library you will use.

3.3. Plasmids for Prey: The Library

With commercially available kits come commercially available libraries. Undoubtedly, purchasing a library is easier than making your own. It is not the purpose of this protocol to explain how to make a cDNA plasmid library; however, we felt we would be remiss if we did not at least mention this part of the protocol. Your library choice will be dependent on your experimental system of interest and individual hypothesis.

Key features of the library that we will discuss are in regard to the vector in which such a library will be cloned. The library vector, or "prey," for the two-hybrid screen should have several main features. First, it should contain a selectable marker for growth in yeast. This marker will likely provide autotrophy for a specific amino acid (as other yeast markers described to this point). In our screen, the library was cloned into the pVP16 vector (generated by Stan Hollenberg), which has the *LEU2* gene, allowing growth of auxotrophic yeast on leu⁻ agar plates. Second, it should contain an AAD with a multiple cloning site at the carboxy-terminus for inserting the cDNA library. It also should contain a selectable marker for growth in bacteria and necessary bacterial and yeast origins of replication. Shuttling the vector to *Escherichia coli* is necessary for analysis and maintenance after identification in the screen. Although our system does not have this feature, some systems will include an additional selectable marker in the library (prey) vector that provides a selective advantage in *E. coli* over the bait plasmid. This marker facilitates isolation of the prey plasmid after completion of the screen. The most common is a cyclohexamide selection marker found in commercially available library vectors.

3.4. Yeast Transformation

Similar to bacterial transformation, yeast transformation is the introduction of a DNA plasmid with a selectable marker into yeast. In the case of yeast, the selectable marker usually is an enzyme that allows growth of yeast in the absence of an amino acid or nucleic acid. Here, we list three methods for transformation. The small-scale transformation is for introducing your bait vector prior to initiating the screen and the controls before and after the screen. There also are two methods for large-scale transformation: lithium acetate and electroporation. Electroporation has been the method of choice for us, but others have reported great success with lithium acetate. For this reason we offer both (*see* **Note 6**).

3.4.1. Small-Scale Transformation by Lithium Acetate

This protocol (adapted from **ref. *13***) is used to introduce the bait plasmid into yeast in preparation for introduction of the library. Generating a strain containing the bait plasmid increases number of yeast transformed with both the bait plasmid and a library plasmid. However, this protocol can be used to introduce any vectors into yeast and is useful in several other aspects of this protocol as well (i.e., determining specificity of interaction between bait and prey, determining if a bait is appropriate for the assay).

1. Grow yeast cells to an optical density (OD) at a wavelength of 595 (OD_{595}) = 1.0 in YPAD (ODs from 0.7 to 3.0 have been used but cells must be in log phase growth). For three transformations, grow a 40-mL culture.
2. Pellet cells via centrifugation at 1500*g*, discard supernatant, and resuspend yeast cells in freshly made 1X TE (pH 7.5), 0.1 *M* LiOAc solution to one-fourth of the original culture volume. Fresh solutions are ideal.
3. Pellet cells and resuspend to a density of 2×10^9 cells/mL in 1X TE, 0.1 *M* LiOAc. For a 40-mL culture, this equals 600 µL.
4. Add 200 µL of cells to a microcentrifuge tube containing 150 µg of salmon sperm DNA and 200 to 500 ng of transforming DNA (i.e., BTM116-bait) and mix thoroughly.

5. Add 700 µL of 1X TE, 0.1 *M* LiOAc, 40% PEG 3350, and mix thoroughly with pipet (do not vortex). Incubate for 30 min at 30°C with gentle shaking (roller drum).
6. Heat shock for 15 min at 42°C.
7. Spin for 30 s in a microcentrifuge at 14,000*g* and aspirate all but 100 µL of the supernatant.

Resuspend cell pellet in 100 µL with P1000 pipet tip and plate. If you are using greater than 1.0 µg of DNA and spreading over several plates, you may want to spread the cells directly after heat shock, i.e., no pelleting.

3.4.2. Large-Scale Transformation by Lithium Acetate

Before introducing the library vector, the bait plasmid is first introduced into the yeast strain by small-scale lithium acetate yeast transformation (adapted from **ref. 7**) and selected for on appropriate agar plates (in our case SD/trp⁻). The resulting strain is used to transform the library.

1. Grow a 2-mL culture of the yeast strain with bait overnight at 30°C in media selecting for the bait plasmid and any other features of your yeast strain (i.e., SD/trp⁻ and ura⁻).
2. Inoculate 100 mL of the same medium with an aliquot of the overnight culture from above and grow at 30°C overnight. The goal is to find a dilution that places the 100-mL culture at mid-log phase (OD_{600} = 1) the next day. Yeast double approximately once an hour, so dilute back to an OD that will yield a culture with an OD between 1 and 3 when you arrive the next morning.
3. Inoculate 1 L of YPAD (prewarmed to 30°C) with the culture from **step 2** (OD at 600 nm should be approx 0.3). Grow at 30°C for 3 h. If the cells were in mid-log phase and the bait does not inhibit growth, they will more than double during this period.
4. Pellet cells at 1500*g* for 5 min at room temperature. (A fixed angle rotor gives better recovery.) Decant supernatant.
5. Resuspend pellet in 500 mL of TE. Respin.
6. Resuspend pellet in 20 mL of 100 m*M* LiAc/ 0.5X TE.
7. Add DNA mixture: 1.0 mL of 10 mg/mL denatured salmon sperm DNA and 500 µg of library plasmid DNA and mix. (Mini-prep DNA that has been phenol-extracted works well as yeast transform better if the DNA is not perfectly "clean." Therefore, RNase treatment of the library DNA is not necessary or desirable.)
8. Add 140 mL of 100 m*M* LiAc/40% PEG-3350/1X TE. Mix well. Incubate 30 min at 30°C.
9. Place into a sterile 2-L beaker and replace foil cover. Add 17.6 mL of dimethyl sulfoxide and swirl to mix. Heat shock at 42°C in a water bath for 6 min, with occasional swirling to facilitate heat transfer. Immediately dilute with 400 mL of YPA (no glucose) and rapidly cool to room temperature in a water bath.
10. Pellet cells at 1500*g* for 5 min at room temperature.
11. Wash cells with 500 mL of YPA.
12. Resuspend pellet in 1.0 L of prewarmed YPAD. Incubate at 30°C for 1 h with gentle shaking. This step gives about a threefold increase in transformation efficiency.

13. Remove 1.0 mL to a microcentrifuge tube. Pellet cells, remove supernatant, and resuspend in 1.0 mL of media selecting for the bait and prey plasmids, but not interaction (SD/trp⁻, ura⁻, and leu⁻). Plate 10 µL and 1 µL (= 1/10⁵ and 1/10⁶ of total) on media selecting for both plasmids, but not the interaction (SD/trp⁻,ura⁻, and leu⁻). This measures the primary transformation efficiency. This protocol should give 10 to 100 million transformants.

14. Pellet the liter of cells as in **step 10**. Resuspend pellet in 500 mL of media selecting for the bait and prey plasmids (SD/trp⁻,ura⁻, and leu⁻). Re-spin and resuspend pellet in 1 L of prewarmed media, again selecting for the bait and prey plasmids. Shake at 30°C for 4 to 16 h. The plating efficiency for activating pairs on the screen selection marker plates appears to dramatically increase some time during this recovery period. At 4 h, the plating efficiency was measured at approx 5% and increased to 50% at 16 h. A shorter recovery period is more desirable and will be sufficient if the primary transformation efficiency is high enough.

15. Pellet cells at 1500*g* for 5 min at room temperature.

16. Wash cells twice with full-selective media (SD/trp⁻, leu⁻, ura⁻, lys⁻, and his⁻). Resuspend final pellet in 10 mL of full-selective media (SD/trp⁻, leu⁻, ura⁻, lys⁻, and his⁻).

17. Plate aliquots on full-selection agar plates (SD/trp⁻, leu⁻, ura⁻, lys⁻, and his⁻). If the density is too high, the true-positive clones will grow poorly. Therefore, the optimal plating volume is dependent on the recovery time. With a 4-h recovery, plate 50 or more plates at 100 µL/plate. With a longer recovery time in SD/ura⁻, trp⁻, and leu⁻), the cells will have divided a number of times, therefore plate decreasing volumes, such as 100, 50, 25, 10, and 5 µL (10 plates for each volume). Typically, approx 10 to 100 µL will give good colony outgrowth, although this number is dependent on the number of doublings during the overnight incubation and the overall transformation efficiency. Also plate 1/10⁶ and 1/10⁷ of the total transformation reaction on plates selecting for just the bait and prey plasmids (SD/trp⁻, leu⁻, and ura⁻) as in **step 13**. Compare the number of transformants on these plates with the number of primary transformants from **step 13**. This allows a calculation of the number of doublings and the percentage of His⁺ colonies which should be screened to roughly cover the number of primary transformants, although this must be corrected for plating efficiency.

3.4.3. Transformation by Electroporation

This protocol (adapted from **ref. 14**) resulted in the best transformation efficiency in the most time efficient manner. Thus, this is our method of choice. However, you should determine which protocol gives you the best efficiency (*see* **Note 6**).

1. Transform the yeast strain with the "bait" plasmid of interest (BTM116-bait) using the small-scale lithiumacetate method described previously (i.e., **Subheading 3.4.1.**).

2. Innoculate 500 mL of YPAD with the two-hybrid yeast strain (L40) with the bait vector and grow to an OD₅₉₅ of 1.3 to 1.5.

3. Divide the tube into two 250-mL centrifuge bottles and spin at 3000g for 5 min at 4°C. Discard the supernatant.

4. Resuspend each pellet in a total of 250 mL of ice-cold sterile water.

5. Centrifuge at 3000g for 5 min at 4°C.

6. Resuspend each pellet in 100 mL of ice-cold sterile water, combine in one bottle and fill to 250 mL.

7. Centrifuge at 3000g for 5 min at 4°C as described previously and discard supernatant.

8. Resuspend in 20 mL of ice-cold, 1 M sterile, filtered sorbitol and transfer to smaller (i.e., 50 mL oak ridge) centrifuge tube.

9. Centrifuge at 3000g for 5 min at 4°C. Discard the supernatant.

10. Resuspend yeast pellet by pipetting in 500 μL of sorbitol (keep on ice). The final volume varies from 1 to 1.5 mL.

11. In separate microcentrifuge tubes (there are usually enough competent cells to do 45 electroporations), add 40 μL of the competent cell mixture. (Be sure to mix each time before aliquoting the cells. The mixture is very dense and settles quickly.) Add 200 ng of library DNA to microcentrifuge tubes. Leave one tube without DNA for a "no DNA" control. Allow to sit on ice 10 min.

12. Transfer tube contents to 0.2-cm sterile electroporation cuvet. Tap to bottom.

13. Pulse at 1.5 kV, 25 uF, 200 Ω.

14. Immediately add 500 μL of ice-cold, 1 M sorbitol to the electroporation cuvet and transfer back to Eppendorf tube. For large-scale electroporations, you can combine all the electroporations into a single 50-mL conical (**Note:** Remember to keep a "no DNA" transformation control in a separate tube). This step gives an average transformation efficiency without having to plate a transformation efficiency control for each sample.

3.5. Plating on Selective Media

1. To determine the transformation efficiency for extrapolation to total cDNAs screened, plate 10 μL and 50 μL of transformed yeast on plates containing the Bait:Prey interaction marker (i.e., SD/trp⁻, leu⁻, and ura⁻).

2. Spread the remaining cells on plates lacking the interaction marker (SD/trp⁻, leu⁻, ura⁻, lys⁻, and his⁻), 200 μL /10-cm plate or 500 μL /15-cm plate. These plates are your actual screen. (**Note:** For electroporated libraries: although the original electroporation protocol suggests including 1 M sorbitol in the plates, we have obtained 10-fold better efficiency without sorbitol using the L40 yeast strains and plasmids. However, this efficiency may be dependent on the strain used. If transformation efficiency is poor, try including 1 M sorbitol in the plate recipe for transformation by electroporation. Expect approx 6000 transformants from 200 ng of DNA.)

3.6. Screening Positive Clones

Once the initial screen is complete, it is necessary to verify that isolated clones are true positives. There are four phases to this:

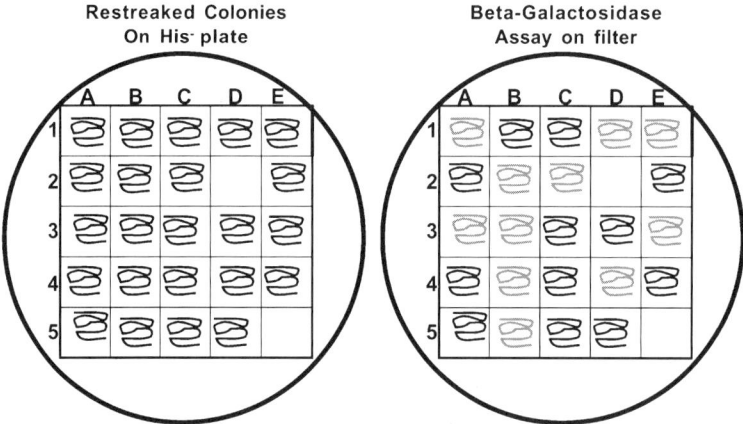

Fig. 4. β-Galactosidase screening of HIS3-positive clones from the yeast two-hybrid screen. HIS+ colonies are streaked to a grid on agar plates lacking histidine **(left panel)**. Positive yeast will grow and yield white colonies. Streaked yeast are lifted onto a nitrocellulose filter and assayed for LAC Z activity. Yeast streaks that turn blue (depicted in gray) should be considered for further analysis **(right panel)**.

1. Restreaking identified colonies on selective plates and assaying for β-galactosidase activity.
2. Retransforming isolated prey vectors with the bait vector into fresh yeast.
3. Introducing the identified prey vector with the bait vector lacking the gene of interest or containing an unrelated gene to verify that the prey is not interacting with the DBD independent of the protein of interest.
4. Introducing the identified prey vector with no bait vector to determine if the prey binds the promoter of the reporter independent of the DBD.

3.6.1. Replating on Selective Media

Positive colonies from an initial screen are streaked in a grid to new selective plates (SD/trp⁻, leu⁻, ura⁻, lys⁻, and his⁻; *see* **Fig. 4**). Plates are grown at 30ºC until the yeast are visible (usually 2–3 d), which serves two purposes. The first is to weed out false-positives caused by overplating. If yeast cells are plated to densely, there will sometimes be growth even though the yeast are not truly autotrophic for the marker. This grid also serves as a template to assay for the second reporter for the yeast system, Lac Z. Colonies that grow are assayed for β-galactosidase activity as described in the following section.

3.6.2. β-Galactosidase Assay

The β-galactosidase assay is used as a secondary screen to amino acid autotrophy. This assay was included in the screen to remove false-positives resulting from prey containing a DNA binding domain interacting with the promoter independent of the bait. It weeds out a fair amount of these false-positives and is worth doing. The filter assay is recommended for mass screening. We also include a quantitative assay for determining relative affinity of the inter- action, which can aid in determining which clones to pursue following the screen.

3.6.2.1. β-GALACTOSIDASE COLONY ASSAY FOR LARGE-SCALE SCREENING (FILTER ASSAY)

This assay can be used to assay the grid of positives restreaked to selective media in the previous step. If there are a large number of positive colonies that make restreaking prohibitive, nitrocellulose filters can be applied directly to the screen plates and assayed for β-galactosidase activity. When using this approach, remember to mark the plate and nitrocellulose for orientation so you can go back and isolate the original colony. We do this by poking three asymmetric holes through the filter and agar using a syringe needle dipped in black India ink, which will limit the number of colonies that need to be restreaked to selective media.

1. Place a dry circular nitrocellulose filter onto yeast colonies grown in plates. This filter should be the same size and shape as the plate. In our case, we use 10-cm diameter circular filters for our 10-cm Petri dishes.
2. Remove filter and place colony side up on a precooled aluminum boat floating in a sea of liquid nitrogen. After 30 s, immerse boat and filter for 5 s. Remove filter and place at room temperature, colony side up, until thawed.
3. Prepare a Petri dish for the reaction. In the lid of a 10-cm dish, place 3 mL of Z-buffer containing 50 μL of 50 mg/mL X-gal. Place two no. 1 Whatman filter circles in the Z-buffer, followed by the nitrocellulose filter, colonies facing up. Try to avoid air bubbles. Cover with the bottom of the dish and place at 30°C. For longer incubations, the Petri dish should be placed in a humidified chamber. Strong interactions yield detectable color in less than 30 min.
4. The original agar plates should be incubated at 30°C overnight to regrow the colonies lifted by the filter for the assay. Recover viable cells from Lac Z-positive colonies by lining up the filter with the original agar plate. We have found that placing asymmetric holes in the filter paper with a syringe needle dipped in India ink aids in realigning the colonies with the filter paper.

3.6.2.2. Quantitative β-Galactosidase Activity Assay

This is an in vitro assay for β-galactosidase activity in yeast (adapted from **ref. 15**) via preparation of crude extract. Activity is then normalized to the amount of protein assayed. This quantitative approach allows determination of relative interaction affinity. It is also useful for determining endogenous transactivation potential of the bait (*see* **Subheading 3.2.**). It should be mentioned that commercial kits also are available for generating extracts for this assay (i.e., Y-Per from Pierce). Although we have not used these kits for this purpose, it is likely they will work as well as the method described here.

1. Grow a 5-mL culture of cells to 1 to 2×10^7 cell/mL in media selective for both the bait and prey vectors.
2. Chill cells on ice and harvest by centrifugation (between 1000 and 1500g for 5 min in a clinical centrifuge is adequate). Keep cells on ice from this point on.
3. Resuspend in 250 µL of breaking buffer. Cells can be frozen at –20°C and assayed at a later date. All of the following steps can be performed in a 1.5-mL microfuge tube.
4. If cells have been frozen, thaw on ice. Add glass beads (0.45–0.5 mm) to fill liquid to the meniscus. Add 125 µM of phenylmethylsulfonylfluoride to a final concentration of 1.25 µM .
5. Vortex at top speed, six times for 15 s. Chill on ice between bursts.
6. Add 250 µL of breaking buffer to the beads. Mix well and withdraw the liquid extract by plunging the tip of a 1000-µL pipetter to the bottom of the tube. Place extracted solution into a new tube.
7. Clarify the extract by 15-min centrifugation in a microcentrifuge at 14,000g.
8. The assay is performed by adding 10 to 100 µL of extract to 0.9 mL of Z-buffer with 0.03 M β-mercaptoethanol (added fresh and brought to pH 7.0 with NaOH). Bring volume to 1 mL with breaking buffer. The tube containing the mixture is incubated at 28°C in a water bath for 5 min and the reaction is initiated by addition of 0.2 mL of ONPG (4 mg/mL in H_2O, sterile filtered and stored at –20°C). Carefully note the time of addition. Incubate at 28°C until the mixture has acquired a pale yellow color. Terminate the reaction by adding 0.5 mL of 1 M Na_2CO_3 and note time. Measure the OD at 420 nm.
9. Measure the protein concentration in the extract using the dye-binding assay of Bradford (*16*). Dilute the Bradford reagent fivefold. Filter the diluted reagent through Whatman 540 paper or equivalent. To 1 mL of diluted reagent, add 10 to 20 µL of extract and mix. Measure the blue color formed at 595 nm. Use disposable plastic cuvets to prevent the formation of a blue film. Prepare a standard curve using several dilutions (0.1 to 1 mg/mL) of bovine serum albumin dissolved in breaking buffer. Typical extracts prepared in this fashion contain 0.5 to 1 mg per milliliter of protein.
10. Express the specific activity of the extract according to the following formula:

$$OD_{420} \times 1.7/0.0045 \times [\text{protein}] \times \text{extract volume} \times \text{time}$$

where OD_{420} is the OD of the product o-nitrophenol at 420 nm. The factor 1.7 corrects for the reaction volume. The factor 0.0045 is OD of a 1 nmol/mL solution of o-nitrophenol. Protein concentration is expressed as milligrams per milliliter. Extract volume is the volume in milliliters; time is in minutes. Specific activity is expressed as nanomoles per minute per milligram of protein.

3.6.3. Isolation of Plasmids From Yeast

For the rest of the false-positive controls, the library plasmid DNA needs to be extracted from yeast and transferred to *E. coli* (*see* **Note 7**). We recommend the following procedure nicknamed the "smash and grab" (adapted from **ref. *17***).

1. Grow a 5-mL culture overnight in media selecting only for the library (prey) plasmid (SD/leu⁻). By using media that selects for the prey plasmid while not selecting for the bait plasmid, some of the yeast will spontaneously lose the bait vector in some of the yeast cells. Spread 50 µL of a 1:10,000 dilution of this culture on SD/leu⁻ agar plates. After growth, replica plate to SD/trp⁻ plate and then to a fresh SD/leu⁻ plate. Pick a colony that grows on the SD/leu⁻ plate, but not the SD/trp⁻ plate to 5 mL of SD/leu⁻ media. Such a colony will have lost the bait vector and now allow for isolation of the library vector only.
2. Pellet the yeast at 1500g at room temperature for 5 min
3. Resuspend pellet in 0.3 mL of lysis buffer and transfer to 1.5-mL tube. Add approx 150 µL of glass beads (0.45–0.50 mm) and 0.3 mL of phenol/chloroform. (Remove any beads adhering near the top of the tube as they will provide a channel of escape for phenol during the next step). Vortex the tubes vigorously for 1 min.
4. Pellet debris in a microcentrifuge for 1 min.
5. Transfer aqueous phase to a new tube. Precipitate DNA by adding 2 volumes of 200 proof ethanol and one-tenth volume of 3 M sodium acetate, storing at –20°C for 40 min, and microcentrifuging at top setting at 4°C for 20 min. Wash with 70% ethanol.
6. Resuspend DNA pellet in 25 µL of TE and electroporate *E. coli* cells with 1–2 µL. (**Notes:** The yeast miniprep also can function as a template for polymerase chain reaction amplification of cDNA inserts. Include 0.1% Tween-20 in the reaction mix. Amplification of cDNA inserts using surrounding vector sequence will allow subcloning into another vector useful for subsequent analysis (i.e., glutathione-*S*-transferase fusion vector).
7. Plate on luria broth plates with appropriate bacterial selection marker (ampicillin is most common).
8. Make bacterial stocks of clones (add 15% glycerol to growing culture and freeze at –80°C. Isolate DNA from bacteria via miniprep for reintroduction into yeast to screen for false-positives.

3.6.4. Reintroducing Positive Clones and Ruling Out False-Positives

Once the plasmids have been isolated from yeast and stored in bacteria, it is time to rule out the potential false-positives outlined in **Subheading 3.2.**, **items 2–4**.

Yeast cells that are auxotrophic for a marker can revert to autotrophy as the result of recombination at the reporter locus. To rule out false-positives caused by reversion, we recommend transforming fresh yeast cells with the bait and each identified prey vector, which has been transferred to *E. coli* and purified by DNA miniprep. Using the small-scale lithium acetate transformation protocol (**Subheading 3.4.1.**), transform the yeast with the two vectors simultaneously and plate half on plates lacking only the amino acids for the vector selectable markers (SD/trp⁻ and leu⁻) and half on plates also lacking the amino acid for which the interaction reporters renders the yeast autotrophic (SD/trp⁻, leu⁻, ura⁻, lys⁻, and his⁻). Only prey that yield colonies positive on both plates should be pursued further.

Simultaneously with the previous transformation, include a control for interaction between prey and the DBD. In this case, the isolated prey vector is transformed into yeast cells with the bait vector lacking the actual bait (BTM116) or containing an unrelated bait (i.e., BTM116-daughterless). Transformed yeast should not grow on plates lacking his. If it does grow, it indicates that the prey is interacting with the DBD or the fusion domain and not the protein of interest. It may also be interesting at this stage to include bait-deletion mutants generated at the beginning of the protocol to determine whether the prey interacts specifically with a domain within the bait.

Finally, the prey vector should be transformed into L40 yeast by itself. Half of the transformed cells should be plated on media selecting for the prey vector alone (SD/leu⁻) and the remaining half should be plated on media lacking the amino acid to select for both the vector and the interaction reporter (SD/leu⁻ and his⁻). This step will determine whether the prey is activating transcription by directly binding the reporter promoter. Again, this control can be performed at the same time as the two previous screens to rule out false-positives but must be plated on different selective media.

3.6.5. Analysis of Positive Interactors

Prey that meet all the aforementioned outlined criteria are strong candidates for true interaction (*see* **Note 8**). If the screen is working well, you should expect between 0 and 100 true-positives. This level provides a workable number of proteins to follow up. At this point, the cDNA should be identified by sequencing of the prey plasmid now stored in *E. coli*. Further experimentation should be performed to verify the interaction biochemically and functionally. Also, some logic should be applied to determining authenticity of a clone, as well as empirical determinations. Based on known protein abundance (*see* **Note 9**), known function (*see* **Note 10**), or strength of interaction (*see* **Note 11**) some clones determined to be true-positive may need to be reconsidered. These criteria will aid in determining which interacting proteins to pursue first.

If this version of the protocol does not yield positive clones (*see* **Note 12**) or yields too many positive clones (*see* **Note 9**), it is possible that another yeast screen is more appropriate for your protein of interest, or that another approach will be necessary.

4. Notes

1. It is important to consider the features of the bait plasmid you plan to use in the screen. There are now numerous such plasmids available each with a unique feature making it more or less desirable for use. The most basic divergence in vectors is the choice of DBD used for producing the fusion protein. By far the most common DBDs used are the *E. coli* Lex A DBD and the yeast Gal4 DBDs. We feel the Lex A DBD has several advantages over the Gal4. First, Lex A is not

conserved or normally expressed in yeast. Thus, there is no need to remove the Lex A gene from the yeast genome. The Gal 4 domain must be used in Gal4⁻ yeast, which do not grow as well as their wild-type counter parts. There are few, if any, yeast proteins that will affect Lex A function in yeast, unlike Gal 4. Finally, because Lex A also is not conserved in higher eukaryotes, it has very few spurious interactions with the expressed library proteins (indeed only one has been reported to date). Thus, Lex A provides a low background for the screen with fewer concerns for yeast growth.

2. Another problem may be that the bait does not enter the nucleus. This step is especially crucial for proteins that are normally cytoplasmic or membrane bound. The first suggestion is to use a bait vector that has a NLS. Because the NLS functions through protein interactions, it will increase the numbers of clones that are not specific to the protein of interest. However, these will be easily detected as prey that gives a positive with the bait vector lacking the protein of interest (*see* **Subheading 3.6.4.**). Another reason your protein of interest may not get into the nucleus is the presence of domains targeting your protein to membranes. If there are known transmembrane- or membrane-binding domains, it is recommended that these be removed. Not only will they interfere with nuclear entry, but they will create a high background in the screen.

3. Whether using Lex A or Gal4, addition of your protein of interest to the fusion creates a "fusion domain." This protein sequence, which occurs at the juncture between the DBD and the bait protein, has been reported to create false-positives in the screen. To begin to address this, we suggest making deletion constructs containing smaller parts of the protein. If the interaction is owing to spurious interaction at the "fusion domain," it should not bind to any of the deletion mutants created. We also recommend verifying interaction in an independent system, such as gluthathion-*S*-transferase fusion protein affinity column pull-down assays, coimmunoprecipitation, or verification in a yeast two-hybrid system using a different DBD.

4. Low protein stability can also be a problem, especially for large proteins. If a protein contains a domain known to decrease stability, this may be deleted. If not, simply using a shorter stretch of the protein will increase stability.

5. Whether you suspect your protein to have hydrophobic or nonspecific interactions, we feel it is a good idea to do a pilot library screen first (described in **Subheading 3.1.**). This will give you an idea of how many positive clones to expect in the real screen and determine if the bait you have chosen will give a manageable number of positive interactors.

6. Before doing an all-out large-scale library screen, we recommend practicing both library transformation protocols on a small scale. This will help you determine which protocol is best for you without wasting library DNA. It will also help you determine whether you need to scale up your transformation to screen the library fully (you should aim for a transformation protocol that yields 10,000 to 100,000 transformants/μg DNA).

7. New screens use mating strategies to remove the time-consuming shuttling of the prey vector into *E. coli*. These strategies are not available for the screen dis-

cussed here, but should be considered if using a system in which they are available. However, in the end, you will want any clones of interest shuttled to *E. coli*, so this aspect of the protocol is still necessary.

8. Once interacting prey cDNAs have been shuttled to *E. coli*, the inserts can be characterized by enzyme restriction mapping, which will give an idea of how many duplicate clones you have identified prior to sequencing.

9. If your protein of interest binds to an abundant protein (such as cytoskeletal components), these clones may swamp out rare and lower affinity interactions. In this case, the screen may be of less use to you. However, if among the 100 clones you identify from screening 1 million yeast colonies, only one clone is actin, it is unlikely that this is a true interaction. Abundant proteins should be identified more frequently among the true-positives than rare proteins.

10. Because yeast are eukaryotes, there are a fair number of proteins conserved with mammalian proteins. If your screen is not working, it is possible that yeast proteins are interfering with your bait (i.e., sequestration, modification, degradation, etc.). Some of these may make your protein unsuitable for use in the yeast two-hybrid screen.

11. After transformation of the library, the number of days plates are left in the incubator will determine how many positive clones you get. Normally, yeast colonies will appear 2 to 3 d after transformation. If left longer, the more colonies will grow over time. It may be useful to pick early-positives and delineate between them and late-positives. Late-positives may indicate weaker interaction. It is up to you and the volume of positives you have as to how many clones you are willing to analyze.

12. Because yeasts are a single-cell organism, it is also possible that modifications needed for interaction are not present. A yeast tri-hybrid (tribrid) screen is now available for interactions that require a third party or modification, such as phosphorylation, not endogenously available in yeast.

References

1. Fields, S. and Song, O. (1989) A novel genetic system to detect protein-protein interactions. *Nature* **340,** 245–246.
2. Brent, R. and Ptashne, M. (1985) A eukaryotic transcriptional activator bearing the DNA specificity of a prokaryotic repressor. *Cell* **43,** 729–736.
3. Ma, J. and Ptashne, M. (1988) Converting a eukaryotic transcriptional inhibitor into an activator. *Cell* **55,** 443–446.
4. Chien, C. T., Bartel, P. L., Sternglanz, R., and Fields, S. (1991) The two-hybrid system: a method to identify and clone genes for proteins that interact with a protein of interest. *Proc. Natl. Acad. Sci. USA* **88,** 9578–9582.
5. Gyuris, J., Golemis, E., Chertkov, H., and Brent, R. (1993) Cdi1, a human G1 and S phase protein phosphatase that associates with Cdk2. *Cell* **75,** 791–803.
6. Durfee, T., Becherer, K., Chen, P. L., et al. (1993) The retinoblastoma protein associates with the protein phosphatase type 1 catalytic subunit. *Genes Dev.* **7,** 555–569.

7. Vojtek, A. B., Hollenberg, S. M., and Cooper, J. A. (1993) Mammalian Ras interacts directly with the serine/threonine kinase Raf. *Cell* **74**, 205–214.
8. Jordan, K. L., Evans, D. L., Steelman, S., and Hall, D. J. (1996) Isolation of two novel cDNAs whose products associate with the amino terminus of the E2F1 transcription factor. *Biochemistry* **35**, 12,320–12,328.
9. Jordan-Sciutto, K. L., Dragich, J. M., Caltagarone, J., Hall, D. J., and Bowser, R. (2000) Fetal Alz-50 clone 1 (FAC1) protein interacts with the Myc-associated zinc finger protein (ZF87/MAZ) and alters its transcriptional activity. *Biochemistry* **39**, 3206–3215.
10. Strachan, G. D., Morgan, K. L., Otis, L. L., et al. (2004) Fetal Alz-50 Clone 1 interacts with the human orthologue of the Kelch-like Ech-associated protein. *Biochemistry* **43**, 12,113–12,122.
11. Bartel, P. L., Chien, C. T., Sternglanz, R., and Fields, S. (1993) In: *Cellular Interactions in Development: A Practical Approach* (Hartley, D. A., ed.). IRL Press at Oxford University Press, Oxford, UK, pp. xviii.
12. Brent, R. and Ptashne, M. (1984) A bacterial repressor protein or a yeast transcriptional terminator can block upstream activation of a yeast gene. *Nature* **312**, 612–615.
13. Elble, R. (1992) A simple and efficient procedure for transformation of yeasts. *Biotechniques* **13**, 18–20.
14. Becker, D. M. and Guarente, L. (1991) High-efficiency transformation of yeast by electroporation. *Methods Enzymol.* **194**, 182–187.
15. Rose, M. and Botstein, D. (1983) Construction and use of gene fusions to lac Z (beta-galactosidase) that are expressed in yeast. *Methods Enzymol.* **101**, 167–180.
16. Bradford, M. M. (1976) A rapid and sensitive method for the quantitation of microgram quantities of protein utilizing the principle of protein-dye binding. *Anal. Biochem.* **72**, 248–254.
17. Ward, A. C. (1990) Single-step purification of shuttle vectors from yeast for high frequency back-transformation into *E. coli. Nucleic Acids Research* **18**, 5319.

13

Analysis of PDZ Domain Interactions Using Yeast Two-Hybrid and Coimmunoprecipitation Assays

Hyun Woo Lee, Jaewon Ko, and Eunjoon Kim

Summary

The PDZ domain is a protein–protein interaction module that interacts with a C-terminal short peptide motif in its binding partners. A variety of methods have been used to study PDZ domain interactions. This chapter details the two methods most commonly used in the analysis of PDZ interactions: yeast two-hybrid and coimmunoprecipitation assays. In addition, we discuss the features that must be considered for an efficient analysis of PDZ interactions.

Key Words: PDZ; yeast two-hybrid; coimmunoprecipitation.

1. Introduction

The PDZ (PSD-95/Dlg/ZO-1) domain is a protein–protein interaction module that recognizes a C-terminal short peptide motif in the target proteins (*1,2*). The PDZ domain is an approx 90-amino-acid region that folds into a globular structure containing six antiparallel β-strands and two α-helices. C-terminal peptides bind to a groove in the PDZ domain. On the basis of the C-terminal amino acid sequences of their peptide ligands, PDZ domains fall into three classes: class I binds to peptides ending with -X-S/T-X-Φ (X, any amino acid residue; Φ, hydrophobic residues); class II to peptides ending with -X-Φ-X-Φ; and class III to peptides ending with -X-D/E-X-Φ (*2*).

PDZ domain-containing proteins are implicated in the organization of macromolecular protein complexes containing membrane, signaling, and cytoskeletal proteins at sites of cellular junctions, such as neuronal synapses (*1–7*). This chapter describes two major and complementary methods for studying PDZ domain interactions: yeast two-hybrid and coimmunoprecipitation. In addition to experimental details, we also describe the features that must be considered for an efficient analysis of PDZ interactions.

From: *Methods in Molecular Biology, vol. 332: Transmembrane Signaling Protocols, Second Edition*
Edited by: H. Ali and B. Haribabu © Humana Press Inc., Totowa, NJ

2. Materials

2.1. Yeast Two-Hybrid Assay

1. Yeast two-hybrid vectors (**Table 1**).
2. Yeast strains (**Table 2**).
3. Salmon sperm DNA (10 mg/mL solution; Sigma, St. Louis, MO).
4. 1X TE (10 mM Tris-HCl + 1 mM ethylene diamine tetraacetic acid [EDTA], pH 7.5).
5. 1X LiAc solution (100 mM lithium acetate, pH 7.5).
6. 50% Polyethylene glycol (PEG; Sigma). PEG-3000 or PEG-3500 is more efficient than PEG-8000.
7. Dimethyl sulfoxide (Sigma).
8. Yeast peptone dextrose (YPD) medium: 20 g/L Bacto-peptone (BD, Sparks, MD), 10 g/L yeast extract (USB, Cleveland, OH), 2% glucose (Sigma). For the preparation of YPD plates, add 15 g/L agar before autoclaving.
9. –LT plate: 6.7 g nitrogen base, 100 mL of 10X Trp-Leu-His dropout mixture (*see* **step 10**), 40 mL of 50% glucose, 10 mL of 200 mg/mL histidine, and 15 g/L agar.
10. –HLT plate: omit histidine from –LT and add 3-amino-1,2,4-triazole (3-AT) to a final concentration of 2.5 mM. It is important to cool the media down to approx 55°C before adding 3-AT because it is destroyed at high temperatures.
11. 10X Trp-Leu-His drop-out mixture: 300 mg of L-isoleucine, 1500 mg of L-valine, 200 mg of L-adenine, 200 mg of L-arginine HCl, 300 mg of L-lysine HCl, 200 mg of L-methionine, 500 mg of L-phenylalanine, 2000 mg of L-threonine, 300 mg of L-tyrosine, and 200 mg of L-uracil per 1 L.
12. 1 M 3-AT.
13. X-Gal solution (Sigma): dissolve 5-bromo-4-chloro-3-indolyl-D-galactopyranoside (X-gal) in N,N-dimethylformamide at 40 mg/mL. Store in the dark at –20°C.
14. Z-buffer: 16.1 g/L $Na_2HPO_4 \cdot 7H_2O$, 5.5 g/L $NaH_2PO_4 \cdot H_2O$, 0.75 g/L KCl, and 0.246 g/L $MgSO_4 \cdot 7H_2O$. Sterilize the solution by autoclave or filtration.
15. Whatman filter paper no. 1 (125 mm) (Whatman Ltd.; Maidstone, UK).

2.2. Coimmunoprecipitation

1. HEK293 T-cells: a human embryonic kidney cell line, American Type Culture Collection CRL-1573.
2. 2.5 M $CaCl_2$.
3. 2X HEPES-buffered saline: 50 mM HEPES, 280 mM NaCl, and 1.5 mM Na_2HPO_4. The final pH should be 7.1. Store at 4°C.
4. Cell lysis buffer: phosphate-buffered saline (PBS) containing 1% Triton X-100 (PBST).
5. Protease inhibitors: 1 mM phenylmethylsulfonyl fluoride, 2 mg/mL aprotinin, 2 mg/mL leupeptin, 2 mg/mL pepstatin, and 2 mg/mL benzamidine.
6. Primary antibodies against target proteins.
7. Protein A-Sepharose (Amersham Pharmacia; Uppsala, Sweden).
8. Equipment for sodium dodecyl sulfate-polyacrylamide gel electrophoresis equipment.

Table 1
List of Commonly Used Yeast Two-Hybrid Vectors

Vector	Selection marker	Functional domain	Promoter
GAL4-based			
pAS1	TRP1	GAL4DB + HA	ADH1 (full length)
pAS2	TRP1	GAL4DB + HA	ADH1 (full length) CYH2
pAS2-1	TRP1	GAL4DB	ADH1 (full length) CYH2
pGBT9	TRP1	GAL4DB	ADH1 (truncated)
pMA424	HIS3	GAL4DB	
pGAD2F	LEU2	GAL4AD	
pGAD424	LEU2	GAL4AD	ADH1 (truncated)
pGAD10	LEU2	GAL4AD	ADH1 (truncated)
pGAD-GL	LEU2	GAL4AD	ADH1 (truncated)
pGAD-GH	LEU2	GAL4AD	ADH1 (full length)
pGAD1318	LEU2	GAL4AD	ADH1 (full length)
pSE1107	LEU2	VP16AD	
pSD-10	URA3	GAL4AD	
pACT1	LEU2	GAL4AD	
pACT2	LEU2	GAL4AD + HA	ADH1 (truncated)
LexA-based			
pBHA	TRP1	LexA	ADH1 (truncated)
pBTM116	TRP1	LexA	ADH1 (truncated)
pLexA	HIS3	LexA	ADH1 (full length)
pB42AD	TRP1	B42 + SV40 NLS + HA	GAL1 (full length), inducible promoter
pHybLex/Zeo	Zeocin	LexA	ADH1 (truncated)
pYESTrp	TRP1	V5 epitope + SV40 NLS + B42	GAL1 (full length), inducible promoter
pGilda	HIS3	LexA	GAL1 (full length), inducible promoter

235

Table 2
List of Commonly Used Yeast Strains

		Expression of reporter	
Strain	Reporter genes	Uninduced	Induced
H7Fc	*LacZ,*	—	Low
	HIS3	—	High
YRG-2	*LacZ,*	—	Low
	HIS3	—	High
SFY526	*LacZ*	—	High
Y187	*LacZ*	—	High
Y190	*LacZ,*	—	High
	HIS3	Low	High
CG-1945	*LacZ,*	—	Low
	HIS3	Very low	High
L40	*LacZ,*		
	HIS3		

3. Methods

3.1. Yeast Two-Hybrid Assay

3.1.1. Yeast Two-Hybrid Vectors

In the yeast two-hybrid system, two separate proteins are fused to the DNA-binding or the activation domains of a transcription factor. When the two proteins interact, the DNA-binding and -activation domains are brought together, reconstituting a bipartite transcription factor and causing the transactivation of reporter genes, such as *lacZ* (encoding β-galactosidase) or *his3* (encoding imidazole acetol phosphate transaminase; **Fig. 1**).

There are several vectors and host strains that can be used for yeast two-hybrid screening (**Tables 1** and **2**). In this chapter, we describe the system using pBHA (LexA DNA-binding domain [DBD]) as the bait vector, pGAD10 (Gal4 activation domain) as the prey vector, and the L40 strain. pBHA and pGAD10 are a shuttle vectors that replicate autonomously in both *Escherichia coli* and *Saccharomyces cerevisiae* and carry the *bla* gene, which confers ampicillin resistance to *E. coli*. pBHA contains the *TRP1* nutritional gene that allows yeast auxotroph cells to grow on synthetic media lacking tryptophan (**Fig. 2**). pGAD10 contains the *LEU2* nutritional marker gene, which allows yeast cells carrying pGAD10 to survive on limiting media lacking leucine (*see* manuals from Clontech for more details on this vector). Expression of the DBD and activation domain fusion proteins from pBHA and pGAD10 vectors, respectively, in yeast cells is directed by the constitutive *ADH1* promoter. The nuclear localization signal in Gal4-activation domain fusion proteins from

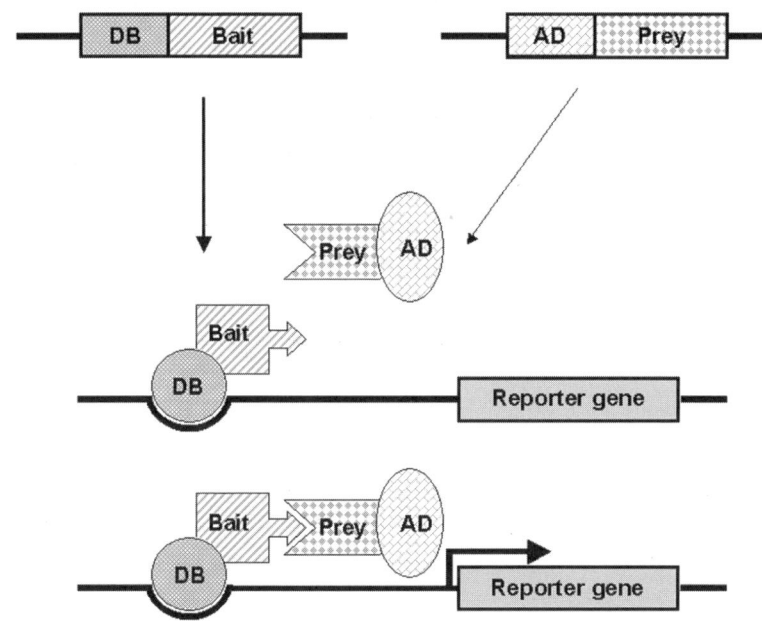

Fig. 1. Principles of the yeast two-hybrid system. Two chimeric fusion proteins, the bait containing the DNA-binding domain and the prey containing the activation domain, are expressed in the same yeast cell. The interaction of the two proteins reconstitutes a functional transcription activator, which leads to transcriptional activation of the reporter gene. DB, DNA-binding domain; AD, activation domain.

pGAD10 assists their nuclear targeting, whereas LexA–DBD fusion proteins from pBHA do not have a nuclear localization signal and are distributed in both the nucleus and cytosol.

3.1.2. Generation of Bait and Prey Constructs

PDZ interactions involve two binding partners: a C-terminal PDZ-binding peptide and the PDZ domain. We usually subclone the C-terminal peptide of a protein into the pBHA bait vector and the PDZ domain into the pGAD10 prey vector because we often find that the PDZ domain in pBHA activates reporter gene expression in the absence of a PDZ interaction through "self-activation" (*see* **Note 1**). For generation of bait and prey constructs, standard subcloning protocols can be used. In brief, C-terminal peptides and PDZ domains are subcloned in frame into the multiple cloning sites of pBHA and pGAD10, respectively (*see* **Note 2**). Sometimes short peptides (approx 7 amino acid residues) are sufficient to mediate PDZ interactions. In this case, we anneal two synthetic oligonucleotides, instead of amplifying inserts by polymerase chain reaction, and subclone them into pBHA vector.

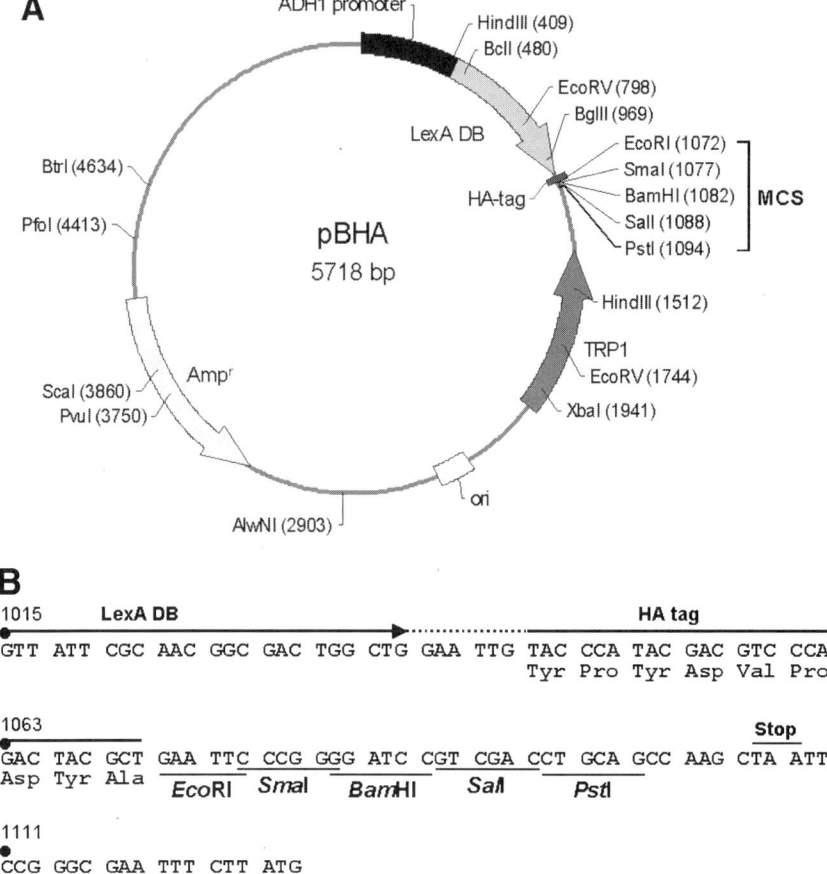

Fig. 2. Schematic diagram of the restriction map (**A**) and multiple cloning sites (**B**) of the pBHA bait vector. Zeocin is a glycopeptide antibiotic of the bleomycin family. pHybLex/Zeo carries the zoeocin resistance gene. The SV40 nuclear localization signal is the simian virus 40 large T-antigen nuclear localization signal (-PKKKRVE-). B42 is an *E. coli* protein that acts as a transcription activation domain. The 1500-bp full-length ADH1 promoter, which normally drives the expression of the metabolic enzyme alcohol dehydrogenase 1, leads to high-level expression of sequences under its control. In contrast to this full-length promoter, expression from a truncated 410-bp ADH1 promoter leads to low or very low levels of fusion protein expression.

Because the free carboxylate group at the C-terminus of the peptide is important for the PDZ interaction, care should be taken to add a stop codon at the end of the peptide. Indeed, an extra amino acid residue at the peptide C-terminus has been shown to abolish the PDZ interaction. Also, point mutations in peptides and PDZ domains are often used as negative controls (*see* **Note 3**).

3.1.3. Transformation of Yeast Cells

1. Pick a colony of L40 yeast and use it to inoculate 30 to 50 mL of YPD broth. Grow overnight at 30°C under aerobic conditions with vigorous shaking (250 rpm). On the next day (the optical density $[OD]_{600}$ of the culture is usually >1.5), dilute samples of the culture into 10 mL of fresh YPD broth so that the OD_{600} is approx 0.15, and grow at 30°C with vigorous shaking (250 rpm) until the OD_{600} reaches 0.4 to 0.6 (approx 4 h). The L40 yeast colonies are maintained on YPD plates at 4°C, and old colonies, which turn pink, should be transferred to fresh plates every 3 to 4 wk.

2. Mix bait DNA (500 ng; approx 2 µL of miniprep-purified DNA), prey DNA (500 ng), and salmon sperm DNA (100 µg; 10 µL of 10 mg/mL solution) in a microcentrifuge tube. Because salmon sperm DNA is very viscous, if necessary, boil it for 5 to 10 min so that it can be pipetted. We recommend the use of both negative (i.e., bait DNA + empty prey vector for self-activation check) and positive (known interactions) controls.

3. Harvest yeast cells by centrifugation in a table-top centrifuge (1000g) at room temperature and resuspend the cell pellet in one-fifth volume of sterile water. Centrifuge the cells at 1000g and resuspend the pellets in 1X TE/1X LiAc solution (100 µL per sample).

4. Add 100 µL of resuspended yeast cells to the DNA mixtures in each microcentrifuge tube.

5. Add 600 µL of 40% PEG solution in 1X TE/1X LiAc to each tube. Because PEG is very viscous, cutting off the end of a 1-mL pipet tip can help pipetting.

6. Incubate the samples at 30°C for 30 min in a shaking water bath (250 rpm). Invert the tubes every 10 to 15 min because the yeast cells tend to sediment.

7. Add one-tenth volume (70 µL) of dimethylsulfoxide and vortex briefly or invert several times.

8. Heat shock the cells by incubating the tubes in a water bath at 42°C for 15 min and then chill the tubes on ice for 1 to 2 min.

9. Collect the cells by centrifugation for 10 s in a microcentrifuge at approx 15,000g and resuspend the pellet in 1 mL of YPD.

10. Grow the resuspended cells in a shaking water bath at 30°C for 2 h, and invert the microcentrifuge tubes every 30 min.

11. Collect the cells by centrifugation in a microcentrifuge at approx 15,000g, and resuspend the pellet in 50 µL of YPD.

12. Spread the cells on –LT and –HLT plates (25 µL per plate).

13. Incubate the plates for 2 to 3 d in a 30°C incubator.

3.1.4. HIS3 Growth Counting and β-Gal Assay

As mentioned previously, pBHA and pGAD10 plasmids carry *TRP1* and *LEU2* as auxotrophic markers, respectively. Therefore, yeast transformants containing both plasmids should survive on the –LT plate. In addition, yeast transformants in which bait and prey fusion proteins interact should induce *HIS3* reporter gene expression and overcome the histidine-lacking environment of the –HLT plate (*see* **Notes 4** and **5**). HIS3 growth is measured by comparing the number of yeast colonies on the –LT and –HLT plates. Semiquantitative measurements of HIS3 growth are often reported (i.e., +++, >60%; ++, 31 to 60%; +, 10 to 30%; –, no significant growth *[8,9]*).

β-Galactosidase activity is measured by soaking yeast colonies with X-gal and checking how long it takes them to turn blue. For this assay, yeast colonies on –LT plates (rather than those on –HLT plates) are used because the size of the yeast colonies on –HLT plates varies depending on the levels of HIS3 expression. β-Galactosidase activities can also be semiquantitatively assessed (i.e., +++, <45 min; ++, 45 to 90 min; +, 91 to 240 min; –, no significant β-galactosidase activity *[8,9]*). Experimental details of β-galactosidase are described herein.

1. Place a Whatman no. 1 filter on top of the yeast colonies on the –LT plate.
2. As soon as the filter gets wet, carefully lift it off the plate with forceps. Make sure that most of the colonies are transferred to the filter.
3. Submerge the filter in liquid nitrogen for 10 to 15 s.
4. Take the frozen filters out of liquid nitrogen and thaw them at room temperature for approx 30 s.
5. Carefully place the filter colony side up onto a Whatman no. 1 filter soaked with X-gal solution in Z-buffer (17 μL of the X-gal stock solution in 1 mL of Z-buffer). Avoid trapping bubbles between the two filter papers.
6. Incubate the filter at room temperature and check how long it takes for the yeast colonies to turn blue.

3.2. Coimmunoprecipitation

The coimmunoprecipitation assay detects the formation of a complex between two proteins in vivo or in heterologous cells. This chapter describes coimmunoprecipitation in heterologous cells. In contrast to the yeast two-hybrid assay, which usually detects the PDZ interaction between two partial proteins (C-terminal peptide and the PDZ domain), the coimmunoprecipitation assay usually involves complex formation between two full-length proteins. The interaction also occurs in a mammalian cellular environment where appropriate posttranslational modifications can occur. It should be noted, however, that the association of proteins in coimmunoprecipitation assays could be indirect and should therefore be confirmed by other complementary assays, such as the yeast two-hybrid assay.

3.2.1. Transfection

1. Transfect 60 to 70% confluent HEK293 T-cells in a 60-mm culture dish with mammalian expression plasmids using the calcium phosphate precipitation method. LipofectAMINE can also be used as an alternative reagent for transfection. Use the appropriate negative controls (*see* **Note 6**).

2. Three hours before the transfection, replace the culture medium with Dulbecco's modified Eagle's medium.
3. Dilute 3 µg of expression constructs (for co-transfections, 3 µg each per construct) in sterile water to a volume of 216 µL. Add 24 µL of 2.5 M $CaCl_2$. Add the DNA–$CaCl_2$ mixture drop by drop (approximately one drop per second) to 240 µL of 2X HEPES-buffered saline while gently vortexing.
4. Incubate the mixture at room temperature for 30 min to allow the DNA and $CaCl_2$ to aggregate.
5. Mix by vortexing and gently add 480 µL of the DNA–$CaCl_2$ aggregates to the cells (add a few drops at a time, and mix by swirling the plate).
6. Incubate the plate for 4 to 6 h at 37°C in a 5% CO_2 incubator.
7. After the incubation, replace culture media with fresh Dulbecco's modified Eagle's medium.

3.2.2. Preparation of Cell Lysates

1. Remove media, and wash HEK293 T-cells co-expressing a PDZ domain protein and its binding partner twice with 1 mL of cold PBS.
2. Add 400 µL of cold (4°C) PBST supplemented with protease inhibitors.
3. Scrape adherent cells off the plate with a rubber policeman and transfer the lysates to a new microcentrifuge tube.
4. Briefly sonicate the cells at low power, and rock the tubes for 30 min at 4°C.
5. Centrifuge the samples for 20 to 30 min at 15,000g in a microcentrifuge at 4°C.
6. Transfer the supernatant to a new tube. Typically, 90% of the cell lysate is used for immunoprecipitation assay, and 10% is set aside as the input sample for immunoblotting.

3.2.3. Immunoprecipitation

1. Dilute 200 µL of the supernatant with 300 µL of PBST containing protease inhibitors, add 2 µg of primary antibodies (*see* **Note 7**), and incubate the samples for 90 min at 4°C on a rocking platform.
2. Add 30 µL of buffer-equilibrated protein A-Sepharose (50% slurry), and incubate the samples for 90 min at 4°C. The use of resin-coupled primary antibodies usually gives better results than a soluble antibody followed by protein A-Sepharose.
3. Wash the resin three times with at least 10 times the bed volume of PBST supplemented with 1 mM EDTA and 1 mM EGTA (protease inhibitors not needed) and gently invert the tubes. Collect the resin by centrifugation for 10 s at approx 500g in a microcentrifuge. Be careful to avoid removing any of the resin while removing the supernatant.
4. After the final washing, remove the residual buffer in the bed by using a Pasteur pipet with a very fine tip.
5. Add 20 µL of 2X sodium dodecyl sulfate-polyacrylamide gel electrophoresis loading buffer to the resin and boil for 5 min.
6. Sediment the beads by centrifugation for 10 s at 500g in a microcentrifuge and remove the supernatant for further analysis.

7. Analyze the immunoprecipitates by Western blotting. Try to also analyze the input samples (*see* **step 6** in **Subheading 3.2.2.**) and immunoblot for both the PDZ domain protein and the potential associated proteins.

4. Notes

1. The problem of self-activation caused by the PDZ domain in pBHA can be solved by using baits containing additional regions flanking the PDZ domain. Alternatively, 3-AT, which lowers the biosynthesis of histidine by inhibiting an imidazole glycerolphosphate dehydratase *(10)*, can be used. Recommended 3-AT concentrations depend on the type of yeast strains (e.g., 1–5 mM for L40). Excess 3-AT may reduce the survival of yeast cells.

2. Although the PDZ domain is approx 90 amino acid residues, domain prediction programs often predict PDZ domains smaller than their actual size and, therefore, are non-functional in experiments. Whenever possible, the boundary of target PDZ domains should be determined by carefully comparing them to known PDZ domains. Furthermore, it is common to use approx 10 additional amino acid residues at both ends of the PDZ domain during their construction.

3. Point mutations in the C-terminal peptide or in the PDZ domain can be used to generate negative control constructs. Commonly mutated residues in the C-terminal peptides include the last hydrophobic residue and the residue at the –2 position (Ser/Thr for class I PDZ-binding peptides and hydrophobic residues for class II PDZ-binding peptides). In the PDZ domain, the carboxylate-binding loop (R/K-XXX-GLGF), which is required for binding to the C-terminal carboxylate oxygens of peptides, is highly conserved and point mutated in negative controls *(11,12)*. In addition to these point mutations, irrelevant PDZ domains are ideal as additional negative controls to demonstrate the specificity of the PDZ interaction under investigation.

4. Negative results in the yeast two-hybrid assay do not necessarily mean that the two proteins do not interact because the fusion proteins expressed in yeast cells may not be functional. A way to test whether they are functional is to use positive controls, such as a known binding partner. Alternatively, the problem may be solved by reciprocal transfer of the inserts, that is, by moving an insert from pBHA to the pGAD10 vector. This swapping of inserts might improve improper folding and steric hindrance of the fusion proteins.

5. Additional features should be considered when studying PDZ interactions. Sometimes the PDZ domain by itself is not sufficient for the interaction with the peptide and needs the presence of adjacent regions/domains (often another PDZ domain), which may be attributable to a need for structural stabilization, as shown in the GRIP and syntenin PDZ proteins *(11,13,14)*. In addition, phosphorylation either in the PDZ-binding peptides or in PDZ domains regulates their interaction *(1,2,15–19)*.

6. A commonly used negative control in the coimmunoprecipitation assay is to use lysates from cells transfected with only a single construct. More relevant negative controls for PDZ coimmunoprecipitation are, as in the yeast two-hybrid assay, point mutations in the C-terminal PDZ-binding peptide or in the PDZ

domain. Other commonly used negative controls include the deletion of the last three or four residues of the C-terminal peptide or the deletion of the whole PDZ domain. Sometimes these mutations or deletions do not disrupt the interaction. This suggests that the protein association under investigation is mediated by domains or motifs other than the PDZ–peptide interaction, as in the interaction between the Shank PDZ protein and the bPIX guanine nucleotide exchange factor (*9*).

7. Because the PDZ domain interaction involves the C-terminal peptide in the binding protein, a protein that is epitope-tagged at its C-terminus cannot be used for PDZ coimmunoprecipitation analysis. For the same reason, primary antibodies that are raised against the C-terminal peptide may not efficiently precipitate the complex. In this case, coimmunoprecipitation should be attempted in both orientations (i.e., immunoprecipitations using antibodies to both the PDZ protein and the putative PDZ-binding protein).

References

1. Kim, E. and Sheng, M. (2004) PDZ domain proteins of synapses (review). *Nat. Rev. Neurosci.* **5,** 771–781.
2. Sheng, M. and Sala, C. (2001) PDZ domains and the organization of supramolecular complexes. *Annu. Rev. Neurosci.* **24,** 1–29.
3. Montgomery, J. M., Zamorano, P. L., and Garner, C. C. (2004) MAGUKs in synapse assembly and function: an emerging view. *Cell Mol. Life Sci.* **61,** 911–929.
4. Scannevin, R. H. and Huganir, R. L. (2000) Postsynaptic organization and regulation of excitatory synapses. *Nat. Rev. Neurosci.* **1,** 133–141.
5. McGee, A. W. and Bredt, D. S. (2003) Assembly and plasticity of the glutamatergic postsynaptic specialization. *Curr. Opin. Neurobiol.* **13,** 111–118.
6. Li, Z. and Sheng, M. (2003) Some assembly required: the development of neuronal synapses. *Nat. Rev. Mol. Cell Biol.* **4,** 833–841.
7. Kennedy, M. B. (2000) Signal-processing machines at the postsynaptic density. *Science* **290,** 750–754.
8. Kim, E., Niethammer, M., Rothschild, A., Jan, Y. N., and Sheng, M. (1995) Clustering of Shaker-type K+ channels by interaction with a family of membrane-associated guanylate kinases. *Nature* **378,** 85–88.
9. Park, E., Na, M., Choi, J., et al. (2003) The Shank family of postsynaptic density proteins interacts with and promotes synaptic accumulation of the beta PIX guanine nucleotide exchange factor for Rac1 and Cdc42. *J. Biol. Chem.* 278, 19,220–19,229.
10. Braus, G. H., Grundmann, O., Bruckner, S., and Mosch, H. U. (2003) Amino acid starvation and Gcn4p regulate adhesive growth and FLO11 gene expression in *Saccharomyces cerevisiae. Mol. Biol. Cell* **14,** 4272–4284.
11. Koroll, M., Rathjen, F. G., and Volkmer, H. (2001) The neural cell recognition molecule neurofascin interacts with syntenin-1 but not with syntenin-2, both of which reveal self-associating activity. *J. Biol. Chem.* **276,** 10,646–10,654.
12. Grootjans, J. J., Reekmans, G., Ceulemans, H., and David, G. (2000) Syntenin-syndecan binding requires syndecan-synteny and the co-operation of both PDZ domains of syntenin. *J. Biol. Chem.* **275,** 19,933–19,941.

13. Feng, W., Shi, Y., Li, M., and Zhang, M. (2003) Tandem PDZ repeats in glutamate receptor-interacting proteins have a novel mode of PDZ domain-mediated target binding. *Nat. Struct. Biol.* **10,** 972–978.

14. Zhang, Q., Fan, J. S., and Zhang, M. (2001) Interdomain chaperoning between PSD-95, Dlg, and Zo-1 (PDZ) domains of glutamate receptor-interacting proteins. *J. Biol. Chem.* **276,** 43,216–43,220.

15. Cohen, N. A., Brenman, J. E., Snyder, S. H., and Bredt, D. S. (1996) Binding of the inward rectifier K+ channel Kir 2.3 to PSD-95 is regulated by protein kinase A phosphorylation. *Neuron* **17,** 759–767.

16. Choi, J., Ko, J., Park, E., et al. (2002) Phosphorylation of stargazin by protein kinase A regulates its interaction with PSD-95. *J. Biol. Chem.* **277,** 12,359–12,363.

17. Tanemoto, M., Fujita, A., Higashi, K., and Kurachi, Y. (2002) PSD-95 mediates formation of a functional homomeric Kir5.1 channel in the brain. *Neuron* **34,** 387–397.

18. Hu, L. A., Chen, W., Premont, R. T., Cong, M., and Lefkowitz, R. J. (2002) G protein-coupled receptor kinase 5 regulates beta 1-adrenergic receptor association with PSD-95. *J. Biol. Chem.* **277,** 1607–1613.

19. Gardoni, F., Mauceri, D., Fiorentini, C., Bellone, C., Missale, C., Cattabeni, F., et al. (2003) CaMKII-dependent phosphorylation regulates SAP97/NR2A interaction. *J. Biol. Chem.* **278,** 44,745–44,752.

14

Clustering Assay for Studying the Interaction of Membrane Proteins With PDZ Domain Proteins

Jaewon Ko and Eunjoon Kim

Summary

Some membrane proteins must be clustered at target sites to efficiently perform their functions. PDZ domain-containing scaffold proteins bind to the tails of target membrane proteins and promote their localization and clustering on the cell surface. This chapter describes the experimental details of the clustering assay, using the interaction between potassium channels and PSD-95, an abundant PDZ domain protein in neuronal synapses, as a model.

Key Words: Membrane protein; clustering; PSD-95; PDZ; immunocytochemistry.

1. Introduction

For proper function, membrane proteins, including receptors, ion channels, and cell adhesion molecules, must cluster on the membrane surface at their specific target sites. The clustering of membrane proteins is thought to be mediated by interaction of their cytoplasmic regions with scaffolding proteins.

PDZ (PSD-95/Dlg/ZO-1) domain-containing proteins are well-known examples of such scaffolding proteins. The PDZ domain is an approx 90-amino-acid module that mediates protein–protein interaction and is found in more than 400 proteins in the human and mouse *(1–6)*. It is a globular domain that contains a groove and hydrophobic pockets on its surface, through which it binds the C-terminus of its target proteins *(7)*.

One of the best-characterized PDZ domain-containing proteins is PSD-95, a key scaffolding protein at neuronal synapses *(1–6)*. The three PDZ domains of PSD-95 bind the C-termini of various membrane proteins, including potassium channels and *N*-methyl-D-aspartate receptors *(8–10)*. In addition to membrane proteins, PSD-95 interacts with a variety of signaling, scaffolding, and

From: *Methods in Molecular Biology, vol. 332: Transmembrane Signaling Protocols, Second Edition*
Edited by: H. Ali and B. Haribabu © Humana Press Inc., Totowa, NJ

cytoskeletal proteins *(11–14)*, contributing to the assembly of macromolecular protein complexes in excitatory neuronal synapses.

The clustering assay is an in vitro experimental system that can be used to analyze the interaction between membrane proteins and PDZ domain proteins. In the clustering assay, co-expressed membrane and PDZ domain proteins form clusters at the surface membrane, where they are co-localized. In contrast, when expressed alone, membrane and PDZ proteins do not form clusters and are usually diffusely distributed throughout the cell.

PSD-95 was first shown to cluster Kv1.4 potassium channels on the plasma membrane of heterologous cells *(8)*. Subsequently, PSD-95 has been shown to cluster a variety of other membrane proteins, including inward rectifier potassium channels, glutamate receptor subunits, α1-adrenergic receptors, stargazin (α-amino-3-hydroxy-5-methyl-4-isoxazolepropionic acid glutamate receptor-interacting membrane protein), and frizzled (receptor for Wnt proteins *[8,15–22]*). PICK1, another PDZ protein, has been shown to cluster membrane proteins, including ephrins and their receptors, GluR2, mGluR7 glutamate receptor subunits, UNC5H (receptor for netrin-1), and the monoamine plasma membrane transporter *(23–27)*. Other PDZ proteins shown to cluster membrane proteins include GRIP, Shank/ProSAP, and S-SCAM *(28–30)*. Thus, the interaction of membrane proteins with PDZ proteins appears to be a mechanism by which membrane proteins are clustered at their target membranes.

Although clustering of membrane proteins is an interesting phenomenon, its underlying mechanism and physiological significance are still largely elusive. In the case of the well-known clustering of potassium channels by PSD-95, important molecular mechanisms underlying the clustering include the direct interaction between the potassium channel and PSD-95, the multimeric nature of both potassium channels and PSD-95, the palmitoylation of PSD-95, and the intramolecular interaction between the Src homology (SH)3 and guanylate kinase domains of PSD-95 *(31–36)*. A current model for the clustering of potassium channels by PSD-95 is that the clusters represent two-dimensional lattice-like structures in which multimeric channels and PSD-95 form a macromolecular complex. Unlike PSD-95, SAP97, a member of the PSD-95 family, has been shown to cause the formation of intracellular clusters of potassium channels in heterologous cells *(35,37)*, suggesting that clustering might have functions at sites other than the plasma membrane.

Although further studies must be performed to understand the mechanisms and roles of the clustering phenomenon, the clustering assay, together with other in vitro methods, can be used to help investigate the interaction of membrane proteins with their cytosolic scaffolds. In this chapter, we describe experimental details of the clustering assay using the interaction between the Kv1.4 potassium channel and PSD-95 in heterologous cells as a model.

2. Materials

2.1. Cell Culture and Transfection

1. COS-7 cells or other heterologous cells (American Type Culture Collection, Rockville, MD; *see* **Note 1**).
2. Dulbecco's modified Eagle's medium (DMEM; Invitrogen, Life Technologies, Rockville, MD).
3. Heat-inactivated fetal bovine serum (Invitrogen, Life Technologies).
4. Gentamicin (Invitrogen, Life Technologies).
5. LipofectAMINE (Invitrogen, Life Technologies).
6. Opti-MEM I reduced serum medium (Invitrogen, Life Technologies).
7. Sterile round microscope glass cover slips (13 or 18 mm diameter; Fisher Scientific, Pittsburgh, PA).
8. 12-Well tissue culture plates (Nunc, Kamstrupvej, Danmark).
9. Microscope slides (25 × 75 × 1 mm; Fisher Scientific, Pittburgh, PA).
10. Poly-D-lysine (Sigma, St. Louis, MO).
11. Porcelain rack (Thomas Scientific, Swedesboro, NJ).
12. Glass chamber for nitric acid washing.
13. Nitric acid (Junsei Company, Japan).

2.2. Immunocytochemistry

1. Phosphate-buffered saline (PBS): 135 mM NaCl, 4.5 mM Na$_2$HPO$_4$, 1.5 mM KH$_2$PO$_4$, pH 7.4.
2. Paraformaldehyde (Sigma).
3. 100% Cold (−20°C) methanol (Merck, Darmstadt, Germany; absolute grade).
4. Triton X-100 (Amresco, Santa Cruz, CA).
5. Blocking buffer: 3% horse serum, 0.1% crystalline-grade bovine serum albumin in PBS.
6. Store at 4°C or −20°C (for long-term storage).
7. Vectashield mounting solution (Vector Laboratories, Burlingame, CA).
8. Primary antibodies against the proteins expressed in heterologous cells. Store in aliquots at 4°C or −70°C (for long-term storage).
10. Secondary antibodies conjugated to fluorescent dyes, such as Cy3 (red) or fluorescein isothiocyanate (FITC) (green; Jackson Research Laboratory, West Grove, PA).

2.3. Fluorescence Microscopy

Confocal laser scanning or conventional fluorescence microscope with appropriate detection filters and water or oil objectives with high numerical apertures (×40 or ×63 for most applications).

3. Methods

3.1. Cleaning Cover Slips

1. Place round cover slips in a porcelain rack and, in a hood, submerge the rack for 36 to 48 h in a glass chamber containing nitric acid.

2. Rinse the cover slips with MilliQ water twice for 1 h and twice more for 30 min.
3. Cover the rack containing the washed cover slips with aluminum foil.
4. Bake the rack in a furnace for 6 h at 225°C.

3.2. Coating Cover Slips

1. Place round cover slips in 12-well plates (one cover slip in each well).
2. Place and spread approx 100 µL (or enough to cover the surface) of poly-D-lysine (1 µg/mL in H₂O) on the surface of each cover slip.
3. Close the lid of the 12-well plate and incubate the plate overnight at room temperature.
4. Rinse the cover slips three times with sterile water.

3.3. Transfecting Cells

1. Spread cells onto coated cover slips in a 12-well plate at low density (15 to 20% confluence).
2. Incubate the plate overnight in a 5% CO_2 incubator at 37°C. This step will allow the attachment of the cells to the surface of coated cover slips.
3. Transfect the cells with mammalian expression plasmids using LipofectAMINE according to the manufacturer's protocols (*see* **Note 2**). Briefly mix 2.4 µL of LipofectAMINE with 40 µL of Opti-MEM. Meanwhile, incubate 0.4 µg of expression constructs (for co-transfections, 0.2 µg each) in 40 µL of Opti-MEM. Mix the DNA with LipofectAMINE, and incubate for 30 min at room temperature.
4. Add 320 µL of Opti-MEM to the DNA + lipid mixture (total volume 400 µL).
5. Wash the cells once with Opti-MEM.
6. Place all of the of the final mixture (400 µL) onto each well and incubate the plate for 4 to 6 h in a 5% CO_2 incubator at 37°C.
7. After the incubation, replace Opti-MEM with 1 mL of prewarmed (37°C) DMEM supplemented with 5% fetal bovine serum.
8. After 24 h of incubation in a 5% CO_2 incubator at 37°C, examine the cells with a microscope. If cell debris is observed, exchange the old media with fresh, prewarmed (37°C) DMEM.

3.4. Immunocytochemistry

1. Forty-eight hours after transfection, wash the cover slips three times for 5 min each with PBS (with gentle shaking). This step and the following steps in this section can be performed in new 12-well plates.
2. Fix the cells for 5 min with 4% paraformaldehyde in PBS (*see* **Note 3**) or with cold methanol (–20°C; *see* **Note 4**).
3. Wash the cells three times for 5 min each with PBS.
4. Permeabilize the cells for 2 min with 0.1% Triton X-100 in PBS.
5. Wash the cells three times for 5 min each with PBS.
6. Incubate the cells for 30 min with blocking buffer at room temperature with gentle shaking.

7. Incubate the cells for 1 h at room temperature with relevant primary antibodies in blocking buffer. The primary antibody concentration and incubation time can be adjusted if necessary (*see* **Note 5**).
8. Wash the cells three times for 10 min each with PBS.
9. Incubate the cells for 30 min at room temperature with fluorescence dye-conjugated secondary antibodies (Cy3, Cy5, or FITC) in blocking buffer. The incubation time can be extended to 1 h if necessary.
10. Wash the cells three times for 10 min each with PBS.
11. Place approx 50 μL of Vectashield solution on a microscope slide, and carefully place the stained cover slip face (cell-attached side down) onto the wetted area of the slide. Try to avoid forming bubbles.

3.5. Image Acquisition and Data Analysis

Fluorescence images are acquired using a confocal or conventional fluorescence microscope usually using ×40 or ×63 water or oil objectives. Acquire the images of co-transfected cells using both channels (e.g., Cy3 and FITC channels; **Fig. 1**; *see* **Note 6**). Co-clustering efficiency is quantified by measuring the number of cells showing co-clusters out of the total number of co-transfected cells.

4. Notes

1. COS-7 cells are flatter than other heterologous cells, such as HEK293T-cells. Because co-clusters are more easily observed in flatter cells than in other less flat cells, COS-7 cells are preferred for the co-clustering assay.
2. In our experience, high expression levels of transfected proteins are important for successful co-clustering experiments. Thus, we recommend the use of mammalian expression vectors with strong promoters, such as cytomegalovirus. Single transfections are useful negative controls for the clustering assay. In addition, mutant proteins that cannot interact with PDZ domain proteins, including point mutants in the C-terminus or deletion mutants lacking the last three or four amino acid residues, can also be used as negative controls.
3. High-quality paraformaldehydes are recommended for immunocytochemistry experiments.
4. Methanol often is preferred for the fixation of cytoskeletal proteins. When methanol is used as a fixative, permeabilization can be omitted because methanol also permeabilizes the cells.The use of high-quality methanol is recommended.
5. The standard concentration for primary antibodies is 1 μg/mL. If low expression levels of the transfected proteins are expected, use a higher concentration of primary antibodies and an extended incubation time (as long as 2 h).
6. The presence of co-clusters on the surface of cells can be supported by Z-stack optical sectioning of the cells in confocal laser scanning microscopy (**Fig. 1**).

Kv1.4 WT (ETDV)   PSD-95   Kv1.4 mut (ETDA)   PSD-95

Kv1.4 WT (ETDV)   SAP97   Kv1.4 mut (ETDA)   SAP97

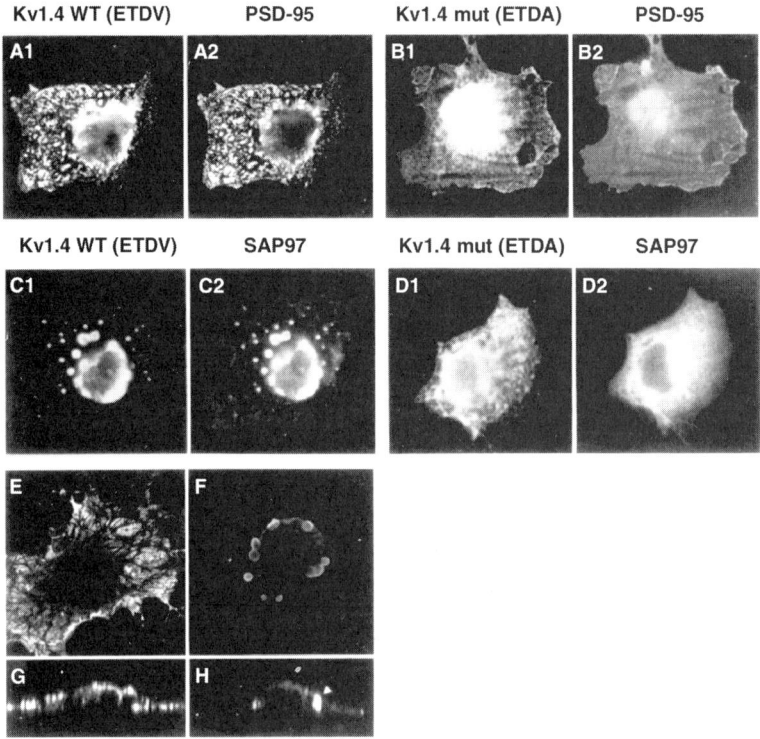

Fig. 1. Different patterns of Kv1.4 co-clustering with PSD-95 and SAP97. COS-7 cells were co-transfected as follows: (A,E,G) wild-type Kv1.4 and PSD-95; (B) Kv1.4 C-terminal -ETDA mutant and PSD-95; (C,F,H) wild-type Kv1.4 and SAP97; (D) Kv1.4 C (-ETDA) mutant and SAP97. Each pair of pictures (A1,A2–D1,D2) represents the same co-transfected cell visualized through different immunofluorescence filter channels. As indicated, the left half of each pair (A1,B1,C1,D1) shows the distribution of the Kv1.4 or Kv1.4 mutant, visualized with Cy3-labeled secondary antibodies; the right half (A2,B2,C2,D2) shows the distribution of PSD-95 or SAP97, labeled with fluorescein isothiocyanatesecondary antibodies. Only the wild-type Kv1.4 shows co-clustering with PSD-95 and SAP97. In cells co-transfected with Kv1.4 (-ETDA) mutant and PSD-95 or SAP97, both proteins are diffusely distributed in a pattern similar to singly transfected cells. (E–H) Confocal microscope images show that PSD-95/Kv1.4 co-clusters are on, or very close to, the surface, whereas SAP97 forms coaggregates with Kv1.4 that are intracellular and concentrated in the perinuclear region. (E) and (F) are horizontal confocal sections and (G) and (H) are vertical sections (only Kv1.4 immunofluorescence is shown). Arrowhead in H indicates an intracellular aggregate of Kv1.4 in a cell co-transfected with SAP97. In (G), two layers of Kv1.4 clusters can be discerned toward the thicker middle part of the cell, presumably associated with adherent and nonadherent surfaces of the cell. This figure was reprinted from **Fig. 1** of **ref. 37**.

References

1. Kim, E. and Sheng, M. (2004) PDZ domain proteins of synapses. *Nat. Rev. Neurosci.* **10**, 771–781.
2. Sheng, M. and Sala, C. (2001) PDZ domains and the organization of supramolecular complexes. *Annu. Rev. Neurosci.* **24**, 1–29.
3. McGee, A. W. and Bredt, D. S. (2003) Assembly and plasticity of the glutamatergic postsynaptic specialization. *Curr. Opin. Neurobiol.* **13**, 111–118.
4. Montgomery, J. M., Zamorano, P. L., and Garner, C. C. (2004) MAGUKs in synapse assembly and function: an emerging view. *Cell Mol. Life Sci.* **61**, 911–929.
5. Garner, C. C., Nash, J., and Huganir, R. L. (2000) PDZ domains in synapse assembly and signalling. *Trends Cell Biol.* **10**, 274–280.
6. Kennedy, M. B. (2000) Signal-processing machines at the postsynaptic density. *Science* **290**, 750–754.
7. Doyle, D. A., Lee, A., Lewis, J., Kim, E., Sheng, M., and MacKinnon, R. (1996) Crystal structures of a complexed and peptide-free membrane protein- binding domain: molecular basis of peptide recognition by PDZ. *Cell* **85**, 1067–1076.
8. Kim, E., Niethammer, M., Rothschild, A., Jan, Y. N., and Sheng, M. (1995) Clustering of Shaker-type K+ channels by interaction with a family of membrane-associated guanylate kinases. *Nature* **378**, 85–88.
9. Kornau, H. C., Schenker, L. T., Kennedy, M. B., and Seeburg, P. H. (1995) Domain interaction between NMDA receptor subunits and the postsynaptic density protein PSD-95. *Science* **269**, 1737–1740.
10. Niethammer, M., Kim, E., and Sheng, M. (1996) Interaction between the C terminus of NMDA receptor subunits and multiple members of the PSD-95 family of membrane-associated guanylate kinases. *J. Neurosci.* **16**, 2157–2163.
11. Brenman, J. E., Chao, D. S., Gee, S. H., et al. (1996) Interaction of nitric oxide synthase with the postsynaptic density protein PSD-95 and alpha1-syntrophin mediated by PDZ domains. *Cell* **84**, 757–767.
12. Chen, H. J., Rojas-Soto, M., Oguni, A., and Kennedy, M. B. (1998) A synaptic Ras-GTPase activating protein (p135 SynGAP) inhibited by CaM kinase II. *Neuron* **20**, 895–904.
13. Kim, J. H., Liao, D., Lau, L. F., and Huganir, R. L. (1998) SynGAP: a synaptic RasGAP that associates with the PSD-95/SAP90 protein family. *Neuron* **20**, 683–691.
14. Kim, E., Naisbitt, S., Hsueh, Y. P., et al. (1997) GKAP, a novel synaptic protein that interacts with the guanylate kinase- like domain of the PSD-95/SAP90 family of channel clustering molecules. *J. Cell Biol.* **136**, 669–678.
15. Garcia, E. P., Mehta, S., Blair, L. A., et al. (1998) SAP90 binds and clusters kainate receptors causing incomplete desensitization. *Neuron* **21**, 727–739.
16. Roche, K. W., Ly, C. D., Petralia, R. S., et al. (1999) Postsynaptic Density-93 Interacts with the delta2 Glutamate Receptor Subunit at Parallel Fiber Synapses. *J. Neurosci.* **19**, 3926–3934.

17. Hu, L. A., Tang, Y., Miller, W. E., et al. (2000) Beta 1-adrenergic receptor association with PSD-95. Inhibition of receptor internalization and facilitation of beta 1-adrenergic receptor interaction with N-methyl-D-aspartate receptors. *J. Biol. Chem.* **275,** 38,659–38,666.

18. Wong, W., Newell, E. W., Jugloff, D. G., Jones, O. T., and Schlichter, L. C. (2002) Cell surface targeting and clustering interactions between heterologously expressed PSD-95 and the Shal voltage-gated potassium channel, Kv4.2. *J. Biol. Chem.* **277,** 20,423–20,430.

19. Nehring, R. B., Wischmeyer, E., Doring, F., Veh, R. W., Sheng, M., and Karschin, A. (2000) Neuronal inwardly rectifying K(+) channels differentially couple to PDZ proteins of the PSD-95/SAP90 family. *J. Neurosci.* **20,** 156–162.

20. Chen, L., Chetkovich, D. M., Petralia, R. S., et al. (2000) Stargazin regulates synaptic targeting of AMPA receptors by two distinct mechanisms. *Nature* **408,** 936–943.

21. Hering, H. and Sheng, M. (2002) Direct interaction of Frizzled-1, -2, -4, and -7 with PDZ domains of PSD-95. *FEBS Lett.* **521,** 185–189.

22. Kim, E., Cho, K. O., Rothschild, A., and Sheng, M. (1996) Heteromultimerization and NMDA receptor-clustering activity of Chapsyn-110, a member of the PSD-95 family of proteins. *Neuron* **17,** 103–113.

23. Xia, J., Zhang, X., Staudinger, J., and Huganir, R. L. (1999) Clustering of AMPA receptors by the synaptic PDZ domain-containing protein PICK1. *Neuron* **22,** 179–187.

24. Williams, M. E., Wu, S. C., McKenna, W. L., and Hinck, L. (2003) Surface expression of the netrin receptor UNC5H1 is regulated through a protein kinase C-interacting protein/protein kinase-dependent mechanism. *J. Neurosci.* **23,** 11,279–11,288.

25. Boudin, H., Doan, A., Xia, J., et al. (2000) Presynaptic clustering of mGluR7a requires the PICK1 PDZ domain binding site. *Neuron* **28,** 485–497.

26. Torres, G. E., Yao, W. D., Mohn, A. R., et al. (2001) Functional interaction between monoamine plasma membrane transporters and the synaptic PDZ domain-containing protein PICK1. *Neuron* **30,** 121–134.

27. Torres, R., Firestein, B. L., Dong, H., et al (1998) PDZ proteins bind, cluster, and synaptically colocalize with Eph receptors and their ephrin ligands. *Neuron* **21,** 1453–1463.

28. Hirao, K., Hata, Y., Yao, I., et al. (2000) Three isoforms of synaptic scaffolding molecule and their characterization. Multimerization between the isoforms and their interaction with N-methyl-D-aspartate receptors and SAP90/PSD-95-associated protein. *J. Biol. Chem.* **275,** 2966–2972.

29. Tobaben, S., Sudhof, T. C., and Stahl, B. (2000) The G protein-coupled receptor CL1 interacts directly with proteins of the Shank family. *J. Biol. Chem.* **275,** 36,204–36,210.

30. Bruckner, K., Pablo Labrador, J., Scheiffele, P., Herb, A., Seeburg, P. H., and Klein, R. (1999) EphrinB ligands recruit GRIP family PDZ adaptor proteins into raft membrane microdomains. *Neuron* **22,** 511–524.

31. Hsueh, Y. P., Kim, E., and Sheng, M. (1997) Disulfide-linked head-to-head multimerization in the mechanism of ion channel clustering by PSD-95. *Neuron* **18,** 803–814.

32. El-Husseini, A. E., Topinka, J. R., Lehrer-Graiwer, J. E., et al. (2000) Ion channel clustering by membrane-associated guanylate kinases. Differential regulation by N-terminal lipid and metal binding motifs. *J. Biol. Chem.* **275,** 23,904–23,910.

33. Christopherson, K. S., Sweeney, N. T., Craven, S. E., Kang, R., El-Husseini Ael, D., and Bredt, D. S. (2003) Lipid- and protein-mediated multimerization of PSD-95: implications for receptor clustering and assembly of synaptic protein networks. *J. Cell Sci.* **116,** 3213–3219.

34. Shin, H., Hsueh, Y. P., Yang, F. C., Kim, E., and Sheng, M. (2000) An intramolecular interaction between Src homology 3 domain and guanylate kinase-like domain required for channel clustering by postsynaptic density-95/SAP90. *J. Neurosci.* **20,** 3580–3587.

35. Tiffany, A. M., Manganas, L. N., Kim, E., Hsueh, Y. P., Sheng, M., and Trimmer, J. S. (2000) PSD-95 and SAP97 exhibit distinct mechanisms for regulating K(+) channel surface expression and clustering. *J. Cell Biol.* **148,** 147–158.

36. Hsueh, Y. P. and Sheng, M. (1999) Requirement of N-terminal cysteines of PSD-95 for PSD-95 multimerization and ternary complex formation, but not for binding to potassium channel Kv1.4. *J. Biol. Chem.* **274,** 532–536.

37. Kim, E. and Sheng, M. (1996) Differential K$^+$ channel clustering activity of PSD-95 and SAP97, two related membrane-associated putative guanylate kinases. *Neuropharmacology* **35,** 993–1000.

II

SPECIFIC TOPICS

D. Rho Guanosine Triphosphatases and Reconstitution of Signaling Complexes

15

Mammalian Cell Microinjection Assay to Study the Function of Rho Family Guanosine Triphosphatases

Ritu Garg and Anne J. Ridley

Summary

Microinjection is an excellent technique for studying the acute responses of cells to proteins and can be used to investigate the effects of mutations in proteins on their activity. It has been used widely to study the responses to Rho family guanosine triphosphatases and is particularly useful for cell types that are difficult to transfect. Here, we describe the procedure for microinjecting cells with purified recombinant proteins or with expression vectors encoding proteins, and for analyzing the cells after injection.

Key Words: Rho GTPase; microinjection; actin cytoskeleton; growth factors; fibroblasts; MDCK cells.

1. Introduction

Microinjection has been used widely as a technique to introduce proteins and DNA into mammalian cells. A major advantage of microinjection over transfection approaches is that it is possible to analyze very early responses to proteins: responses to microinjected proteins can be detected within minutes, and expression of proteins encoded by microinjected DNA can often be detected within 2 h. In addition, most cells, including primary cells, are microinjectable, whereas many cell types are not readily transfectable. Analysis of responses in microinjected cells usually is based on immunocytochemical approaches because, in general, it is not possible to inject sufficient numbers of cells to conduct biochemical studies. In some cases, however, microinjection has been used to analyze changes in protein phosphorylation, for example, after the injection of fibroblasts with cyclic adenosine monophosphate (cAMP)-dependent protein kinase (1).

From: *Methods in Molecular Biology, vol. 332: Transmembrane Signaling Protocols, Second Edition*
Edited by: H. Ali and B. Haribabu © Humana Press Inc., Totowa, NJ

Microinjection approaches have been important in defining the early responses of cells to a number of small Ras-related guanosine triphosphate (GTP)-binding proteins. Injection of recombinant Ras protein showed that it stimulated DNA synthesis, morphological transformation, and membrane ruffling *(2,3)*. Injection studies have been used widely to establish the roles of members of the Rho family of Ras-related proteins in regulating actin organization. By microinjecting recombinant proteins, RhoA was shown to regulate actin stress fiber formation, whereas Rac1 regulates membrane ruffling and the formation of lamellipodia, and Cdc42 regulates filopodium formation *(4–8)*. In addition, microinjection of Ras, Rac, and Rho proteins into Madin-Darby canine kidney (MDCK) epithelial cells has shown that Ras and Rac are required for motility responses of these cells to hepatocyte growth factor/scatter factor *(9)*.

Although recombinant proteins provide an ideal approach for investigating cytoskeletal reorganization induced by Rho GTPases, there are situations in which this is not possible. Some Rho GTPases cannot be purified in adequate quantities from *Escherichia coli*, and alternative approaches, such as purification from insect cells, are very long and labor intensive. Many investigators have instead introduced expression vectors encoding Rho GTPases into cells *(8,10–12)*. DNA can be introduced into cells by many different methods, including calcium phosphate-mediated transfection, lipofection, diethylaminoethyl (DEAE)-dextran-mediated transfection, and electroporation. Microinjection has the advantage that changes to the cytoskeleton can be analyzed early after protein expression, without the accumulation of long-term changes in gene expression that may indirectly affect cell morphology. Microinjection of DNA also provides a rapid means of assessing the localization of Rho GTPases in cells *(10,11)*, which is not possible by protein injections. By expressing a protein tagged with an epitope, it is possible to follow its localization independently of endogenous proteins *(11)*.

Here, the method used to microinject recombinant Rho proteins and DNA expression vectors encoding Rho proteins into Swiss 3T3 cells and MDCK cells is presented (*see* **Note 1**). Many cell types have been used to study Rho protein function, including macrophages, many cancer cell lines, and endothelial cells. Generally, we find that it is useful to compare responses in at least two different cell types to get a good overview of a protein's function. The microinjection technique was initially described in detail by Graessmann and Graessmann *(13)*. Protein or DNA solution is loaded into glass pipets, which have been pulled to a fine point at one end of approx 0.5- to 1-μm diameter. A micromanipulator is used to position the point of the glass pipet very close to the cells to be injected. The other end of the pipet is attached via tubing to a pressure regulator. Air pressure applied to this end of the pipet forces the solution out of the pointed end of the pipet. The pipet is manipulated so that it transiently pierces the plasma membrane of a cell, allowing the solution in the pipet to enter the cell (*see* **Note 2**). The pipet remains within the cell for only a very short period (<0.5 s) and then is removed, allowing the membrane to reseal. The volume of solution introduced into cells is between 5 and 10% of their total volume, or approx 10^{-14} L.

In microinjection experiments with Rho GTPases, DNA, or recombinant protein is injected into 100 to 150 cells over a period of 10 to 20 min. The cells subsequently are incubated for varying lengths of time with or without addition of growth factors, then fixed, permeabilized, and stained to show injected cells together with either phalloidin to show actin filaments, or with various other antibodies to detect, for example, focal adhesion proteins.

2. Materials

2.1. Cell Culture (see Note 3)

1. Dulbecco's modified Eagle's medium (DMEM) containing 0.11 g/L sodium pyruvate, 4.5 g/L glucose can be purchased from Life Sciences. Antibiotics are stored in aliquots at –20°C and added to a final concentration of 100 U/mL penicillin and 100 μg/mL streptomycin. Medium is stored at 4°C.
2. Fetal calf serum (FCS) is batch tested and selected from various sources (*see* **Note 4**). It is stored in 50-mL aliquots at –20°C.
3. Phosphate-buffered saline (PBS)-A is 137 mM NaCl, 2.7 mM KCl, 8.1 mM Na_2PO_4, 1.47 mM KH_2PO_4.
4. 13-mm Diameter glass cover slips (Chance Propper, No. 1½) are cleaned by washing first with nitric acid, then extensively with distilled water, and finally with ethanol. They are then baked before use.

2.2. Microinjection

1. Goat, rat, or rabbit immunoglobulin (Ig)G (10 mg/mL) is stored at 4°C, and can be obtained from Pierce. The choice of species depends on which antibodies are being used to stain cells after microinjection.
2. Protein injection buffers:
   a. 10 mM Tris-HCl, pH 7.5, 150 mM NaCl, 5 mM $MgCl_2$.
   b. 20 mM HEPES-Cl, pH 7.2, 100 mM KCl, 5 mM $MgCl_2$.Buffer a usually is used for recombinant Rho GTPases because it is compatible with biochemical assays carried out with these proteins. Buffer b is closer in composition to the cytoplasm of cells and, therefore, is preferable where cells or responses monitored may be highly sensitive to changes in, for example, levels of sodium. A variety of different microinjection buffer compositions has been used by others and are documented in the literature.
3. DNA injection buffers:
   a. PBS-A (*see* **Subheading 2.1.**).
   b. 10 mM Tris-HCl, pH 7.5, 150 mM NaCl. Mg^{2+} ions are omitted from DNA injection buffers because DNases require Mg^{2+}.

4. The programmable pipet puller (model no. 773) used is obtained from Campden Instruments.
5. Glass pipets are 1.2-mm bore; they can be obtained from Clark Electroinstruments, Reading, UK.
6. The microinjection station consists of an inverted phase-contrast microscope fitted with a heated stage and an enclosed Perspex chamber. The temperature and CO_2 concentration in the chamber are maintained by the temperature regulator TRZ3700 and CTI controller 3700, obtained from Zeiss. Humidity is provided by placing a Perspex dish containing sterile distilled water in the chamber. Cells are injected using an Eppendorf microinjector (model no. 5246) and micromanipulator (model no. 5171).

2.3. Fixing and Staining Cells

1. Formaldehyde can be obtained from BDH as a 40% solution containing 9 to 11% methanol. It is toxic by inhalation, therefore to minimize exposure, fixing cells are placed in a fume cupboard and formaldehyde is disposed of in the fume cupboard outlet. Dilute fresh 1:10 (v/v) in PBS immediately before use. Formaldehyde fixation is adequate for analyzing F-actin using phalloidin.
2. 6% (w/v) paraformaldehyde is prepared from solid paraformaldehyde by dissolving an appropriate amount (approx 1 g) in PBS-A with heating in a fume cupboard. It dissolves at approx 80°C. Dilute 1:2 in PBS (with Ca^{2+} and Mg^{2+}) to give final 3% solution. Paraformaldehyde is best when made fresh, but can be stored at –20°C. Once thawed, do not refreeze. Paraformaldehyde is used for most antibody staining.
3. PBS is PBS-A (see **Subheading 2.1., step 3**) containing 0.9 mM $CaCl_2$ and 0.5 mM $MgCl_2$. It can be obtained from Gibco as a 10X stock solution and diluted with sterile distilled water.
4. Blocking solution (PBS/bovine serum albumin [BSA]) is PBS containing 1% BSA. It can be stored for several weeks at 4°C provided 0.02% azide is added to prevent the growth of micro-organisms. For some cell types or antibodies other blocking solutions are needed. For example, for macrophages, we generally use 10% goat serum to block IgG receptors on the macrophage surface.
5. 0.2% Triton X-100/PBS. A stock solution of 10% Triton X-100 is used.
6. Tetramethylrhodamine isothiocyanate (TRITC)-labeled phalloidin (Sigma) is toxic in high quantities, but not at the levels used here for staining cells. Dissolve in sterile distilled water at a concentration of 50 μg/mL and store in small aliquots at –20°C in a light-sealed container.
7. Fluorescein isothiocyanate-labeled anti-rat/rabbit/goat IgG can be obtained from Jackson Research Laboratories.
8. Mountant: 0.1% p-phenylenediamine (antiquench), 10% (w/v) Mowiol (obtained from Calbiochem), 25% (w/v) glycerol, 100 mM Tris-HCl pH 8.5. Mowiol is stored at 4°C without p-phenylenediamine, but once the p-phenylenediamine is added, it is stored in 100-μL aliquots at –70°C. Once thawed, these aliquots can be kept wrapped in aluminium foil at –20°C for approx 1 wk; however, the p-phenylenediamine is sensitive to light and temperature.
9. 1- To 1.2-mm-thick glass slides can be obtained from Chance Propper.

3. Methods

3.1. Preparation of Cells for Microinjection

1. Grow Swiss 3T3 cells (*see* **Note 3**) and MDCK cells in DMEM containing 10% FCS in a humidified incubator at 37°C with 10% (v/v) CO_2.

2. Passage Swiss 3T3 cells and MDCK cells every 3 to 4 d by washing with PBS-A, then incubating for 2 to 3 min with 0.05% trypsin, 0.02% ethylenediaminetetraacetic acid (EDTA). Swiss 3T3 cells are seeded in 80-cm^2 flasks at a density of 3×10^5 cells per flask. MDCK cells are seeded in 25-cm^2 flasks at a density of 1 to 2×10^5 cells per flask.

3. Prepare cover slips by drawing a cross with a diamond-tipped marker pen. This facilitates localization of injected cells. Sterilize by dipping in 100% ethanol and flaming. Place in 18-mm diameter wells in four-well dishes.

4. For microinjection of Swiss 3T3 cells, seed at a density of 3×10^4 per 18-mm well. At this density, Swiss 3T3 cells reach confluence in 3 d. After 5 to 7 d, remove medium and replace with DMEM (no FCS) for approx 16 h. Transfer each cover slip to a separate 35-mm dish containing 2 mL of DMEM, using fine forceps and a 21-gage needle, bent at the end to facilitate lifting the cover slip.

5. For microinjection of subconfluent MDCK cells, seed at a density of 10^4 cells per well. They are microinjected 3 d after seeding, when the majority of cells are in colonies of between 16 and 80 cells. Alternatively, to analyze confluent cells, they are seeded at a greater density and injected 4 to 5 d after seeding. Transfer each cover slip to a 35-mm dishes containg 2 mL of DMEM/5% FCS approx 1 h before microinjection.

6. Keep cells in an incubator close to the microinjector to minimize changes in temperature and medium pH during transfer to and from the microinjector.

3.2. Injection of Proteins

1. The methods for purifying recombinant proteins for microinjection have been previously described in detail *(6,14)*. Proteins are expressed as glutathione *S*-transferase fusion proteins in *E. coli*. In general, from a 1-L culture of *E. coli*, approx 100 µL of concentrated protein is obtained. Proteins are stored in 10-µL aliquots in liquid nitrogen, and the activity of each protein preparation is determined by GTP-/guanosine diphosphate-binding assay after thawing an aliquot *(14)*.

2. Thaw protein aliquots on ice. After thawing, the protein can be used for several days provided it is kept at 4°C (*see* **Note 5**). It should not be refrozen because freezing results in loss of activity.

3. Turn on the temperature regulator and CO_2 controller at least 20 min before beginning microinjection to allow the temperature to reach 37°C and CO_2 levels to reach 10%.

4. Pull pipets on a pipet puller according to the manufacturer's instructions (*see* **Note 6**). Pipets can be stored by pressing the middle of each pipet onto a strip of Blu-Tak adhesive, in a 150-mm diameter plastic dish with a lid.

5. Dilute rat IgG to 1 mg/mL in protein injection buffer (*see* **Note 5**). Centrifuge the proteins, protein injection buffer, and diluted rat IgG for 5 min at 4°C, 13,000*g*,

to pellet small particles that will block up the microinjection pipets. Mix proteins, buffer, and Rat IgG in sterile 600-μL microfuge tubes to give the required concentrations of proteins and final concentration of 0.5 mg/mL rat IgG. Store proteins on ice until adding to the microinjection needle. Proteins are normally injected at concentrations between 5 and 500 μg/mL.

6. If inhibitors (e.g., kinase inhibitors) are to be tested for their effects on the response of cells to a protein, add them to the cells before microinjection and leave the cells in the incubator for the appropriate length of time needed for the inhibitor to enter cells (*see* **Note 7**).

7. Take a dish containing cells on a cover slip from the incubator. Gently press down the cover slip at the edge onto the dish with a yellow tip, to exclude air bubbles and prevent the cover slip moving during microinjection. Place the dish on the microscope stage and localize the etched cross using a low power objective.

8. Load approx 1 μL of protein or DNA solution into an Eppendorf microloader tip and then load this into a glass pipet. Care should be taken to ensure that bubbles are not present in the solution in the pipet.

9. Insert the pipet into the holder, then using the joystick, move it to the center of the cover slip, looking from above the stage. Subsequently, looking down the microscope, bring the pipet down so that it is nearly in focus above the cells. A bright spot, representing the meniscus, should appear first. On higher power, again bring the pipet to be nearly in focus but just above the surface of the cells.

10. Cells normally are injected in manual mode (*see* **Note 2**) using a ×32/0.4 NA objective and ×10 eyepieces. Clear the pipet at high pressure (approx 7000 hPa) briefly (<5 s) before injecting cells at a working pressure of around 1000 hPa (*see* **Note 5**). If the aim is to inject all cells in a given area, use the photoframe as a guide to work round all the cells in view. Between 100 and 150 cells are normally injected over the course of 10 to 20 min, then the dish is returned to the incubator.

11. To determine the effects of injected proteins on growth factor responses, add growth factors to the medium 15 to 30 min after finishing injections, and mix gently.

3.3. Injection of DNA

Optimal expression of DNA in cells depends on the enhancer/promotor in the expression vector. Generally, simian virus 40- or cytomegalovirus-based promotor/enhancer systems work well in most, although not all, cell types. In quiescent cells or primary cells, it is worth experimenting with different expression vectors to find one that expresses efficiently in the cells of interest, and then subclone all complementary DNAs into this vector.

1. Purification of DNA. For microinjection, we have found that it is important to have highly purified DNA, preferably purified by $CsCl_2$ density gradient centrifugation, to minimize the levels of toxic contaminants from *E. coli* that affect expression of plasmid-encoded genes, such that far fewer injected cells effectively express the protein of interest. DNA purified with a commercially avail-

able plasmid purification kit (for example, those supplied by Qiagen), using endotoxin-free reagents can also provide high-quality DNA. With highly purified DNA, the percentage of microinjected cells that express protein from the microinjected plasmid can approach 100%.

2. Prepare cells for microinjection as described for protein injection (*see* **Subheading 3.1.**).

3. Immediately before injection, dilute DNA to an appropriate concentration (1 to 100 ng/µL) in DNA microinjection buffer. The concentration of DNA required for microinjection can vary greatly depending on the vector, insert, cell type, and conditions. To minimize toxic side effects owing to overexpression of a protein, the DNA should be titrated down to find the lowest level which still gives detectable, high-efficiency protein expression.

4. Centrifuge the diluted DNA at 13,000*g* for 5 min at 4°C, then carefully transfer approx 5 µL to another microfuge tube (*see* **Note 5**).

5. Load DNA into a micropipet and inject (*see* **Subheading 3.2.**) into the nucleus of cells.

6. To determine the effects of expressed proteins on growth factor responses, add growth factors between 3 and 4 h after the injection.

7. The time after DNA injection at which cells are analyzed should be kept to a minimum because increased accumulation of protein over time can be toxic (*see* **Note 8**) With Rho GTPases, protein expression and cellular responses can be detected within two hours of microinjecting expression vectors. If left for 16 h, however, most injected cells die. With some larger proteins, it may be necessary to incubate cells for longer than 2 h, but in our experience 4 h is always sufficient.

3.4. Fixing and Staining Cells

1. For analysis of responses to proteins, fix cells at time points after injection ranging from 5 min to 24 h (*see* **Note 9**). For DNA, fix cells 1 to 24 h after injection. At the appropriate time point, wash the cells with PBS (*see* **Note 10**), then fix in 4% paraformaldehyde/PBS or 3% paraformaldehyde/PBS for at least 15 min. Cells can be left in fixing solution for up to 2 h without detrimental effects on the staining with phalloidin (*see* **Note 11**).

2. After fixation, transfer cover slips to 18-mm diameter wells containing PBS, then wash six times with PBS. An optional incubation step with 50 m*M* ammonium chloride in PBS (10 min) can be included to quench residual formaldehyde.

3. Permeabilize for 5 min with 0.2% Triton X-100 in PBS, then wash two times with PBS.

4. To stain for protein-injected cells containing IgG and for actin filaments, incubate each cover slip with 200 µL of a 1:400 dilution of fluorescein isothiocyanate-labeled anti-rat/rabbit/goat IgG together with 0.1 µg/mL TRITC-phalloidin in PBS for 30 to 60 min. During this incubation, place dishes on a rocker at low speed. When incubating with TRITC-phalloidin, keep the dishes in the dark by covering with aluminum foil (*see* **Note 12**).

5. To stain for DNA-injected cells, incubate cover slips in blocking solution for 30 min, then with primary antibodies to the protein expressed by the DNA (usually epitope-tagged with myc, HA, or FLAG epitopes) in PBS for 30 to 60 min at room temperature. Wash cover slips six times with PBS. This is followed by incubation with fluorophore-labeled secondary antibodies and TRITC-phalloidin (*see* **step 4**).

6. To stain with primary antibodies (e.g., to focal adhesion proteins, such as vinculin) where antibody stocks are limiting, remove cover slips from the wells using fine forceps and a 21-gage needle bent at the end. Immediately invert onto a 15-µL drop of antibody solution (in blocking solution) on parafilm. Place the parafilm on top of a dish in a sandwich box containing a small amount of distilled water to maintain humidity.

7. Wash cover slips in multiwell dishes six times with PBS and place on a rocker for a final wash in PBS for 5 min. Mount cover slips on slides with Mowiol solution containing *p*-phenylenediamine as antiquench. This mountant takes about 1 h to set permanently at room temperature. Before this, cover slips cannot be viewed using oil immersion objectives.

8. Store slides at 4°C, in a light-tight slide container. Cells are viewed and photographed on a conventional epifluorescence microscope or on a confocal microscope. Locate the etched cross under phase-contrast microscopy at low power, and subsequently locate microinjected cells using epifluorescence. It is advisable to make images of cells within 1 wk of staining, as nonspecific background fluorescence increases gradually over time.

4. Notes

1. Microinjection is a technique which requires demonstration from an experienced person. Companies that supply microinjection equipment, such as Zeiss, often run training courses. Intensive workshops are also run occasionally by various organizations, for example, the European Molecular Biology Organization.

2. Most microinjection setups allow the researcher to use either a semiautomatic or manual mode of injection. Some setups are completely automatic. Our experience is that injecting on manual mode is by far the preferred mode. Although it takes more practice to learn the muscle coordination, the user learns to inject each cell according to its morphology and take up of protein, and the survival rate is far better than on semiautomatic mode. For injections into confluent, serum-starved Swiss 3T3 cells, no other method is appropriate, as it is essential that as few cells as possible are killed. If many cells are killed, the cells are no longer confluent.

3. Swiss 3T3 cells change in morphology during passaging in culture. They gradually lose their contact inhibition and grow to greater densities, preventing the accurate analysis of the actin cytoskeleton for which they have been favored. Eventually, spontaneously transformed cells will multiply more rapidly and take over the culture. It is important to monitor their growth very carefully. We routinely only passage the cells about 8 to 10 times after thawing.

4. Batch testing of FCS is crucial for the successful maintenance of Swiss 3T3 cells and MDCK cells. For Swiss 3T3 cells, some batches inhibit growth almost completely, whereas others stimulate very rapid proliferation. To maintain the cells for up to 10 passages in culture, it is important to have a batch of FCS that is intermediate, that is, does not support the most rapid proliferation, as this leads more rapidly to loss of contact inhibition and a more transformed phenotype. For MDCK cells, some batches of serum promote a more "scattered" phenotype, so that the analysis of scatter factor-induced scattering is not as tight. We routinely test six batches of serum from various sources, every 15 to 18 mo, and select one from these.

5. Difficulties in getting the protein or DNA to flow out of the pipet may have several causes:

   a. The DNA, protein or IgG may be contaminated. The IgG should be aliquoted and stored at 4°C. Aliquots of Rac and Rho proteins can be used for several days after thawing, but no longer than 1 wk. DNA should be injected within a few hours after diluting in injection buffer.

   b. The microfuge tube used to make up the final injection mix contains some dust/particulate matter. Often just respinning the protein solutions and mixing the components again in a fresh tube can solve the problem.

   c. At high concentrations, it is difficult to inject protein or DNA. For protein, difficulties occur at above approx 5 mg/mL, whereas for DNA, concentrations of 0.5 mg/mL and greater can be problematic.

6. The optimal program for pipet pulling has to be determined by trial and error. With the Campden Instruments programmable pipet puller, each instrument behaves differently and must be individually programmed to obtain a certain shape of pipet. It is advisable to optimize the program with pipets containing a solution of IgG, rather than buffer alone, as buffer flows more easily than protein.

7. During microinjection, the cells are exposed to a strong light source for 10 to 15 min, which must be taken into account when analyzing the effects of added drugs, for example, tyrosine kinase inhibitors, on the responses to Rho/Rac proteins. Controls are performed where the effects of the drugs on growth factor responses are tested on the microinjection microscope with the light source on.

8. If no or few protein-expressing cells are detected after DNA injection, this may be the result of one of the following reasons:

   a. The DNA is degraded. DNA should be diluted into injection buffer as close as possible to the time of injection.

   b. The promotor/enhancer of the vector expresses poorly in the cell type being injected.

   c. The protein is toxic to the cells or induces cell detachment from the substratum. Cells should be analyzed at earlier time points after injection. To ensure that cells are being successfully injected, include IgG (as for protein injections) in the injection buffer.

9. Rho and Rac effects on the actin cytoskeleton can be detected within 5 min of microinjecting the proteins. The extent and timescale of the response will depend on the concentration of protein injected *(8,9,12,15)*.

10. It is important to use PBS with Mg^{2+} and Ca^{2+} for fixing and staining because Ca^{2+} is required for many cell–cell and cell–extracellular matrix interactions, and Mg^{2+} is required to maintain cytoskeletal organization.

11. The methods for staining cells described here work for localization of actin filaments and injected cells. Many variations exist for immunocytochemical staining techniques, and when using other antibodies it is necessary to test different blocking steps and different dilutions to obtain optimal results. Methanol fixation cannot be used with phalloidin, so for antibodies where methanol fixation is required, we use anti-actin antibodies to visualize actin.

12. We have found that TRITC-phalloidin (obtained from Sigma) is light-sensitive, so weak actin filament staining can be the result of excessive exposure to light. We only freeze–thaw aliquots a maximum of three times. During incubation of cover slips with TRITC-phalloidin, the dishes are wrapped in aluminum foil.

References

1. Lamb, N. J. C, Fernandez, A., Conti, M. A., et al. (1988) Regulation of actin microfilament integrity in living nonmuscle cells by the cAMP-dependent protein kinase and the myosin light chain kinase. *J. Cell Biol.* **106,** 1955–1971.

2. Feramisco, J. R., Gross, M., Kamata, T., Rosenberg, M., and Sweet, R.W. (1984) Microinjection of the oncogene form of the human H-ras (t-24) protein results in rapid proliferation of quiescent cells. *Cell* **39,** 109–117.

3. Bar-Sagi, D. and Feramisco, J. R. (1986) Induction of membrane ruffling and fluid-phase pinocytosis in quiescent fibroblasts by *ras* proteins. *Science* **233,** 1061–1068.

4. Paterson, H. F., Self, A. J., Garrett, M. D., Just, I., Aktories, K., and Hall, A. (1990) Microinjection of recombinant p21*rho* induces rapid changes in cell morphology. *J. Cell Biol.* **111,** 1001–1007.

5. Ridley, A. J. and Hall, A. (1992) The small GTP-binding protein rho regulates the assembly of focal adhesions and actin stress fibers in response to growth factors. *Cell* **70,** 389–399.

6. Ridley, A. J., Paterson, H. F., Johnston, C. L., Diekmann, D., and Hall, A. (1992) The small GTP-binding protein rac regulates growth factor-induced membrane ruffling. *Cell* **70,** 401–410.

7. Nobes, C. and Hall, A. (1995) Rho, Rac, and Cdc42 GTPases regulate the assembly of multimolecular focal complexes associated with actin stress fibres, lamellipodia, and filopodia. *Cell* **81,** 53–62.

8. Kozma, R., Ahmed, S., Best, A., and Lim, L. (1995) The Ras-related protein Cdc42Hs and bradykinin promote formation of peripheral actin microspikes and filopodia in Swiss 3T3 fibroblasts. *Mol. Cell. Biol.* **15,** 1942–1952.

9. Ridley, A. J., Comoglio, P. M., and Hall, A. (1995) Regulation of scatter factor/ hepatocyte growth factor responses by Ras, Rac and Rho proteins in MDCK cells. *Mol. Cell. Biol.* **15,** 1110–1122.

10. Adamson, P., Paterson, H. F., and Hall, A. (1992) Intracellular localization of the p21rho proteins. *J. Cell Biol.* **119,** 617–627.
11. Paterson, H., Adamson, P., and Robertson, D. (1995) Microinjection of epitope-tagged Rho family cDNAs and analysis by immunolabeling. *Methods Enzymol.* **256,** 162–173.
12. Self, A. J., Paterson, H. F., and Hall, A. (1993) Different structural organization of Ras and Rho effector domains. *Oncogene* **8,** 655–661.
13. Graessmann, M. and Graessmann, A. (1983) Microinjection of tissue culture cells. *Methods Enzymol.* **101,** 482.
14. Self, A. J. and Hall, A. (1995) Purification of recombinant Rho/Rac/G25K from *Escherichia coli. Methods Enzymol.* **256,** 3–10.
15. Ridley, A. J. (1995) Microinjection of Rho and Rac into quiescent Swiss 3T3 cells. *Methods Enzymol.* **256,** 313–320.

16

Affinity-Based Assay of Rho Guanosine Triphosphatase Activation

Mary Stofega, Celine DerMardirossian, and Gary M. Bokoch

Summary

The recognition that Rho guanosine triphosphatases (GTPases) (Rho, Rac, and Cdc42) play important regulatory roles in many areas of cell biology has made the ability to measure their activity in cells an important biological tool. Because Rho GTPases become activated by conversion from guanosine diphosphate-bound states to guanosine triphosphate (GTP)-bound forms, affinity-based methods to detect the formation of GTP-Rho GTPases have been developed and are widely used for the purpose of assessing Rho GTPase activities in biological studies.

Key Words: Rho; Rac; Cdc42; Rho GTPases; affinity-based activation assay; PBD assay; RBD assay; pulldown assay.

1. Introduction

With the recognition of the many biological roles of Rho guanosine triphosphatases (GTPases) (including members of the Rho, Rac, and Cdc42 subfamilies), the ability to directly measure their activity in cell samples has become an important biochemical tool for the cell biologist. GTPases cycle from inactive (guanosine diphosphate [GDP]-bound) forms to active (guanosine triphosphate [GTP]-bound) forms that interact with and regulate components of intracellular signaling pathways. The identification of binding domains in these effector protein targets that specifically recognize the active, GTP-bound form of the upstream Rho GTPase has provided the basis for affinity-based assays of Rho GTPase activation (**Fig. 1**).

The Rac- and Cdc42-regulated p21-activated kinase 1 (Pak1) contains in the N-terminal regulatory region a specific site for interaction with the active GTP forms of these two GTPases. This region, referred to as the CRIB domain (*1*)

From: *Methods in Molecular Biology, vol. 332: Transmembrane Signaling Protocols, Second Edition*
Edited by: H. Ali and B. Haribabu © Humana Press Inc., Totowa, NJ

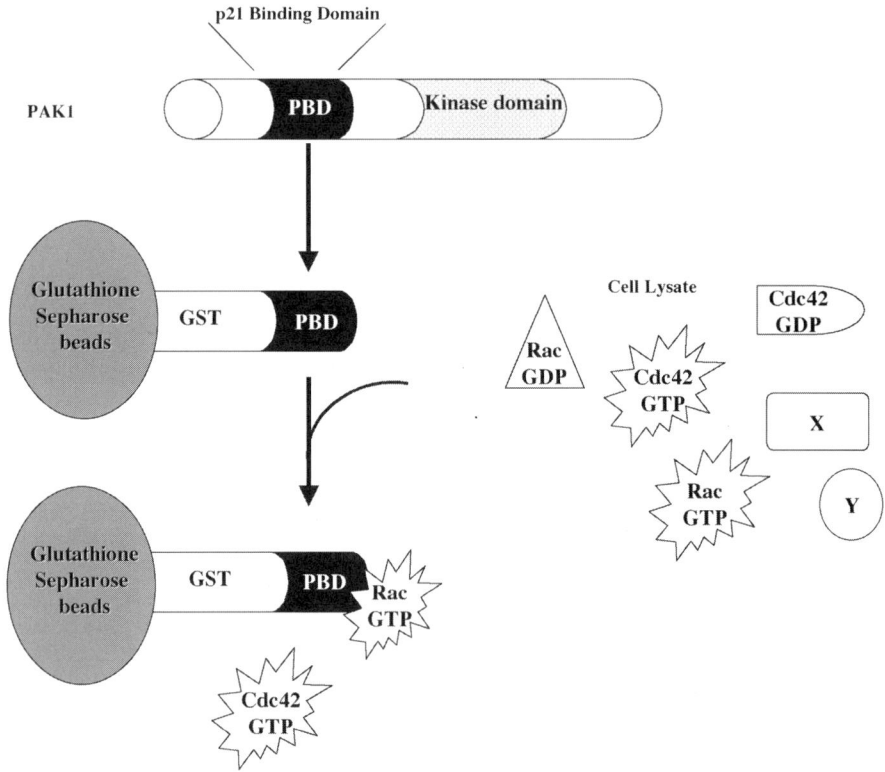

Fig. 1. Principle of the affinity precipitation (pulldown) assay to detect active Rac and Cdc42 using the glutathione-*S*-transferase–Pak1 PBD-binding domain. **Note:** X and Y represent nonrelevant proteins in the cell lysate.

(Cdc42/Rac interactive binding domain) or p21-binding domain (PBD), has a minimal sequence required for specific GTPase binding consisting of amino acids 74 to 89 *(2)*. A homologous but Cdc42-selective CRIB domain is found in Wiscott Aldrich syndrome protein (WASP) amino acids 235 to 268 *(3)*. Binding affinities of the Pak1 PBD range from 20 n*M* to 1 µ*M* depending on the length of the peptide encompassing the minimal CRIB domain *(2)*. A Rho binding site is contained within residues 7 to 89 of Rhotekin *(4)*. Using such probes, it is possible to selectively bind GTP-Rho GTPases because this active form is generated during cell activation (**Fig. 1**). The isolated GTP–GTPases are then detected through the use of specific antibodies for the particular Rho GTPase being assayed. We describe here in detail the methods for performing Rac/Cdc42 activation assays based on the use of the binding domain from p21-activated kinase (PBD assay *[5–10]*) and RhoA activation assays based on the use of the binding domain from Rhotekin (RBD assay *[11]*).

2. Materials
2.1. Pak1 PBD Assay for Rac and/or Cdc42

1. Complementary DNA (cDNA) encoding amino acids 67 to 150 of human PAK1 cloned into pGEX2T.
2. Luria broth (LB).
3. Ampicillin.
4. Isopropyl-β-D-thiogalactopyranoside (IPTG).
5. Bacterial lysis buffer: 50 mM Tris-HCl, pH 7.5, 150 mM NaCl, 5 mM MgCl$_2$, 1 mM dithiothreitol (DTT), freshly added; 1 mM ethylene diamine tetraacetic acid (EDTA); 1 mM phenylmethylsulfonyl fluoride (PMSF), freshly added.
6. Wash buffer: 50 mM Tris-HCl, pH 8.0, 150 mM NaCl, 5 mM MgCl$_2$, 1 mM DTT, freshly added, 1 mM PMSF, freshly added, and 1 µg/mL aprotinin, freshly added.
7. Glutathione Sepharose 4B beads.
8. Cell lysis buffer: 50 mM Tris-HCl, pH 7.5, 200 mM NaCl, 5 mm MgCl$_2$, 1 mM DTT, 1% NP-40, 10% glycerol, 1 mM PMSF, freshly added, 10 µg/mL aprotinin, freshly added, and 10 µg/mL leupeptin, freshly added.
9. PBD binding buffer: 25 mM Tris-HCl, pH 7.5, 40 mM NaCl, 30 mm MgCl$_2$, 1 mM DTT, 1% NP-40, 1 mM PMSF, freshly added, 10 µg/mL aprotinin, freshly added, 10 µg/mL leupeptin, freshly added.
10. Anti-Cdc42 and/or anti-Rac antibodies. Multiple commercial sources are available. We suggest polyclonal (5087) from Santa Cruz for Cdc42, and monoclonal (23A8) from Upstate Biotechnology for Rac1.
11. GTPγS and GDP.

2.2. Rhotekin Rho Binding Domain-Based Assay for RhoA

1. cDNA encoding amino acids 7 to 89 of human Rhotekin cloned into pGEX2T.
2. LB.
3. Ampicillin.
4. IPTG.
5. Bacterial lysis buffer: 50 mM Tris HCl, pH 7.5, 150 mM NaCl, 5 mM MgCl$_2$, 1 mM EDTA, 1 mM PMSF, freshly added, 1 mM DTT, freshly added.
6. Wash buffer: 25 mM Tris-HCl, pH 7.5, 5 mM MgCl$_2$, and 1 mM EDTA, freshly added, 1 mM PMSF, freshly added, and 1 µg/mL aprotinin, freshly added.
7. Glutathione sepharose 4B beads.
8. Rho binding domain (RBD) lysis buffer: 50 mM Tris-HCl, pH 7.2, 1% Triton X-100, 0.5% sodium deoxycholate, 0.1% sodium dodecyl sulfate (SDS), 500 mM NaCl, 10 mM MgCl$_2$, 1 mM PMSF, freshly added, 10 µg/mL aprotinin, freshly added, and 10 µg/mL leupeptin, freshly added.
9. RBD wash buffer: 50 mM Tris-HCl, pH 7.2, 150 mM NaCl, 10 mM MgCl$_2$, 1 mM DTT, freshly added, 1% NP-40, 1 mM PMSF, freshly added, and 10 µg/mL aprotinin and 10 µg/mL leupeptin, both freshly added.
10. Anti-Rho antibody. We have used the monoclonal RhoA antibody from Upstate Biotechnology.
11. Aluminum fluoride (AlF$_4$)$^-$: 10 mM sodium fluoride, 20 µM aluminum, 10 mM MgCl$_2$.

3. Methods

3.1. PBD Assay for Rac and/or Cdc42

3.1.1. Construction of Glutathione-S-Transferase–PBD Fusion Protein

cDNA for amino acids 67 to 150 of Pak1 PBD is amplified by polymerase chain reaction (PCR) and cloned in to the pGEX 2T vector at the *Bam*H1-*Eco*R1 sites and transformed into DH10B *Escherichia coli*. Transformed bacteria are plated out in 100 µg/mL ampicillin plates overnight at 37°C, and single colonies are picked and grown overnight at 37°C with 100 µg/mL ampicillin. Plasmid DNA is isolated and checked for proper orientation and sequencing of the PBD insert. Transformed bacteria containing the PBD insert are stored at –80°C in 20% glycerol in LB (*see* **Note 1**).

3.1.2. Preparation of Glutathione-S-Transferase–PBD Protein

3.1.2.1. Inoculation and Induction of Glutathione-*S*-Transferase–PBD Transcription

1. Prepare 1 L of LB with 100 µg/mL ampicillin.
2. Inoculate 40 mL of LB with ampicillin with glutathione-*S*-transferase (GST)–PBD glycerol stock and grow overnight at 37°C with shaking.
3. Inoculate a 1-L flask with overnight culture.
4. Grow at 37°C until optical density (OD) of 600 nm is 0.7 to 0.8 (approx 3 h). Save 50 µL of bacteria culture (this is the uninduced sample).
5. Induce transcription with 0.8 m*M* IPTG and grow for 3 h at 30°C. Save approx 50 µL (this is the induced sample).

3.1.3. Preparation of Glutathione Beads

1. Take equivalent of 1 mL of dry beads (Glutathione Sepharose 4B) and centrifuge for 5 min, 4°C at 2000*g*.
2. Wash beads twice in 10 mL of H$_2$O, twice in 10 mL of bacterial lysis buffer, and once in 10 mL of bacterial lysis buffer +1 µg/mL aprotinin.

3.1.4. Harvest and Sonication of E. coli

1. Spin down *E. coli* culture 10 min at 2000*g* at 4°C (*see* **Note 2**).
2. Resuspend pellet in 10 mL of bacterial lysis buffer with 1 mg/mL lysozyme, 20 µg/mL DNase I, and 1 µg/mL aprotinin.
3. Incubate on ice for 30 min.
4. Sonicate bacteria on ice, incubate for another 15 min.
5. Centrifuge 10 min, 2000*g* at 4°C.
6. Collect supernatant and save 50 µL of supernatant sample (this is the GST–PBD sample).

3.1.5. Incubation of E. coli Cytosol With Glutathione Beads

1. Combine Glutathione Sepharose 4B beads with *E. coli* supernatant from **Subheading 3.1.4.**, **step 6**, and incubate either 2 h or overnight at 4°C while inverting.

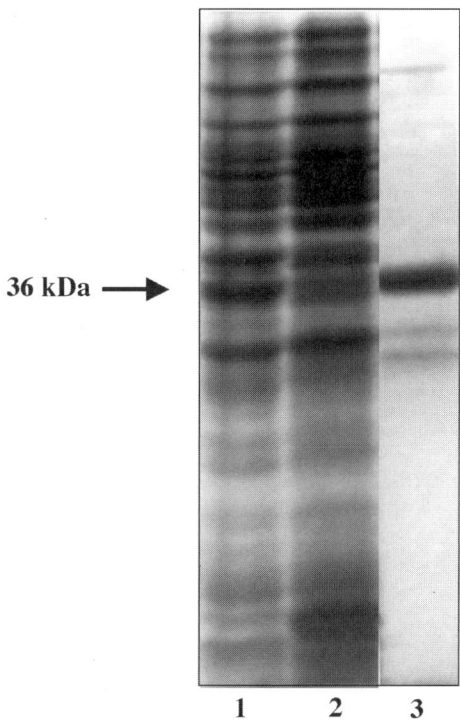

36 kDa ➔

| 1 | 2 | 3 |

1 *E.coli* **supernatant after induction**
2 **unbound proteins in *E.coli* supernatant**
3 **GST-RBD bound to glutathione sepharose beads**

Fig. 2. Analysis of glutathione-*S*-transferase–Rho-binding domain preparation on a 12% (w/v) sodium dodecyl sulfate-polyacrylamide gel stained with Coomassie blue protein dye.

2. Centrifuge for 5 min at 2000*g* at 4°C, and save 50 μL of supernatant as unbound protein sample.
3. Wash beads 5× 10 mL in wash buffer.
4. Aliquot beads in washing buffer with 10% v/v glycerol and store at –80°C. Determine protein concentration of GST–PBD.

3.1.6. Analysis of GST–PBD Purification Process by SDS-Polyacrylamide Gel Electrophoresis (see **Note 3**)

1. Analyze approx 5 μL of samples from uninduced, induced, GST–PBD unbound, and final purified GST–PBD aliquots by SDS-polyacrylamide gel electrophoresis (PAGE) and Coomassie staining to check protein induction, levels, and purity of the final GST–Pak PBD (*see* **Fig. 2**).
2. Determine final protein concentration of product.

3.1.7. Preparation of Cell Extracts for Pak1 PBD Assay

1. Wash cells twice in ice-cold PBS and lyse cells for 30 min on ice in cell lysis buffer (*see* **Note 4**). Centrifuge cell lysates for 10 min at 2000*g* at 4°C to clarify lysates.
2. Determine the protein concentration for each sample.
3. Normalize total cell protein for each sample. The maximal final volume is 500 μL, and, if required, the samples are diluted in PBD binding buffer.
4. Add approx 10 μg of purified GST–PBD beads to samples and incubate for 1 h at 4°C while inverting.
5. Centrifuge the samples at 2000*g* for 2 min, aspirate supernatant, and wash three times in PBD binding buffer.
6. Add Laemmli sample buffer, heat at 100°C for 5 min, and perform SDS-PAGE on a 12% SDS-PAGE gel.
7. Transfer to nitrocellulose and perform Western blot analysis with appropriate Rac and/or Cdc42 antibodies. Typical growth factor-induced stimulation of Rac1 GTP formation is shown in **Fig. 3B**.

Positive and negative controls for the assay should be performed with lysates in which the endogenous Rho GTPases are loaded with either GDP (negative control) or with GTPγS (positive control; *see* **Subheading 3.1.8.**). In addition, lysates from cells transiently overexpressing cDNAs for constitutively active Cdc42- or Rac1-Q61L, or dominant-negative Cdc42- or Rac1-T17N can be used as controls (*see* also **Note 9**).

3.1.8. Nucleotide Loading of Cell Lysates

As a positive control, add EDTA to a final concentration of 10 m*M* in cell lysates prepared in cell lysis buffer, and then add GTPγS to a final concentration of 100 μ*M*. Incubate for 15 min at 30°C and add MgCl$_2$ to 60 m*M* to stop nucleotide exchange (*see* **Note 5**).

As a negative control, add EDTA to 10 m*M* in cell lysates prepared in cell lysis buffer, and add GDP to 1 m*M*. Incubate for 15 min at 30°C and add MgCl$_2$ to 60 m*M* to stop nucleotide exchange. The GTPγS- or GDP-loaded lysates can be used in the PBD assay as described in **Subheading 3.1.** Typical results are shown in **Fig. 3A**.

3.2. RBD Assay for RhoA

3.2.1. Construction of GST–RBD Fusion Protein

cDNA for amino acids 7 to 89 of Rhotekin containing the RhoA binding domain is amplified by PCR and cloned in to the pGEX 2T vector at the *Bam*H1–*Eco*R1 sites, and transformed into DH10a *E. coli*. Transformed bacteria are plated out in 100 μg/mL ampicillin plates overnight at 37°C and single colonies are picked and grown overnight at 37°C with 100 μg/mL ampicillin; Plasmid DNA is isolated and checked for proper orientation and sequencing of

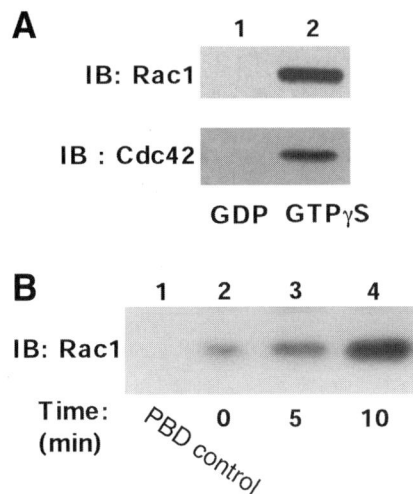

Fig. 3. Affinity precipitation of activated Rac1 and Cdc42 with glutathione-*S*-transferase (GST)–PDZ-binding domain (PBD). **(A)** SK-BR-3 breast carcinoma cells were lysed in cell lysis buffer and cell lysates were loaded with guanosine diphosphate or guanosine triphosphate (GTP)γS. Nucleotide-loaded lysates were incubated with GST fusion protein containing the p21-binding domain of Pak1 (GST–PBD) to precipitate activated Rho GTPases. Activated Rac1 **(top panel)** or Cdc42 **(bottom panel)** were detected by Western blot analysis of precipitated proteins with monoclonal anti-Rac1 antibody or polyclonal anti-Cdc42 antiserum, respectively. Similar amounts of Rac1 or Cdc42 were detected in cell lysates by Western blot analysis with anti-Rac1 or anti-Cdc42 antibodies (data not shown). **(B)** HeLa cells were stimulated with 100 ng/mL recombinant human epidermal growth factor for 0, 5, or 10 min. Cells were lysed in cell lysis buffer and were incubated with GST–PBD to affinity precipitate activated Rho GTPases. As a control (Lane 1), GST–PBD was not incubated with cell lysate. Activated Rac1 was detected by Western blot analysis of precipitated proteins with monoclonal anti-Rac1 antibody. There were equal amounts of Rac1 protein in each sample, as determined by Western blot.

the RBD insert. Transformed bacteria containing the RBD insert are stored at –80°C in 20% glycerol in LB.

3.2.2. Preparation of GST–RBD

3.2.2.1. INOCULATION AND INDUCTION OF GST–RBD TRANSCRIPTION

1. Prepare 1 L of LB with 100 μg/mL ampicillin.
2. Inoculate 40 mL of LB with ampicillin with GST–RBD glycerol stock and grow overnight at 30°C with shaking.
3. Inoculate 1-L flask with overnight culture.
4. Grow at 30°C until OD at 600 nm is 0.6–0.8. Save 50 L of bacteria culture (this is the uninduced sample).

5. Induce transcription with 0.2 m*M* IPTG and grow for 3 h at 30°C. Save approx 50 μL (this is the induced sample).

3.2.3. Preparation of Glutathione Beads

1. Take equivalent of 1 mL of dry beads (Glutathione Sepharose 4B) and centrifuge for 5 min, 4°C at 2000*g*.
2. Wash beads 2X 10 mL of H$_2$O, 2X 10 mL of bacterial lysis buffer, and 1X 10 mL of bacterial lysis buffer +1 μg/mL aprotinin.

3.2.4. Harvest and Sonication of E. coli

1. Spin down *E. coli* culture 10 min at 2000*g* at 4°C.
2. Resuspend pellet in 10 mL of bacterial lysis buffer with 1 mg/mL lysozyme, 20 μg/mL DNase I, and 1 μg/mL aprotinin.
3. Incubate on ice for 15 min.
4. Sonicate bacteria on ice, then incubate for another 15 min on ice.
5. Centrifuge 20 min at 2000*g* at 4°C.
6. Collect supernatant and save 50 μL of supernatant sample (this is the GST–RBD sample).

3.2.5. Incubation of E. coli Cytosol With Glutathione Beads

1. Combine Glutathione Sepharose 4B beads with *E. coli* supernatant, and incubate 2 h or overnight at 4°C while inverting.
2. Centrifuge for 5 min at 2000*g* at 4°C; save 50 μL of supernatant as unbound protein sample.
3. Wash beads 5× 10 mL in 25 m*M* Tris-HCl, pH 7.5, 1 m*M* EDTA, 5 m*M* MgCl$_2$, and 5% glycerol.
4. Aliquot beads in washing buffer with 10% v/v glycerol and store at –80°C.
5. Determine protein concentration of purified GST–RBD.

3.2.6. Analysis of GST–RBD Purification Process by SDS-PAGE

1. Analyze approx 5 μL of samples from uninduced, induced, GST–Rhotekin RBD unbound, and final purified GST–RBD aliquots by SDS-PAGE and Coomassie staining to check protein induction, integrity, and purity (**Fig. 2**).
2. Determine protein concentration of final product.

3.2.7. Preparation of Cell Extracts for RBD Assay

1. Wash cells twice in ice-cold Tris-buffered saline and lyse cells in cell for 30 min on ice in lysis buffer. Clarify lysates by centrifugation for 10 min at 2000*g*.
2. Determine the protein concentration for each sample.
3. Normalize total cell protein for each sample. The maximal final volume is 500 μL, and, if required, the samples are diluted in RBD lysis buffer.
4. Add approx 20 to 30 μg of purified GST–RBD beads to samples and incubate for 45 to 60 min at 4°C while inverting.
5. Centrifuge the samples at 2000*g* for 2 min, aspirate supernatant, and wash four times in RBD wash buffer.

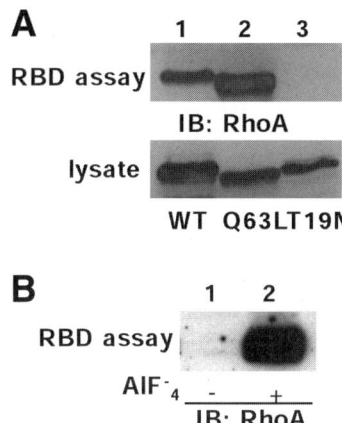

Fig. 4. Affinity precipitation of activated RhoA with glutathione-*S*-transferase (GST)–Rho-binding domain (RBD). **(A)** HeLa cells were transfected with cDNA encoding the indicated RhoA proteins: RhoA Q63L is constitutively guanosine triphosphate-bound, while RhoA T19N is guanosine diphosphate-bound. Cells were lysed in RBD lysis buffer and were incubated with RBD–GST. Activated RhoA was detected by Western blot analysis of precipitated proteins with monoclonal RhoA antibody **(top panel)**. Similar levels of RhoA wild-type or mutant proteins were detected in the HeLa cell lysates by Western blot analysis of whole cell lysates with monoclonal RhoA antibody **(bottom panel)**. **(B)** SK-BR-3 breast carcinoma cells were lysed in RBD lysis buffer and cell lysates were incubated in the presence or absence of AlF_{4-}. Cell lystates were incubated with GST fusion protein containing the Rhotekin binding domain of RhoA (GST–RBD). Activated RhoA was detected by Western blot analysis of precipitated proteins with monoclonal RhoA antibody. Similar amounts of RhoA in SK-BR-3 cell lysates were detected by Western blotting with anti-RhoA antibody (data not shown).

6. Add Laemmli sample buffer, heat at 100°C for 5 min, and perform SDS-PAGE on a 12% SDS-PAGE gel.
7. Transfer to nitrocellulose and perform Western blot analysis with appropriate Rho antibodies (*see* **Note 6**).

A positive control for the assay should be performed with lysates in which the endogenous Rho GTPases are loaded with AlF_{4-} (*see* **Subheading 3.2.8.**). In addition, positive and negative controls for the assay can be performed with lysates from cells transiently overexpressing cDNA for constitutively active RhoA Q63L, or dominant-negative RhoA T19N. Typical results are shown in **Fig. 4** (*see* **Notes 7** and **8**).

3.2.8. Stimulation of Cell Lysates Using AlF_{4-}

As a positive control, add $MgCl_2$ to a final concentration of 10 mM, NaF to a final concentration of 10 mM and, finally, $AlCl_3$ at a final concentration of 20 μ*M* to RBD lysis buffer and RBD wash buffer to generate $AlF_4{}^-$. $AlF_4{}^-$ mimics the activated state of G proteins by binding to the γ-phosphate position in GDP-bound G proteins. Typical results are shown in **Fig. 4B**.

4. Notes

1. Multiple methods based on modern molecular biology can be used in the expression of proteins and in the construction of expression plasmids.
2. Various strains of *E. coli* can be tried to maximize protein expression and purity of the final product. We find that freshly inoculated and induced cultures give best protein yields.
3. Breakdown products of the isolated PBD (or RBD) often are observed (*see* **Fig. 2**). They usually are not detrimental to the assay as long as the intact PBD/RBD is the major product.
4. The choice of the lysate buffer can be varied and optimized for the cells being used. Also for optimal results, the wash conditions for the lysates after binding the PBD or RBD beads should be optimized for the cell type you are working with. Positive controls (GTPγS-loaded or $AlF_4{}^-$-loaded samples) should give a strong clear signal, whereas negative controls (GDP-loaded samples) should give little or no signal.
5. Binding of GTP–GTPase reaches 75% by 30 min. and is maximal by 1 h at 4°C. This time may need to be shortened if GTP hydrolysis in the cell lysate is high. Binding of GTPase to the PBD or RBD inhibits hydrolysis. It is thus sometimes preferable to add the beads to the sample during the lysis step to avoid rapid hydrolysis of GTP to GDP.
6. Sensitivity of the assay will be determined to a large extent by the antibody used for detection. We have recommended some antibodies that have worked for us. However, you should first verify that the antibody you choose to use for detection gives you a strong signal against an aliquot of your cell lysate.
7. We feel it is imperative when performing the PBD or RBD assays on a cell sample for the first time that the investigator carry out the GTPγS or $AlF_4{}^-$ controls, respectively. These allow you to determine the signal you will detect on the immunblots with a maximally activated GTPase. You can then adjust the number of cells you use per sample to get a signal in the detectable range, assuming that the level of endogenously activated GTPase will usually be on the order of 5 to 10% of the maximal activatable GTPase present in the sample. We also conduct these controls routinely to insure that the assay is working properly, and to be able to calculate the percent of total GTPase activated in each sample.
8. Because of differences in their composition, it has been reported that the Rho-binding domain from the Rho effectors Rhotekin, mDia, ROCK, and Citron have differing effectiveness in binding and thus detecting RhoA GTP formation in different biological circumstances *(12)*. Thus, to fully optimize your RhoA assay, it may be useful to test GST–RBDs from these different RhoA effector targets.
9. (General) The PBD assay may be made specific for Cdc42 by using the WASP Cdc42-specific binding domain in place of the Rac/Cdc42-binding sequence from Pak1.

Acknowledgments

The authors thank Ms. Lia Marshall for excellent editorial assistance. The work in our laboratory is supported with grants from the United States Public Health Service/National Institutes of Health.

References

1. Burbelo, P. D., Drechsel, D., and Hall, A. (1995) A conserved binding motif defines numerous candidate target proteins for both Cdc42 and Rac GTPases. *J. Biol. Chem.* **270,** 29,071–29,074.
2. Thompson, G., Owen, D., Chalk, P. A., and Lowe, P. N. (1998) Delineation of the Cdc42/Rac-binding domain of p21-activated kinase. *Biochemistry* **37,** 7885–7891.
3. Abdul-Manan, N., Aghazadeh, B., Liu, G. A., et al. (1999) Structure of Cdc42 in complex with the GTPase-binding domain of the 'Wiskott-Aldrich syndrome' protein. *Nature* **399,** 379–383.
4. Reid, T., Furuyashiki, T., Ishizaki, T., et al. (1996) Rhotekin, a new putative target for Rho bearing homology to a serine/threonine kinase, PKN, and rhophilin in the rho-binding domain. *J. Biol. Chem.* **271,** 13,556–13,560.
5. Benard, V., Bohl, B. P., and Bokoch, G. M. (1999) Characterization of rac and cdc42 activation in chemoattractant-stimulated human neutrophils using a novel assay for active GTPases. *J. Biol. Chem.* **274,** 13,198–13,204.
6. Geijsen, N., van Delft, S., Raaijmakers, J. A., et al. (1999) Regulation of p21rac activation in human neutrophils. *Blood* **94,** 1121–1130.
7. Akasaki, T., Koga, H., and Sumimoto, H. (1999) Phosphoinositide 3-kinase-dependent and -independent activation of the small GTPase Rac2 in human neutrophils. *J. Biol. Chem.* **274,** 18,055–18,059.
8. Benard, V. and Bokoch, G. M. (2002) Assay of Cdc42, Rac, and Rho GTPase activation by affinity methods. *Methods Enzymol.* **345,** 349–359.
9. Bagrodia, S., Taylor, S. J., Jordon, K. A., Van Aelst, L., and Cerione, R. A. (1998) A novel regulator of p21-activated kinases. *J. Biol. Chem.* **273,** 23,633–23,636.
10. Sander, E. E., van Delft, S., ten Klooster, J. P., et al. (1998) Matrix-dependent Tiam1/Rac signaling in epithelial cells promotes either cell-cell adhesion or cell migration and is regulated by phosphatidylinositol 3-kinase. *J. Cell. Biol.* **143,** 1385–1398.
11. Ren, X. D., Kiosses, W. B., and Schwartz, M. A. (1999) Regulation of the small GTP-binding protein Rho by cell adhesion and the cytoskeleton. *EMBO J.* **18,** 578–585.
12. Kimura, K., Tsuji, T., Takada, Y., Miki, T., and Narumiya, S. (2000) Accumulation of GTP-bound RhoA during cytokinesis and a critical role of ECT2 in this accumulation. *J. Biol. Chem.* **275,** 17,233–17,236.

17

Assay of Phospholipase D Activity in Cell-Free Systems

Shankar S. Iyer and David J. Kusner

Summary

Phospholipase D (PLD) enzymes are present in all animal and plant species and have been linked to many critical cellular processes, including proliferation, differentiation, motility, and secretion. The functional significance of PLD derives from its generation of phosphatidic acid, which has both direct signaling properties via activation of numerous kinases, phosphatases, phopspholipases, and other enzymes, as well as via its conversion to diglycerides, the endogenous activators of protein kinase C. The two mammalian PLD isoforms, PLD1 and PLD2, are peripheral membrane proteins that exhibit important physical and functional interactions with the actin cytoskeleton. We outline a cell-free system for the characterization of mammalian PLDs and their activation by physiologic stimuli or pharmacologic agonists for guanine triphosphate-binding proteins. This assay system is used to illustrate the interactions of PLD1 with specific membrane domains and their associated filamentous and monomeric actin components.

Key Words: Phospholipase; signal transduction; enzyme; membrane; phospholipids; phosphatidic acid; macrophage; phagocyte; leukocyte; monocyte; human; inflammation; infection; innate immunity; actin; cytoskeleton; GTP-binding protein; phagocytosis; caveolae; membrane raft.

1. Introduction

Phospholipase D (PLD) enzymes comprise a large class of lipid hydrolases that are present in all organisms from viruses to mammals and in all cells and tissues of metazoan organisms. PLD catalyzes the hydrolysis of membrane phospholipids to yield phosphatidic acid (PA) and the free head group (*1–5*). In the case of phosphatidylcholine (PC), the major substrate of mammalian PLD enzymes and the predominant membrane phospholipid in these cells, choline is the other product of the reaction. The functional importance of PLD derives from its generation of PA a bioactive lipid-signaling molecule. PA

From: *Methods in Molecular Biology, vol. 332: Transmembrane Signaling Protocols, Second Edition*
Edited by: H. Ali and B. Haribabu © Humana Press Inc., Totowa, NJ

directly activates numerous kinases, phosphatases, phospholipases, and other enzymes. PA also serves as a major source of diglycerides, the endogenous activators of protein kinase C, via removal of the phosphate group by phosphatidate phosphohydrolase. In the presence of primary alcohols (e.g., ethanol), PLD enzymes also catalyze a unique "transphosphatidylation" reaction to produce the corresponding phosphatidylalcohol (phosphatidylethanol [PEt]), rather than PA *(1,6,7)*. Thus, PLD activity can be assayed via generation of the natural product, PA, or the phosphatidylalcohol derived from the PLD-specific transphosphatidylation reaction.

PLD activities have been determined in numerous membrane fractions, including the plasma and nuclear membranes, as well as the membranes of organelles (Golgi, lysosomes) and endocytic and exocytic vesicles *(8–12)*. Mammalian PLDs are peripheral membrane proteins that exhibit a prominent physical and functional association with the actin cytoskeleton *(13–19)*. Both mammalian PLD isoforms (PLD1 and PLD2) bind directly to actin with bidirectional functional consequences *(14–16)*. Monomeric G-actin inhibits PLD activity, whereas filmentous F-actin augments it *(14)*. Conversely, stimulation of PLD enzymes has been linked to dynamic rearrangements of the actin cytoskeleton. These unique physical and functional properties of mammalian PLDs have increased the importance of cell-free reconstitution assays. We describe a robust cell-free system for the characterization of several of these critical properties of PLDs, including their physical and functional interactions with the actin cytoskeleton, association with specific membrane domains (e.g., rafts, caveolae), and definition of PLD isoform-specific activities via a novel immunoprecipitation–in vitro assay.

2. Materials

1. Diisopropylfluorophosphate (DFP): 5.8 M stock in anhydrous isopropanol; store at 4°C (*see* **Note 1**; Sigma, St. Louis, MO).
2. Phenylmethylsulfonylfluoride (PMSF): 100 mM stock in ethanol; store at 4°C (Calbiochem, San Diego, CA).
3. Leupeptin: 2 mM stock in H_2O; store at 4°C (Sigma).
4. Ficoll-Hypaque (Sigma).
5. Iscove's medium (BioSource International, Camarillo, CA).
6. RPMI-1640 (Sigma).
7. H/S buffer: 25 mM HEPES, pH 7.4, 125 mM NaCl, 0.7 mM $MgCl_2$, 0.5 mM ethylenebis(oxyethylenenitrilo)tetraacetic acid (EGTA).
8. H/K buffer: 25 mM HEPES, pH 7.4, 100 mM KCl, 3 mM NaCl, 5 mM $MgCl_2$, 1 mM EGTA, 2 µM leupeptin, 0.5 mM PMSF, 1 mM dithiothreitol.
9. Lysis buffer: 0.3% Triton X-100 and 0.5% octyl glucoside in H/K buffer.
10. RIPA-like buffer: 1% Triton X-100, 1% octylglucoside, 1% in H/K buffer.
11. Media for U937 cell line: Iscove's medium, 10% fetal bovine serum, 1% penicillin/streptomycin.

12. Media for THP-1 cell line: RPMI-1640, 10% fetal bovine serum, 0.1 m*M* 2-mercaptoethanol, 1% penicillin/streptomycin.
13. 10% $Mg(NO_3)_2 \cdot 6H_2O$ in ethanol.
14. Chloroform (all organic solvents are high-performance liquid chromatography grade; Fisher, Pittsburgh, PA).
15. Methanol.
16. Ethylacetate.
17. Isooctane.
18. Acetic acid.
19. Guanosine 5'-[γ-thio]triphosphate (GTPγS): 50 m*M* in H_2O, store at –20°C (Roche, Indianapolis, IN).
20. Dipalmitoylphosphatidylcholine (DPPC; all lipids are from Avanti Polar Lipids; Alabaster, AL; *see* **Note 2**).
21. Dipalmitoylphosphatidylethanolamine (PE).
22. Phosphatidylinositol-(4,5)-bisphosphate ($PI(4,5)P_2$).
23. [2-palmitoyl-9,10-³H(N)]-DPPC, specific activity 60–80 Ci/mmol (American Radiolabeled Chemicals, St. Louis, MO).
24. Thin-layer chromatography (TLC) plates, K6 Silica gel plates (250 μm; Fisher).
25. 1,25-dihydroxyvitamin D_3: 0.24 m*M* in dimethyl sulfoxide, protect from light, store at –70°C (Calbiochem).
26. Interferon-γ: 1000 U/μL; store at –70°C (Sigma).
27. Retinoic acid: 1 m*M* in dimethyl sulfoxide, store at –70°C (Sigma).
28. Protein A-Sepharose (Sigma).
29. Rabbit polyclonal anti-PLD1 antibodies (Abs) to the following peptide sequences of PLD1: (1) N-terminus, (PLD1-N), peptide 1–15, (2) Internal sequence, PLD1-I, peptide 525–541, and (3) C-terminus, PLD1-C, peptide 1057–1074.
30. Complement-opsonized zymosan (COZ): Zymosan (Sigma) is washed three times in phosphate-buffered saline (PBS), incubated with 25% serum in PBS for 30 min at 37°C on a rotator, and then washed twice in 4°C PBS.

3. Methods

The methods outlined in this chapter include the following:

1. Cultivation of monocytic cell lines and primary human macrophages.
2. Subcellular fractionation and reconstitution of PLD activity with purified membranes and cytosol.
3. Analysis of PLD activity of the detergent-insoluble cytoskeletal fraction.
4. Characterization of the association of PLD1 with actin.
5. An immunoprecipitation–in vitro PLD assay to determine isoform-specific activity.

3.1. Cultivation of Human Monocytic Cell Lines and Primary Macrophages

Our characterization of the regulation of PLD activity in phagocytic leukocytes uses both human monocytic cell lines (U-937, THP-1) and primary human macrophages derived from blood monocytes. The U937 promonocytic cell line expresses abundant GTPγS-stimulated PLD activity in the undifferentiated state. Thus, this cell line is used for the majority of large-scale biochemical analyses in cell-free assays. THP-1 promonocytic leukemia cells differentiated by cytokine treatment are used when a more mature "macrophage-like" cell line is required, e.g., for studying responses to physiological agonists, and the analysis of macrophage functional responses, such as phagocytosis or generation of a respiratory burst. THP-1 cells also are used for transfection of wild-type and mutant PLD genes by electroporation or lipofection. Primary human monocyte-derived macrophages are used to evaluate the accuracy and physiological relevance of the cell-line models and to study highly differentiated phenotypes, including adhesion and bactericidal activity.

3.1.1. Cultivation of Human Monocytic Cell Lines

1. The U937 human promonocytic leukocyte cell line from American Tissue Type Culture Collection is maintained at 37°C, 7.5% CO_2, at $0.3-1.0 \times 10^6$ cells, and subcultured twice weekly. For large-scale preparations of membranes and cytosol, cells are seeded into 1-L spinner flasks and grown with gentle agitation.
2. The THP-1 human promonocytic leukemia cell line (American Tissue Type Culture Collection) is differentiated to a macrophage-like phenotype (differentiated THP-1 cells [dTHP]-1) by incubation in 1,25-dihydroxyvitamin D_3 (100 nM), interferon-γ (1000 U/mL), and retinoic acid (1 μM) for 48 to 72 h.

3.1.2. Preparation of Primary Human Monocyte-Derived Macrophages

1. Heparinized venous blood is drawn from healthy adult volunteers in accordance with a protocol approved by the human subjects institutional review board.
2. Blood is diluted (1:1) with ice-cold PBS and layered onto 0.5 volume of Ficoll-Hypaque and centrifuged at 500g for 40 min at 25°C with no brake.
3. The upper plasma layer is discarded. The cloudy Ficoll layer, which contains the peripheral blood mononuclear cells (PBMCs), is removed and diluted 1:1 in cold RPMI-1640, and centrifuged at 500g for 10 min at 4°C with full brake.
4. The supernatant is discarded and the cell pellet is resuspended in RPMI-1640 and PBMCs are counted in a hemacytometer.
5. PBMCs are cultured in RPMI-1640 and 20% autologous serum, at a concentration of 2×10^6 cells/mL in Teflon wells for 5 d at 37°C, 5% CO_2.
6. Macrophages are purified by adherence to plastic tissue culture plates for 2 h at 37°C in 5% CO_2. Monolayers are washed three times to remove nonadherent lymphocytes and incubated in RPMI-1640, 2.5% autologous serum, without antibiotics, for use in experiments.
7. Effects of experimental manipulations on macrophage viability are assessed by the exclusion of Trypan blue, and monolayer density is determined by nuclei counting with naphthol blue-black stain. Purity and viability of macrophage preparations are greater than 95%.

3.2. Subcellular Fractionation and Reconstitution of PLD Activity With Purified Membranes and Cytosol

3.2.1. Cell Disruption by Nitrogen Cavitation

1. U937 human promonocytes are grown in a 1-L spinner flask (1.0 to 1.5 × 10⁶ cells/mL) and centrifuged at 600g for 25 min at 4°C.
2. The packed cells are resuspended in 50 mL of H/S buffer containing 1 mg/mL of bovine serum albumin and 10 mM glucose and recentrifuged.
3. The washed cells are suspended in 25 mL of H/S buffer and incubated with 4 mM DFP for 25 min in an ice bucket (*see* **Note 1**). The cells are centrifuged at 800g and the supernatant containing DFP is inactivated in 1 N NaOH.
4. The DFP- treated cells are resuspended in 10 mL of H/K buffer before disruption by N₂ cavitation (450 psi, 25 min, 4°C; *see* **Note 3**). The undisrupted cells and nuclei are removed by centrifugation at 900g for 8 min at 4°C to yield the postnuclear supernatant.

3.2.2. Separation of Membrane and Cytosolic Fractions

1. The postnuclear supernatant is layered over a 1:1 volume of 50% sucrose in water and ultracentrifuged at 150,000g for 60 min at 4°C. The resulting supernatant is recentrifuged at 225,000g for 60 min to remove any contaminating membrane. This second supernatant (cytosol) is filtered through a 0.2-μm filter (*see* **Note 4**). The clarified cytosol is stored as aliquots at –70°C.
2. The membrane fraction at the sucrose interface is pelleted at 225,000g for 60 min at 4°C, resuspended in H/K buffer, and washed by recentrifugation.
3. The washed membrane is suspended in H/K buffer and homogenized with a Tenbroeck tissue grinder.
4. This membrane fraction is enriched in plasma membrane (defined by the presence of virtually all of the human leukocyte antigen (HLA)-Class I antigen in the total cell lysate). Because this fraction also contains the Golgi marker, β-COP, it will be referred to as the "plasma membrane-enriched" or "membrane" fraction for simplicity. The denser fraction, which sediments through the 50% sucrose, is enriched in the lysosomal marker CD63. Protein concentration in the membrane and cytosolic fractions is determined by the method of Bradford *(20)*.

3.2.3. Assay of Membrane-Associated PLD Activity

1. For each reaction condition, 75 μg of the membrane fraction is incubated with 100 μg of cytosol ± GTPγS (1, 10, 100 μM) for 30 min at 37°C in H/K buffer, in a volume of 100 μL.
2. Membranes are re-isolated by centrifugation at 150,000g for 60 min, washed twice with H/K, and resuspended in the same buffer *(9,11,12)*.
3. Washed membranes prepared from incubation with cytosol in the absence of GTPγS are designated M₀, whereas those prepared in the presence of GTPγS are referred to as M$_{GTPγS}$.
4. The mixed vesicle substrate is prepared by addition of 35 nmol DPPC, 540 nmol PE, and 48 nmol PI(4,5)P₂, with 16 μCi of [³H]-DPPC in a 1.5-mL polypropylene

tube, and the solvent slowly evaporated with a stream of N_2 in the hood. 400 μL of H/S buffer without Ca^{2+} or Mg^{2+} is added to the dried lipids, and the suspension is sonicated for 5 min at 25°C with a microtip sonicator probe (Heat Systems, Microson cell disrupter, setting 17) to yield a slightly cloudy solution (*see* **Note 5**).

5. 100 μ*M* GTPγS is included in each sample, and 1.5% ethanol is added to permit detection of the PLD-specific transphosphatidylation product, PEt. Reactions are terminated at 60 min by addition of 500 μL of chloroform:methanol (2:1, v/v).

6. Lipids are extracted, dried under N_2, and analyzed by TLC in an ethylacetate: isooctane:acetic acid (9:5:2, v/v) solvent system *(13,21,22)*. PEt and PA are identified by comigration with purified standards, and [³H]-PEt and [³H]-PA cpm are quantitated by liquid scintillation spectrophotometry, and counts normalized for the total amount of [³H]-labeled phospholipid in each experiment. [³H]-cpm comigrating with PEt are determined for each set of samples in the absence of ethanol, and these background counts are subtracted from each data point.

7. The PLD activity of membranes prepared in the absence of GTPγS (M_o) exhibits very little basal or GTPγS-dependent activity. In contrast, incubation of membranes and cytosol in the presence of GTPγS, followed by re-isolation of membranes ($M_{GTPγS}$) results in a stable association of PLD activity (**Fig. 1**). This latter activity is 21-fold greater than that of M_o and is approximately equal to the maximal level of PLD activity derived from the complete cell-free system of membranes plus cytosol.

3.3. Assay of PLD Activity in the Detergent-Insoluble Cytoskeletal Fraction

The GTPγS-dependent association of PLD activity with the membrane fraction and its stable persistence in the absence of cytosol involves an interaction with the actin-based membrane skeleton and specific membrane domains, i.e., rafts or caveolae. This interaction may be evaluated by determining the stability of the membrane-associated PLD activity to detergent extraction.

3.3.1. Preparation of the Cytosketal Fraction and Determination of Associated PLD Activity

1. Detergent-insoluble cytoskeletal fraction (DIF) is prepared by incubating membranes (M_o and $M_{GTPγS}$) in 0.5% octyl glucoside in H/K buffer for 30 min on ice.

2. The detergent-insoluble pellets are separated by centrifugation at 14,000g for 15 min at 4°C, washed twice in H/K with 0.5% octyl glucoside, and resuspended in H/K buffer without detergent (*see* **Note 6**).

3. The detergent-insoluble fraction prepared from M_o is designated DIF_0, whereas that derived from $M_{GTPγS}$ is referred to as $DIF_{GTPγS}$.

4. The PLD activities of $DIF_{GTPγS}$ and DIF_0 are assayed with [³H]-DPPC vesicles, as noted in **Subheading 3.2.3.**, followed by quantitation of [³H]-PEt and [³H]-PA.

5. DIFs are also prepared from $M_{GTPγS}$ and M_o by extraction with 1% sodium cholate or 1% Triton X-100, and the PLD activities associated with these detergent-insoluble preparations are determined in an identical manner.

6. In separate experiments, the direct effects of each detergent on membrane-associated PLD activity are assessed by incubation of 100 μg of $M_{GTPγS}$ with varying deter-

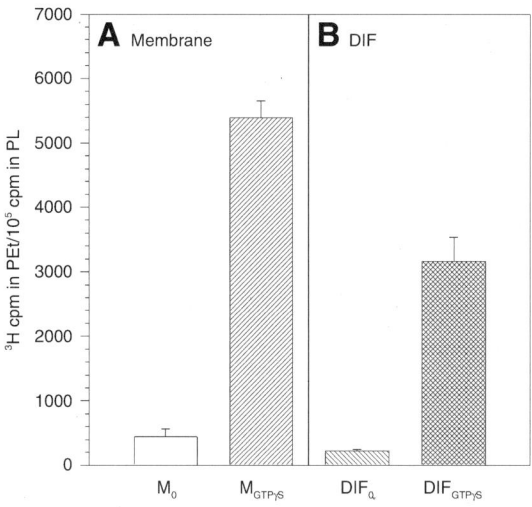

Fig. 1. Stimulation of guanosine triphohsphate (GTP)-binding proteins with guanosine 5'-[γ-thio]triphosphate (GTPγS) induces stable association of phospholipase D (PLD) activity with the plasma membrane and its cytoskeletal fraction. (**A**) Plasma membrane-enriched (75 μg) and cytosolic fractions (100 μg) from U937 cells are incubated in the absence or presence of 100 μ*M* GTPγS for 30 min at 37°C. Membranes (M_o and $M_{GTP\gamma S}$, respectively) are reisolated by centrifugation, washed, and assayed for PLD activity in the presence of 100 μ*M* GTPγS and 1.5% ethanol, via production of [³H]-phosphatidylethanol (PEt) over the course of 60 min. (**B**) The re-isolated M_o and $M_{GTP\gamma S}$ are incubated with 0.5% octyl glucoside for 30 min on ice and the detergent-insoluble fractions, detergent-insoluble cytoskeletal fraction $(DIF)_0$, and $DIF_{GTP\gamma S}$, respectively, are washed twice in H/K buffer containing octyl glucoside, and resuspended in H/K buffer without detergent, for determination of PLD activity in the presence of 100 μ*M* GTPγS. PLD activity is expressed as [³H]-PEt cpm per 10^5 [³H]-cpm in phospholipid. (Reprinted with permission from **ref. *13*.**)

    gent concentrations (0.02–2.0%) in H/K buffer at 37°C for 60 min, followed by quantitation of [³H]-PEt.

7. The detergent-insoluble fraction from $M_{GTP\gamma S}$ ($DIF_{GTP\gamma S}$) exhibits a 20-fold greater activity than that of the control DIF_0 (**Fig. 1**). This activity accounts for about 60% of the total GTPγS-dependent activity of $M_{GTP\gamma S}$.

3.3.2. Analysis of Phospholipid and Protein Content of Membrane and Cytoskeletal Fractions

1. Lipid phosphorus content of membrane and detergent-insoluble fractions is determined by the ashing procedure of Ames (*23*). Briefly, samples are placed in Pyrex test tubes, ashed in $Mg(NO_3)_2$ in ethanol, solubilized in 0.5 *M* HCl, and heated at 100°C for 15 min. The test tubes are covered with marbles to minimize evaporation.

Table 1
Phospholipid and Protein Content of Membrane
and Octyl Glucoside-Insoluble Fractions

Fraction	Lipid phosphorus (nmoles)	Protein (µg)	nmoles P/µg protein
M	18.5 ± 1.50	17.8 ± 0.40	1.04
M_o	17.0 ± 1.50	17.6 ± 0.40	0.97
$M_{GTP\gamma S}$	20.0 ± 0.50	17.1 ± 0.01	1.17
DIF_0	13.5 ± 0.50	6.8 ± 0	1.98
$DIF_{GTP\gamma S}$	17.0 ± 1.00	6.3 ± 0.3	2.70

75 µg of membrane (M) and 100 µg of cytosol from U937 cells are incubated in the absence or presence of 100 µ*M* GTPγS for 30 min at 37°C, followed by washing and re-isolation of the membrane fractions, M_o, and $M_{GTP\gamma S}$, respectively. Membranes are extracted with 0.5% octyl glucoside for 30 min at 4°C, and the detergent-insoluble fractions, DIF_0, and $DIF_{GTP\gamma S}$, are isolated, washed, and suspended in H/K buffer. For each fraction, parallel samples were subjected to analyses of lipid phosphorus and protein content. For analysis of lipid phosphorus, samples are ashed, solubilized in 0.5 *N* HCl, followed by addition of the Ames colorimetric reagent *(23)* and determination of A_{820} nm. Quantitation is performed by reference to a standard curve derived from KH_2PO_4 solutions of known concentration. Protein levels are determined by the Bradford method *(20)*.

2. After cooling to 25°C, Ames colorimetric reagent *(23)* is added, samples are incubated at 37°C for 1 h, cooled to 25°C, and A_{820} nm is determined. Quantitation is performed by reference to a standard curve derived from KH_2PO_4 solutions of known concentration (**Table 1**).
3. Protein levels are determined by the method of Bradford *(20)*.

3.3.3. Determination of the Effects of Detergent Concentration and Physical Properties on Cytoskeletal-Associated PLD Activity

1. A range of concentrations of octylglucoside (0.1–1.0%) is used for extraction of M_o and $M_{GTP\gamma S}$.
2. Membrane pellets are resuspended in detergent-free H/K buffer and the PLD activity is determined, as noted previously. With U937 cells, the level of cytoskeletal-associated PLD activity is inversely proportional to the octylglucoside concentration used for extraction (**Fig. 2A**).
3. To determine whether the significant levels of cytoskeletal PLD activity are the result of a direct activating effect of the detergent, a range of concentrations of octylglucoside is added directly to the activated membrane fraction ($M_{GTP\gamma S}$). In our system, octylglucoside induced a dose-dependent inhibition of the membrane PLD activity (**Fig. 2B**).
4. A range of detergents that differ in charge, chemical structure, and critical micellar concentration may be used for preparation of the cytoskeletal fraction from activated membranes. Use of cholate and Triton X-100 demonstrated that the GTPγS-dependent association of PLD activity with the cytoskeleton is a general

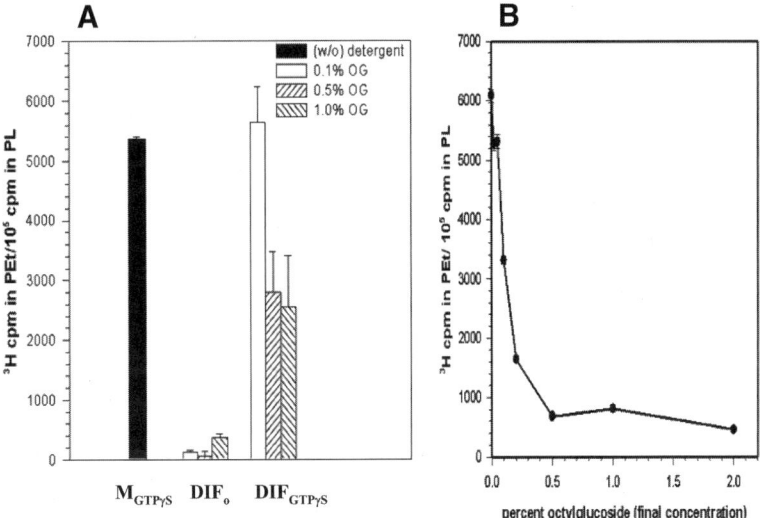

Fig. 2. Determination of the effects of detergent concentration on the phospholipase D activity of membrane and cytoskeletal fractions. The isolated M_o and $M_{GTP\gamma S}$ (as detailed in **Fig. 1**) are incubated with various concentrations of octyl glucoside for 30 min on ice and the cytoskeletal (detergent-insoluble cytoskeletal fraction [DIF]) fractions, DIF_0, and $DIF_{GTP\gamma S}$, respectively, isolated for determination of PLD activity in the presence of 100 µM guanosine 5'-[γ-thio]triphosphate. **(A)** The level of PLD activity associated with the DIF fraction is inversely proportional to the octyl glucoside concentration used for extraction. **(B)** PLD activity of $M_{GTP\gamma S}$ is determined in the presence of externally added octyl glucoside at the indicated final concentrations. Octyl glucoside induces concentration-dependent inhibition of membrane PLD activity. PLD activity is expressed as [^{3}H]-phosphatidylethanol cpm per 10^5 [^{3}H]-cpm in phospholipid.

property of detergents with diverse chemical and physical properties. Of note, extraction of membranes with Triton X-100 was performed according to the flotation method of Lisanti and coworkers, which is used for analysis of caveolae or membrane rafts *(24)*.

3.3.4. Analysis of the Effect of Physiological Stimulation of Intact Cells on the Membrane and Cytoskeletal Association of PLD Activity

To evaluate the potential physiological relevance of PLD association with the actin cytoskeleton, we used intact U937 human promonocytes that were stimulated by a receptor-dependent agonist. Zymosan is a cell wall fragment, consisting of repeated disaccharide moieties, derived from the yeast *Saccharomyces cerevisiae*. Incubation of zymosan in human serum results in its opsonization with fragments of complement component C3 (C3b, C3bi), which are potent agonists for the phagocytic complement receptors, CR1, CR3, and CR4 *(25)*.

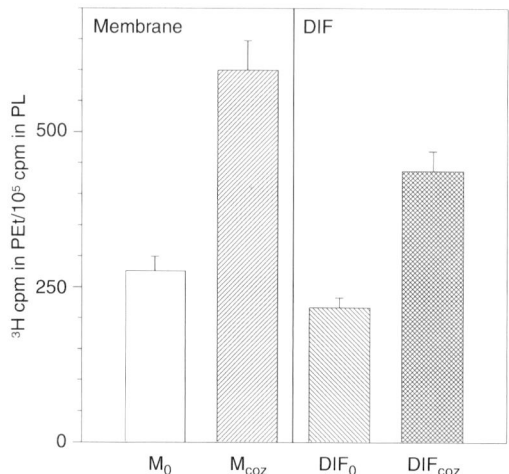

Fig. 3. Phagocytosis by intact U937 cells is accompanied by the stable association of phospholipase D (PLD) activity with the plasma membrane and its cytoskeletal fraction. U937 cells (2×10^8) are incubated with complement-opsonized zymosan (COZ) (at a particle/cell ratio of 10:1) or H/S buffer control for 15 min at 37°C. Membrane fractions from control (M_0) or COZ-treated cells (M_{COZ}) are isolated by N_2 cavitation, followed by density-gradient centrifugation. The detergent-insoluble cytoskeletal fractions ($[DIF]_0$ and DIF_{COZ}, respectively) are prepared by extraction of these membranes with 0.5% octyl glucoside. The PLD activity of each fraction is assayed by the production of [^{3}H]-phosphatidylethanol over the course of 60 min in the presence of 0.5% ethanol and 100 μM guanosine 5'-[γ-thio]triphosphate. Reprinted with permission.

1. 10^8 U937 cells (10^7/mL) in Iscove's medium, 10% FCS are washed and resuspended in H/S buffer.
2. Zymosan is opsonized in 25% serum, as previously described *(25)*.
4. U937 cells are incubated with COZ, at a particle/cell ratio of 10:1, for 15 min at 37°C. Incubations are terminated by centrifugation at 3000g for 1 min at 4°C.
5. The cell pellet is suspended in H/K buffer, disrupted by N_2 cavitation, and the plasma membrane-enriched fraction is isolated, as described in **Subheading 3.2.2.**
6. The PLD activities of the membrane and its octyl glucoside-insoluble fraction derived from COZ-treated cells (M_{COZ} and DIF_{COZ}), are compared with those derived from buffer-treated cell (M_0, DIF_0) by quantitation of [^{3}H]PEt generation from [^{3}H]PC-containing mixed-lipid vesicles. The physiological stimulus COZ, which binds to plasma membrane complement receptors, results in stable association of PLD with actin cytoskeleton (**Fig. 3**).

3.4. Characterization of the Association of PLD1 With Actin

3.4.1. GTPγS Induces Association of PLD1 With the Membrane Cytoskeleton

Low-molecular-weight guanosine triphosphatases (GTPases) of the ARF and Rho families mediate activation of PLD1 by physiological agonists for G

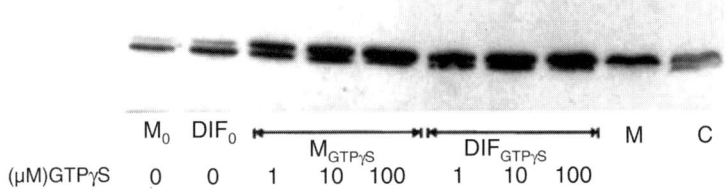

M₀	DIF₀		←—— M_GTPγS ——→	←—— DIF_GTPγS ——→			M	C
(μM)GTPγS	0	0	1	10	100	1	10	100

Fig. 4. Stimulation of guanosine triphosphate (GTP)-binding proteins induces a concentration-dependent association of phospholipase D (PLD)1 with the plasma membrane and its detergent-insoluble cytoskeleton. Plasma membrane (M) and cytosol (C) isolated from U937 cells by nitrogen cavitation and density gradient centrifugation are incubated in the presence of the indicated concentrations of guanosine 5'-[γ-thio]triphosphate (GTPγS) or buffer to prepare $M_{GTPγS}$ (1–100 μM) and M_0. Membrane fractions are extracted with 0.5% octyl glucoside and the respective detergent-insoluble fractions ($[DIF]_{GTPγS}$ [1–100 μM], DIF_0,) isolated by centrifugation. 5×10^6 cell equivalents of each sample are subjected to sodium dodecyl sulfate-polyacrylamide gel electrophoresis on 8% gels, followed by transfer to polyvinylidene difluoride membranes, and Western blotting with polyclonal anti-PLD1 antibody. M_{GTPgS} and $DIF_{GTPγS}$ exhibit increasing amounts of two closely resolved species of PLD1 that are dependent on the concentration of GTPγS. (Reprinted with permission from **ref. *13*.**)

protein-coupled receptors and GTPγS. Rho and ARF are also involved in the regulation of actin cytoskeletal dynamics *(13,26,27)*. The majority of RhoA and ARFs are located in the cytosol of resting U937 cells. It has been shown that guanine nucleotides induce translocation of Rho and ARF GTPases from cytosol to membranes *(13,28,29)*. The level of association of these GTPases with the detergent-insoluble membrane-associated cytoskeleton can be directly determined. Here, we show the association of PLD1 with the membrane-associated cytoskeleton after stimulation with GTPγS.

1. 5×10^6 cell equivalents of the membrane or cytoskeletal fraction, or 100 μg of cytosol, was subjected to sodium dodecyl sulfate-polyacrylamide gel electrophoresis (SDS-PAGE) on 8% gels, as previously described *(13)*. Proteins were transferred to the polyvinylidene difluoride membrane and blocked with 5% nonfat dry milk.
2. Western blotting for PLD1 with polyclonal anti-PLD1 Ab, with detection via horseradish peroxidase-coupled 2° Ab and enhanced chemiluminescence (ECL), was performed as described (**Fig. 4**; *see* **Note 8 *[13]***).

3.4.2. Coimmunoprecipitation of PLD1 and Actin From Purified Membranes

1. To assess the hypothesis that PLD1 and actin physically are associated in membranes from U937 cells, the membrane fraction is solubilized in RIPA-like buffer by incubation on ice for 60 min.
2. Insoluble material is removed by centrifugation at 14,000*g* for 15 min at 4ºC.

IP: anti-PLD1
WB: anti-actin

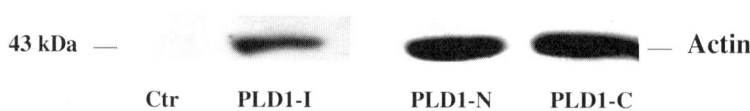

43 kDa —   — Actin

    Ctr   PLD1-I   PLD1-N   PLD1-C

Fig. 5. Actin coimmunoprecipitates with phospholipase D (PLD)1 in membranes from U937 cells. Membranes are incubated in lysis buffer for 1 h on ice and subjected to immunoprecipitation with control irrelevant antibody (Ab) (Ctr) or antipeptide polyclonal Abs generated to an internal (PLD1-I), N-terminal (PLD1-N), or C-terminal (PLD1-C) sequence of PLD1. After washing the immunoprecipitates, we analyzed the samples using sodium dodecyl sulfate-polyacrylamide gel electrophoresis /Western blotting with anti-actin immunoglobulin M monoclonal Ab, with detection by HRP-conjugated secondary Ab and enhanced chemiluminescence. (Reprinted with permission from **ref. *14*.**)

3. The lysate is precleared by incubation with pre-immune serum for 120 min at 4°C followed by a 30 min incubation with 50 µL of a 10% Protein A–Sepharose suspension.

4. After centrifugation at 1000*g* for 5 min at 4°C, the precleared lysate is incubated with control irrelevant polyclonal Ab or one of three polyclonal anti-PLD1 Abs for 5 h at 4°C, followed by an additional 1-h incubation with 50 µL of the 10% Protein A–Sepharose suspension. Three rabbit polyclonal anti-PLD1 Abs to the following peptide sequences of PLD1 were used: (1) PLD1-N, peptide 1–15, (2) Internal sequence, PLD1-I, peptide 525–541, and (3) PLD1-C, peptide 1057–1074.

The resultant immunoprecipitates are washed five times with RIPA-like buffer and subjected to SDS-PAGE on 8% gels. After transfer to PVDF, immunoblotting is performed with an anti-actin immunoglobulin (Ig)M monoclonal Ab, with detection via ECL. Immunoprecipitation with all three anti-PLD1 Abs, but not control irrelevant Ab, demonstrated coimmunoprecipitation of actin (**Fig. 5**).

3.4.3. Determination of PLD1 Binding to Membrane-Associated G-Actin

To assess the binding of PLD1 to the G-actin localized to the plasma membrane, we use a co-sedimentation assay in which DNase 1 is immobilized on Sepharose beads. Because DNase 1 binds specifically to G-actin but not to F-actin *(30,31)*, proteins that bind G-actin are co-sedimented in the DNase I Sepharose pellet *(14)*.

1. 500-µg Aliquots of membranes are solubilized in lysis buffer, as noted in **Subheading 3.4.2.** In select samples, various amounts of purified rabbit skeletal muscle actin are added.

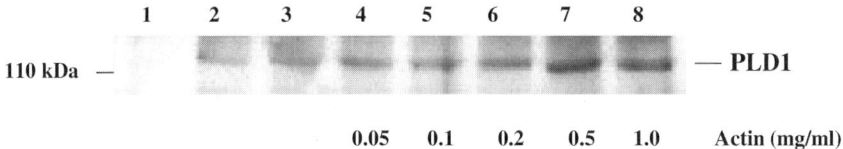

Fig. 6. Phospholipase D (PLD)1 binds membrane-associated G-actin. Purified membranes are solubilized in lysis buffer, and incubated with uncomplexed Sepharose beads (**lane 1**) or DNase I-Sepharose (**lanes 2–8**). The designated amounts of purified α-actin are added to **lanes 4** to **8**, before sedimentation by centrifugation. Sedimented beads are washed in lysis buffer, and associated proteins analyzed by sodium dodecyl sulfate-polyacrylamide gel electrophoresis /Western blotting with anti-PLD1 antibody. (Reprinted with permission from **ref. *14*.**)

2. The lysates are precleared with washed, control Sepharose beads by incubation for 60 min at 4ºC.
3. The precleared lysates are incubated with DNase 1-Sepharose beads for 16 h at 4°C and washed five times with lysis buffer.
4. The washed, pelleted beads are suspended in SDS-PAGE sample buffer, heated to 100°C for 5 min, and subjected to electrophoresis on 8% gels, followed by transfer of the DNase 1-binding proteins to PVDF, and immunoblotting with anti-PLD1 Ab.
5. The results demonstrate that PLD1 is specifically cosedimented by the DNase 1 beads but not by the uncomplexed, control beads (**Fig. 6**). In the presence of increasing concentrations of added G-actin (0.05–1.0 mg/mL), there was a progressive increase in the amount of the co-sedimented PLD1.

3.5. A Novel Immunoprecipitation–In Vitro PLD Assay and Its Application to the Determination of PLD Isoform-Specific Activity

Many cell types express both PLD1 and PLD2, but isoform-specific activities and their relationship to physiological functions remain poorly understood. We developed an immunoprecipitation–in vitro PLD activity assay (*32*) to define the role of the individual PLD isoforms in the regulation of cellular functions, illustrated here for macrophage phagocytosis.

1. Monocyte-derived macrophages or dTHP-1 cells in RPMI-1640 are incubated with COZ (particle:cell ratio, 10:1), 100 nM PMA, or buffer control, for 15 min at 37°C.
2. Incubations are terminated by washing in ice-cold PBS, which also removes nonadherent COZ.
3. Cells are solubilized in lysis buffer by incubation for 1 h on ice (*see* **Note 9**).
4. After centrifugation at 14,000g for 15 min at 4°C, to pellet the insoluble fraction, supernatants are precleared by incubation with Protein A–Sepharose for 120 min at 4°C.

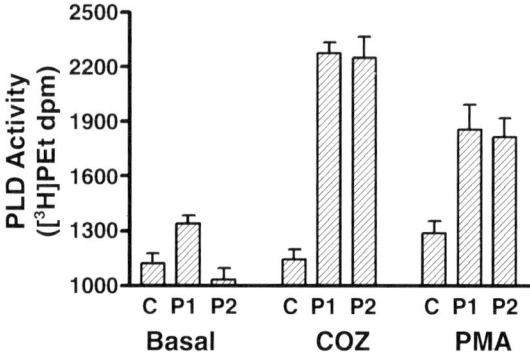

Fig. 7. Phagocytosis is associated with stimulation of both phospholipase D (PLD)1 and PLD2. Differentiated THP-1 macrophages are incubated with buffer (Basal), complement-opsonized zymosan (COZ) (particle:cell, 10:1) or PMA (100 nM) for 30 min. Cells are disrupted in lysis buffer, precleared and then immunoprecipitated with control, pre-immune serum (C) or specific rabbit polyclonal antibodies (Abs) to PLD1 (P1) or PLD2 (P2). After extensive washing of the immunoprecipitates (IPs), PLD activity of each is determined by addition of mixed lipid vesicles containing [^{3}H]-phosphatidylcholine substrate and 1.0% ethanol. The PLD-specific product, [^{3}H]-phosphatidylethanol, is isolated by thin-layer chromatography and quantitated by liquid scintillation spectrometry. Western blotting of the IPs confirmed the specificity of the immunoprecipitating Abs (not shown). Reprinted with permission from **ref. *32***.

5. Precleared lysates are centrifuged at 1000g for 5 min at 4°C, and supernatants incubated with rabbit polyclonal Abs to PLD1, PLD2, or control preimmune serum for 5 h at 4°C, followed by an additional 2-h incubation with 50 µL of 10% Protein A–Sepharose.

6. Immunoprecipitates (IPs) are washed in complete lysis buffer, followed by three washings with 0.5% octyl glucoside in H/K buffer, then five times with H/K buffer without detergents. All treatments and washings are carried out at 4°C.

7. The PLD activity of the IPs is assayed by addition of mixed-lipid substrate vesicles in the presence of 1.0% ethanol, as detailed preously in **Subheading 3.2.3.**

8. Using this approach, we have demonstrated that phagocytosis is associated with stimulation of both PLD1 and PLD2 activities (**Fig. 7**). Stimulation of cells with phorbol myristate acetate (PMA), which is known to activate both PLD1 and PLD2, resulted in increased levels of activity in IPs containing PLD1 (170 ± 14% of control) ,as well as PLD2 (163 ± 7%), supporting the accuracy of this assay (**Fig. 7**).

4. Notes

1. DFP is extremely toxic because of its potent, irreversible inhibition of acetylcholinesterase. DFP should be handled only in the hood with gloves, and a colleague should be present to monitor for toxicity (lightheadedness, fainting). Atropine

injection syringes are available in case of accidental exposure. All equipment contaminated with DFP should be rinsed in 1 N NaOH. After the initial centrifugation of DFP-treated cells, the supernatant is diluted 1:1 in 1 N NaOH and allowed to stand at room temperature for 30 min to inactivate the DFP.

2. The stock solutions of purified lipids are made in chloroform:methanol (9:1). PC and PI(4,5)P$_2$ stocks are 1 mM, whereas the PE stock solution is 2 mM. To increase its solubility, the PE stock is gently warmed in a 37°C water bath. PC and PE stock solutions are stored at –20°C, whereas PI(4,5)P$_2$ is stored as 50-µL aliquots at –70°C. The PE solution, which appears cloudy when removed from the cold, can be rendered clear and colorless in seconds by brief warming in the 37°C bath.

3. Cell disruption is performed by the nitrogen decompression method in the Parr Cell Disruption Bomb. The bomb is precooled to 4°C and left immersed in the ice bucket throughout the disruption process. The cells are pressurized with N$_2$ and sufficient time must be allowed for N$_2$ to dissolve and come to equilibrium with the cells. The N$_2$ pressure and duration of equilibration depend on the number of cells and their surface:volume ratio. Efficiency of cell disruption can be evaluated by monitoring release of cytosolic constituents, such as lactate dehydrogenase, or by microscopic examination. Use of a larger number of cells may require a longer equilibration period. If a high percentage of cells remain undisrupted, a second cavitation step may be performed. Disruption occurs when the pressure in the chamber is released by opening the discharge valve and collecting the homogenate.

4. During the isolation procedure all steps are performed in ice bucket with tubes immersed in ice. Storage of cytosol and membrane fractions should be in frost-free –70°C freezers. The cytosol must be filtered through a 0.2-µm membrane to remove residual contamination with low-density membranes. In experiments in which membrane lipids were labeled by incorporation of radioactive fatty acids, unfiltered cytosols had a 5 to 10% contamination with the membrane fraction.

5. The final lipid vesicle preparation should be translucent, with no flocculent or insoluble material. A few seconds cooling of the tube between the sonic pulses will prevent overheating of the tube. Because 10 µL of this substrate vesicle preparation is used for each sample, this amount is sufficient for at least 35 assays. The substrate vesicles should be used on the day of preparation.

6. After removing the frozen membrane aliquots from the freezer, the protein is thawed quickly at room temperature and the membranes uniformly resuspended using the Tenbroeck homogenizer. A similar approach for resuspension must be followed after every centrifugation step of the membrane fraction. The cytoskeletal fractions are resistant to rapid resuspension and patience must be exercised. All steps should be conducted in the ice bucket. After this protocol, the activities of the membrane and cytoskeletal fractions can be maintained for months at –70°C.

7. The low level of PLD activity obtained with the Triton X-100-insoluble fraction probably represents the significant inhibitory effect of this detergent on PLD activity, and indeed direct addition of this detergent to the PLD assay system exerted greatest inhibitory effect on the activity of M$_{GTP\gamma S}$.

8. Confirmation that the detergent-insoluble fraction derived from purified membranes is consistent with the membrane cytoskeleton can be obtained by Western blotting for the actin-binding proteins, vinculin, α-actinin, talin, and paxillin, as previously described *(9)*.
9. In initial experiments, we used more stringent lysis conditions, including 0.5 to 1.0% Triton X-100 with or without 1% deoxycholate. However, these conditions resulted in very poor recovery of PLD activity, even after detergent exchange into 0.5% octylglucoside followed by extensive washings with detergent-free H/K buffer.

Acknowledgments

We would like to thank James A. Barton for his assistance in the development of several aspects of these PLD assays, and gratefully acknowledge the valuable support and critique of our colleagues in the Inflammation program at the University of Iowa. This work was supported by NIH GM62302 and AI055916 and VA Merit Review grants to D. J. K.

References

1. Brow, H. A. and Sternweis, P. C. (1995) Stimulation of phospholipase D by ADP-ribosylation factor. *Methods Enzymol.* **257,** 313–324.
2. Extol, J. H. (2002) Regulation of phospholipase D. *FEBS Lett.* **531,** 58–61.
3. Exton, J. H. (2002) Phospholipase D-structure, regulation and function. *Rev. Physiol. Biochem. Pharmacol.* **144,** 1–94.
4. Du, G., Morris, A. J., Sciorra, V. A., and Frohman, M. A. (2002) G-protein-coupled receptor regulation of phospholipase D. *Methods Enzymol.* **345,** 265–274.
5. Liscovitch, M., Czarny, M., Fiucci, G., and Tang, X. (2000) Phospholipase D: molecular and cell biology of a novel gene family. *Biochem. J.* **345,** 401–415.
6. Liscovitch, M., Czarny, M., Fiucci, G., and Tang, X. (1999) Localization and possible functions of phospholipase D isozymes. *Biochim. Biophys. Acta.* **1439,** 245–263.
7. Liscovitch, M. (1991) Signal-dependent activation of phosphatidylcholine hydrolysis: role ofphospholipase D. *Biochem. Soc. Trans.* **19,** 402–407.
8. Colley, W. C., Sung, T. C., Roll, R., et al. (1997) Phospholipase D2, a distinct phospholipase D isoform with novel regulatory properties that provokes cytoskeletal reorganization. *Curr. Biol.* **7,** 191–201.
9. Freyberg, Z., Sweeney, D., Siddhanta, A., Bourgoin, S., Frohman, M., and Shields, D. (2001) Intracellular localization of phospholipase D1 in mammalian cells. *Mol. Biol. Cell* **12,** 943–955.
10. Lucocq, J., Manifava, M., Bi, K., Roth, M. G., and Ktistakis, N. T. (2001) Immunolocalisation of phospholipase D1 on tubular vesicular membranes of endocytic and secretory origin. *Eur. J. Cell Biol.* **80,** 508–520.
11. Vitale, N., Caumont, A. S., Chasserot-Golaz, S., et al. (2001) Phospholipase D1: a key factor for the exocytotic machinery in neuroendocrine cells. *EMBO J.* **20,** 2424–2434.

12. Du, G., Altshuller, Y. M., Vitale, N., et al. (2003) Regulation of phospholipase D1 subcellular cycling through coordination of multiple membrane association motifs. *J. Cell Biol.* **162**, 305–315.
13. Iyer, S. S. and Kusner, D. J. (1999) Association of phospholipase D activity with the detergent-insoluble cytoskeleton of U937 promonocytic leukocytes. *J. Biol. Chem.* **274**, 2350–2359.
14. Kusner, D. J., Barton, J. A., Wen, K. K., Wang, X., Rubenstein, P. A., and Iyer, S. S. (2002) Regulation of phospholipase D activity by actin. Actin exerts bidirectional modulation of mammalian phospholipase D activity in a polymerization dependent, isoform-specific manner. *J. Biol. Chem.* **277**, 50,683–50,692.
15. Han, J. M., Kim, Y., Lee, J. S., et al. (2002) Localization of phospholipase D1 to caveolin-enriched membrane via palmitoylation: implications for epidermal growth factor signaling. *Mol. Biol. Cell.* **13**, 3976–3988.
16. Lee, S., Park, J. B., Kim, J. H., et al. (2001) Actin directly interacts with phospholipase D, inhibiting its activity. *J. Biol. Chem.* **276**, 28,252–28,260.
17. Ha, K. S. and Exton, J. H. (1993) Activation of actin polymerization by phosphatidic acid derived from phosphatidylcholine in IIC9 fibroblasts. *J. Cell Biol.* **123**, 1789–1796.
18. Hodgkin, M. N., Clark, J. M., Rose, S., Saqib, K., and Wakelam, M. J. (1999) Characterization of the regulation of phospholipase D activity in the detergent-insoluble fraction of HL60 cells by protein kinase C and small G-proteins. *Biochem. J.* **339**, 87–93.
19. Cross, M. J., Roberts, S., Ridley, A. J., et al. (1996) Stimulation of actin stress fibre formation mediated by activation of phospholipase D. *Curr. Biol.* **6**, 588–597.
20. Bradford, M. M. (1976) A rapid and sensitive method for the quantitation of microgram quantities of protein utilizing the principle of protein-dye binding. *Anal. Biochem.* **72**, 248–254.
21. Xie, M. S. and Dubyak, G. R. (1991) Guanine-nucleotide- and adenine-nucleotide-dependent regulation of phospholipase D in electropermeabilized HL-60 granulocytes. *Biochem. J.* **278**, 81–89.
22. Kusner, D. J. and Dubyak, G. R. (1994) Guanosine 5'-[γ-thio]triphosphate induces membrane localization of cytosol-independent phospholipase D activity in a cell-free system from U937 promonocytic leucocytes. *Biochem. J.* **304**, 485–491.
23. Ames, B. N. (1966) Assay of inorganic phosphate, total phosphate and phosphatases. *Methods Enzymol.* **8**, 115–118.
24. Lisanti, M. P., Scherer, P. E., Vidugiriene, J., et al. (1994) Characterization of caveolin-rich membrane domains isolated from an endothelial-rich source: implications for human disease. *J. Cell Biol.* **126**, 111–126.
25. Kusner, D. J., Hall, C. F., and Schlesinger, L. S. (1996) Activation of phospholipase D is tightly coupled to the phagocytosis of *Mycobacterium tuberculosis* or opsonized zymosan by human macrophages. *J. Exp. Med.* **184**, 585–595.
26. Paterson, H. F., Self, A. J., Garrett, M. D., Just, I., Aktories, K., and Hall A. (1990) Microinjection of recombinant p21rho induces rapid changes in cell morphology. *J. Cell Biol.* **111**, 1001–1007.

27. Zigmond, S. H. (1996) Signal transduction and actin filament organization. *Curr. Opin. Cell Biol.* **8,** 66–73.

28. Houle, M. G., Kahn, R. A., Naccache, P. H., and Bourgoin, S. (1995) ADP-ribosylation factor translocation correlates with potentiation of GTP gamma S-stimulated phospholipase D activity in membrane fractions of HL-60 cells. *J. Biol. Chem.* **270,** 22,795–22,800.

29. Fleming, I. N., Elliott, C. M., and Exton, J. H. (1996) Differential translocation of rho family GTPases by lysophosphatidic acid, endothelin-1, and platelet-derived growth factor. *J. Biol. Chem.* **271,** 33,067–33,073.

30. Shiokawa, D. and Tanuma, S. (2001) Characterization of human DNase I family endonucleases and activation of DNase gamma during apoptosis. *Biochemistry* **40,**143–152.

31. Mori, S., Yasuda, T., Takeshita, H., et al. (2001) Molecular, biochemical and immunological analyses of porcine pancreatic DNase I. *Biochim. Biophys. Acta.* **1547,** 275–287.

32. Iyer, S. S., Barton, J. A., Bourgoin, S., and Kusner, D. J. (2004) Phospholipases D1 and D2 coordinately regulate macrophage phagocytosis. *J. Immunol.* **173,** 2615–2623.

18

Reconstitution System Based on Cytosol-Depleted Cells to Study the Regulation of Phospholipase D

Amanda Fensome-Green and Shamshad Cockcroft

Summary

Phospholipase D (PLD) hydrolyzes phosphatidylcholine to produce the membrane-associated second messenger, phosphatidic acid (PA) and choline. Two phospholipase D enzymes—PLD1 and PLD2—have been identified, although their regulatory mechanisms are yet to be fully understood. To study the regulation of PLD, we established a reconstitution system that allows the study of the PLD enzymes in their native environment while enabling the cytosol to be manipulated. Cells are permeabilized with a bacterial cytolysin (streptolysin O), which produces lesions in the plasma membrane, resulting in the release of cytosolic proteins. With increasing permeabilization times, guanosine 5'-[γ-thio]triphosphate and receptor-activated PLD activity diminishes. Once the conditions for the run-down of the response is established, cellular factors, such as cytosol and purified proteins, can be added to these cells to restore activity. In addition to examining PLD activity, this reconstitution system allows the study of potential cellular targets of PA, such as phosphatidylinositol 4-phosphate (PIP) 5-kinase activity by monitoring PIP_2 synthesis, and also functional outputs, such as exocytosis.

Key Words: ARF; phosphatidylcholine; phosphatidate; streptolysin O; alcohols; permeabilization.

1. Introduction

Cell-surface receptors regulate hydrolysis of cellular phospholipids that are catalyzed by different classes of phospholipases having distinct specificities. Depending on cell type and stimulus, multiple lipid signaling pathways are initiated to allow for the physiological response of the cell to be manifested. Thus, receptors can be coupled to activation of the inositol lipid-specific phospholipase C and the phosphatidylcholine (PC)-specific phospholipase D (PLD) and this gives rise to multiple second messengers.

From: *Methods in Molecular Biology, vol. 332: Transmembrane Signaling Protocols, Second Edition*
Edited by: H. Ali and B. Haribabu © Humana Press Inc., Totowa, NJ

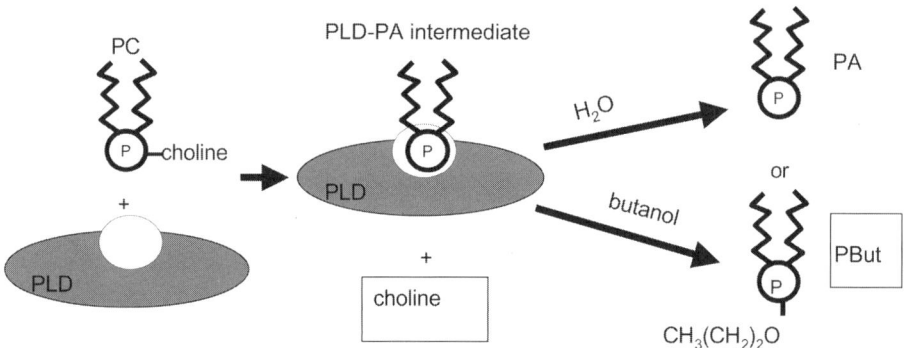

Fig. 1. Hydrolysis of phosphatidylcholine (PC) by phospholipase D (PLD) produces choline and phosphatidic acid (PA). Either product can be assayed as a monitor of PLD activity. However, PA is rapidly metabolized; therefore, it generally is assayed as the stable product, phosphatidylbutanol (PBut), in the presence of 0.2% to 0.5% butanol. Hydrolysis of PC by PLD occurs in two steps. Initially, PLD binds to PC and the first part of the reaction involves the formation of a PA–PLD intermediate by covalent linkage of PA to a histidine residue and the release of the choline headgroup. Either water or a primary alcohol (e.g., butanol) can act as a nucleophile in the second stage of the reaction. In the presence of water, the reaction product is PA, and in the presence of butanol, the reaction production is the metabolically stable product, PBut. PBut has to be separated by thin-layer chromatography, whereas choline, which is water-soluble, can be separated by column chromatography.

PLD catalyses the hydrolysis of PC to produce the lipid-soluble metabolite phosphatidic acid (PA), and the water-soluble headgroup, choline (**Fig. 1**; **refs.** *1–3*). There are two PLD enzymes, PLD1 and PLD2, and the activities of both enzymes are increased when G protein-coupled receptors or receptors that regulate tyrosine kinases are occupied by appropriate agonists. The regulatory mechanisms concerning PLD1 and PLD2 are still being elucidated. The ADP-ribosylation factor (ARF) (ARF1–6) and Rho (Rac, Rho, and Cdc42) family of guanosine triphosphatases (GTPases) have all been identified as regulators of mammalian PLD1. In addition, conventional isoforms of protein kinase C (α, βI, βII, and γ) also can activate PLD1 directly. More importantly, the activation of PLD1 by the three regulators, ARF, Rho, and protein kinase C (PKC), is synergistic, indicating that these activators interact at different sites of the PLD1 molecule. PLD2 also is regulated in a complex manner and oleic acid, ARF proteins, and PKC all can increase PLD2 activity. The activity of PLD1 and PLD2 also can be modulated (inhibitory or excitatory) by many soluble proteins, including G_{M2} activator *(4,5)*, actin *(6)*, α-actinin *(7)*, and synuclein *(8)*.

To study PLD isozymes in their native environment, it is essential to set up cell-based assays where it is possible to study their regulation by cell-surface receptors and their modulation by other cellular factors. An important consideration when studying PLDs is the presentation of the lipid substrate, PC, because the presence of other lipids will profoundly influence the activity. Both PLD1 and PLD2 have a requirement for phosphoinositides with phosphatidylinositol(4,5)bisphosphate being probably the most important in this regard. Both PLD1 and PLD2 contain a PH and a PX domain. These domains commonly are found in many proteins and, where examined, have been shown to bind phosphoinositides *(9)*. PH domains can bind a variety of phosphoinositides, whereas the PX domain can bind PI3P. In addition, a basic region also has been identified as a PIP_2 binding site *(10)*. By using permeabilized cells where the cytosolic compartment can be manipulated, the enzymes can be studied in their native environment because both the lipid substrates and the enzymes (PLD1 and PLD2) remain cell-associated and, hence, in their native state. The use of lipid micelles or vesicles as a source of substrate for the enzyme inevitably leads to loss of many subtle aspects of PLD regulation and is a poor reflection of the cellular environment that these enzymes normally operate in.

An additional advantage of the permeabilized cell system is that many of the functions controlled by the putative second messenger, PA, can also be examined. Thus, the permeabilized system can be used to examine the production of $PI(4,5)P_2$ by the activators of PLD *(11–13)*. Here, we describe the use of a permeabilized cell system that has powerful applications with respect to examination of both the regulators (e.g., ARF proteins *[14]*, PKC *[15]*, oleic acid *[16]*) and modulators of PLD (e.g., G_{M2} activator *[5]*). Permeabilization of cells is used to deplete cytosolic proteins, and this leads to loss of receptor-activated PLD activity. Cytosol or specific proteins can be re-introduced into the cells and this is sufficient to restore receptor-activated PLD activity. For cytosol depletion, cells are permeabilized with streptolysin O, a bacterial (streptococcal) cytolysin, which generates large lesions (approx 15 nm in diameter) in the plasma membrane of cells. The protocol detailed below can be used for studying both G protein-coupled receptors *(17)* and for studying tyrosine kinase receptors *(11)*. In addition, the receptor can be by-passed when G protein-coupled receptors are being studied by using GTPγS, the nonhydrolyzable analog of GTP, to directly activate G proteins of both heterotrimeric, as well as monomeric, GTP-binding proteins of the Ras superfamily *(15)*.

Activation of PLD should be examined under different states of the cell to establish the quality of the restoration with specific proteins (**Fig. 2**). In the first instance, the responsiveness of the system is established by examining "acutely permeabilized" cells, which will determine the extent of the response one is likely to obtain under the most optimal conditions. Here, the receptor agonist (or GTPγS) is added simultaneously with streptolysin O (SLO). Under these conditions, the activation of PLD occurs while the majority of the cellular proteins are still present in the cells. Entry of GTPγS into the permeabilized

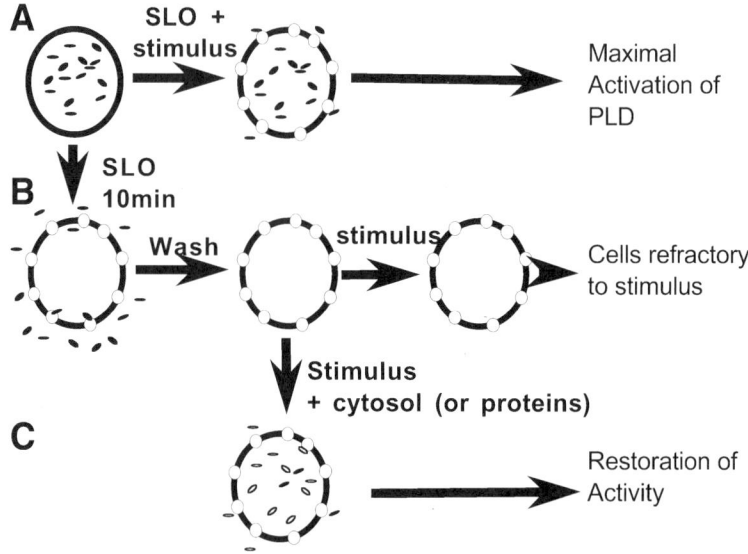

Fig. 2. Illustration of the method used for restoration of G protein-regulated phospholipase D (PLD) activity in permeabilized cell preparations. Three steps are described. (**A**) PLD activity is monitored in acutely permeabilized cells, conditions in which guanosine 5'-[γ-thio]triphosphate (GTPγS) and the permeabilizing agent, streptolysin O (SLO), are added simultaneously. Activation by GTPγS occurs in the presence of cytosolic proteins and maximal stimulation is observed. (**B**) When cells are incubated with SLO first to deplete the cytosolic proteins, the ability of GTPγS to stimulate phospholipase D activity in the "cytosol-depleted" cells is impaired despite the presence of membrane-associated PLD. (**C**) "Cytosol-depleted" cells are reconstituted with addition of exogenous cytosol (or purified proteins, e.g., ARF).

cells occurs within seconds, whereas the loss of the cytosolic proteins from the cells occurs in 5 to 10 min. This activity is simply attributable to the size of the molecules. In addition, GTPγS will retard the loss of ARF proteins by translocating them to membranes *(13)*.

The second step establishes conditions that leads to "run-down" so that the ability of the cells to respond to GTPγS (or a receptor-directed agonist) becomes refractory. This refractory state is achieved by permeabilizing the cells first to allow the leakage of the cytosolic proteins. These cells are referred to as "cytosol-depleted" cells. Conditions for run-down should be empirically determined for each cell type. Some proteins are freely diffusable (e.g., ARF1) and will be released within 5 to 10 min but other proteins are associated with the membrane or cytoskeleton and are only released over a longer period of time (30–45 min), for instance, Rac proteins (*see* **Notes 1** and **2** and ref. *18*).

The third and final step is to restore activation by the re-addition of exogenous cytosol or known proteins that are suspected to be required for PLD signaling. If the identity of the proteins is not known, then the cytosol can be fractionated and the reconstituting factor(s) purified.

2. Materials

1. SLO can be purchased from Sigma (cat. no. S-140). The SLO is supplied in powder form and it is reconstituted in 2 mL of distilled water to give a stock solution of 20 IU/mL. (International units are the manufacturer's arbitrary units.) This solution can be kept at 4°C for 1–2 wk. The solution can get cloudy with time and can be partially clarified on warming at 37°C. However, the cloudiness does not affect permeabilization. Alternatively, the solution can be kept frozen in 50- to 100-µL aliquots at –20°C and used over the course of a 6-mo period.

2. Both primary and cultured cells have been used successfully, including HL60 cells, rat basophilic leukemia mast cells, rat peritoneal mast cells, and human neutrophils. These cells mainly are used as a cell suspension, although attached cells also can be used, but depletion of cytosolic proteins is incomplete compared with cells in suspension. Alternatively, attached cells can be detached by trypsinization or scraping and used in suspension.

3. Permeabilization buffer: 20 mM Na-piperazine-N,N'-bis(2-ethane sulfonic acid) (PIPES), 137 mM NaCl, 2.7 mM KCl, pH 6.8. Stock solution of Na-PIPES (1 M) and a 20X stock of NaCl/KCl are kept at 4°C till required. Two milliliters of stock Na-PIPES and 5 mL of stock NaCl/KCl are diluted to 100 mL and made fresh and the pH adjusted to 6.8. This buffer will be referred to as PIPES. Glucose 1 mg/mL (5.6 mM) and bovine serum albumin 1 mg/mL are added to the PIPES buffer to obtain the permeabilization buffer.

4. Ca/ethylenebis(oxyethylenenitrilo)tetraacetic acid (EGTA) buffers. It is necessary to control the concentration of Ca^{2+} between 10 nM and 10 µM (pCa 8–pCa 5). The resting level of cytosol Ca^{2+} is 100 nM in most cells and increases to micromolar levels upon stimulation; therefore, it is important to clamp the Ca^{2+} concentration to a known value by using Ca–EGTA buffers. The final EGTA concentration is maintained at 3 mM. Stock Ca–EGTA buffers (100 mM) at specific free Ca^{2+} concentrations are prepared and stored at –20°C. Ca^{2+}–EGTA buffer stocks (100 mM) in the range of pCa 8 (10 nM) to pCa 5 (10 µM) are prepared from stock solutions of Ca–EGTA and EGTA of 100 mM concentration, which are then combined in varying proportions to achieve the desired value of free Ca^{2+} (**Table 1**). These values have been obtained using the program CHELATE *(19)*. The two stock solutions are prepared in PIPES buffer (20 mM Na–PIPES, 137 mM NaCl, 2.7 mM KCl, pH 6.8). EGTA is purchased from Fluka because of high purity and CaCl$_2$ is analytical grade obtained from Sigma.

5. Mg adenosine triphosphate (ATP) is made up as a stock solution of 100 mM and can be kept at –20°C for months. ATP is purchased as a disodium dihydrogen salt. To prepare 10 mL of 100 mM stock solution of MgATP, dissolve 605 mg in 10 mL of a solution containing 2 mL of 1 M Tris and 1 mL of 1 M MgCl$_2$. The use

Table 1
Recipe for Ca²⁺ Buffer Solutions

pCa	Vol. (mL) Ca–EGTA	Vol. (mL) EGTA
8	0.112	7.888
7	0.996	7.004
6.5	2.481	5.519
6	4.698	3.302
5.5	6.552	1.448
5	7.501	0.499

Ca–EGTA and EGTA solutions (100 mM) are mixed in the proportions indicated to obtain 8 mL of each buffer stock (100 mM) at the appropriate pCa. In the experiments, the final [EGTA]$_{total}$ is 3 mM. The calculation is based on the assumption that there is 2 mM MgCl$_2$ in the buffer and the buffer pH is 6.8.

of 200 mM Tris effectively results in a neutral solution (pH 7.0). This should be checked with a pH electrode and adjusted accordingly. Freeze-thaw of the solution is not detrimental and 10-mL aliquots can be kept at –20°C and used repeatedly.

6. [*methyl*-3H]choline is purchased from Amersham Biosciences and kept sterile.
7. Medium 199 is purchased from Sigma.
8. Bio-Rex 70 cation exchange resin (sodium form, mesh size 200–400) is obtained from Bio-Rad.
9. GTPγS is obtained as a 10 mM solution from Boehringer.
10. Recombinant ARF1 can be expressed in *Escherichia coli* (15) and cytosol can be prepared from rat brain.

3. Methods

Procedures are described for reconstitution of PLD activity in permeabilized HL60 cells stimulated with GTPγS (**Fig. 2**). Three steps are described: The first step outlines the experimental procedure for the following:

1. Working with acutely permeabilized cells.
2. Conditions to establish run-down.
3. The restoration of PLD activity with ARF proteins. (The method can be adapted to examine any other regulators of PLD activity.)

PLD activity can be measured by two independent methods (**Fig. 1**). The method described here is suitable for working with permeabilized cells, which rely on the release of labeled [³H]choline. The advantage of this method is that many samples can be analyzed in a single experiment, which is essential if you are screening many cytosolic fractions. An alternative method involves labeling of the cells with [³H]myristic acid for 4 h to label the PC pool (12). In the presence of butan-L-ol, phosphatidylbutanol (PBut) is produced at the expense of PA (**Fig. 1**). Transphosphatidylation is a hallmark of PLD activity and provides definitive evidence for PLD activity. It is advisable to confirm the production of PBut in a limited number of experiments, as choline release could occur by other means. Choline release requires simple column chromatography, whereas the measurements of PBut require lipid extraction and separation by thin-layer chromatography.

A typical experiment is illustrated in **Fig. 3**. HL60 cells were permeabilized for different times and subsequently assayed for GTPγS-stimulated PLD activity. As the permeabilization interval increases, the responsiveness of the cells diminishes. The supernatants are analyzed for leaked ARF proteins to demonstrate the depletion of these proteins from the cells. Finally, the cytosol-depleted cells are reconstituted with recombinant ARF proteins to demonstrate that ARF proteins are responsible for PLD activation.

3.1. Assay of PLD Activity in Acutely Permeabilized Cells Using [³H]Choline Release

1. Labeling of cells: to measure PC-hydrolysing PLD activity, the cells are labelled with [*methyl*-³H]choline and release of radiolabeled choline is used as a monitor of activity. HL60 cells normally are grown in RPMI-1640 medium with heat-inactivated 12.5% (v/v) fetal calf serum, 4 mM glutamine, 50 IU/mL penicillin, and 50 μg/mL streptomycin. Radiolabeling of HL60 cells is performed in medium 199 containing 10% fetal calf serum. This medium is used because of its low choline content. HL60 cells are labelled with 0.5 μCi/mL for 48 h. HL60 cells grow to a density of 1–2 × 10⁶ cells/mL and each assay tube has approx 1–2 × 10⁶ cells. A total of 50 mL of cells is therefore sufficient for 50 separate assay conditions. The experiment is always conducted in duplicate. [³H]choline is mainly incorporated into PC (87%) and sphingomyelin (13%). Incorporation into the choline-containing lipids can be determined by extracting the total lipids (*see* **step 8**) and measuring the level of incorporation in the extract. To obtain a good signal, each assay tube should contain approx 10⁵ disintegration per minute (DPM).
2. Labeled HL60 cells (50 mL, 1–2 × 10⁶ cell/mL) are centrifuged at 1000g for 5 min at room temperature. The medium is discarded and the cells resuspended in 40 mL of permeabilization buffer (i.e., PIPES buffer containing albumin and glucose). The cells are pelleted by centrifugation and the process repeated once more. After the final centrifugation, the cells are resuspended in 2 mL of permeabilization buffer. The washed radiolabeled cells are equilibrated at 37°C for 10 to 15 min.
3. 1.5-mL Eppendorf tubes are used for the assays. The final assay volume is 100 μL. 50 μL of reaction mixture is prepared in the Eppendorf tubes containing twice the concentration of SLO (0.4 IU/mL final), MgATP (1 mM final), MgCl₂ (2 mM final), Ca²⁺ buffered with 3 mM EGTA (pCa 5), and GTPγS (10 μM final). The Eppendorf tubes containing the appropriate reagents are prepared at 4°C and put into the water bath at 37°C for 5 min before the addition of the cells.
4. 50-μL aliquots of cells are transferred to the Eppendorf tubes and incubated for 30 min.

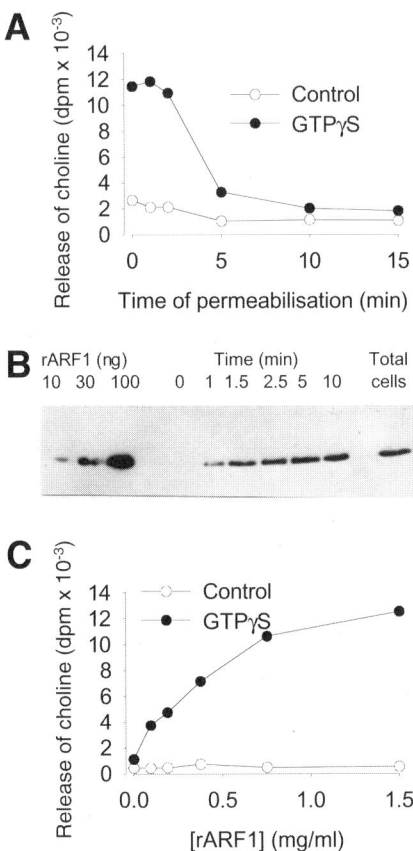

Fig. 3. **(A)** Run-down of guanosine 5'-[γ-thio]triphosphate (GTPγS)-stimulated phospholipase D (PLD) activity. [³H]-choline-labeled HL60 cells were permeabilized and at the indicated times 50-μL aliquots were removed and incubated further in the presence or absence of 10 μ*M* GTPγS. Samples were quenched on ice and then analyzed for [³H]-choline release. The response observed at time 0 is the maximal response observed in acutely permeabilized cells. As the time of permeabilization is increased, the ability of GTPγS to stimulate PLD activity diminishes. **(B)** Reduction of GTPγS-stimulated PLD activity correlates with the release of ARF proteins. HL60 cells were permeabilized for the indicated times, and 1-mL aliquots were removed and centrifuged. The proteins in the supernatants (released proteins) were precipitated with TCA, and redissolved in sample buffer. Samples (10⁶ cell equivalents) were run on a sodium dodecyl sulfate-polyacrylamide gel electrophoresis (SDS-PAGE), transferred, and probed with anti-ARF antibodies. **(C)** Restoration of GTPγS-stimulated PLD activity by rARF1 (nonmyristoylated). Much lower concentrations of ARF proteins can be used if the proteins are myristoylated. Labeled HL60 cells were permeabilized for 10 min and washed. The cells were then incubated at 37°C for 45 min in the presence of the indicated concentrations of rARF1 and GTPγS.

5. The samples are quenched with 375 µL of a mixture of chloroform:methanol (1:2 by vol). The sample is vigorously vortexed and a single phase obtained. A further addition of 125 µL of chloroform, and 125 µL of water is then made to obtain a two-phase system. After vigorous mixing, the samples are centrifuged for 5 min at 1000*g*. The lipids are present in the lower chloroform phase, and the top aqueous phase contains the water-soluble components including free [³H]choline. An aliquot of the top phase is used to analyse the presence of [³H]choline.

6. [³H]choline is separated from glycerophosphocholine and phosphorylcholine by cation chromatography. The aqueous phase containing the choline metabolites are applied to a 1-mL bed volume of Bio-Rex 70 cation exchange resin (sodium form, mesh size 200–400 purchased from Bio-Rad) in a Bio-Rad column. The column is rinsed with 3 mL of water to elute phosphorylated choline metabolites. Radiolabeled choline is quantitatively eluted with 3 mL of 50 m*M* glycine containing 500 m*M* NaCl, pH 3.0, directly into scintillation vials. The Bio-Rex resin is regenerated by extensively washing the resin with 0.5 *M* NaOH, pH 9.0, followed by washing with water. The resin is then washed with 0.1 *M* sodium phosphate, pH 7.0, and finally washed with water.

7. The radioactivity is measured after addition of a scintillation cocktail that is able to accommodate acidic solutions and high salt (e.g., Ultima Gold XR from Canberra Packard).

8. Calculation of data: the increase in labeled choline is expressed as a function of the total radioactivity (DPM) incorporated in the total choline lipids, (which includes both PC and sphingomyelin). The total lipid chloroform extract is carefully removed from the Eppendorf tube and transferred to a clean scintillation vial and the chloroform is allowed to evaporate by leaving it overnight on the bench (or the fume hood). Then, 500 µL of methanol should be added to the dried lipids followed by 2 mL of scintillation cocktail.

3.2. Establishing Conditions for Run-Down of PLD Activity

1. To establish conditions for the run-down of regulated PLD activity, 4 mL (1–2 × 10^7 cells/mL) of washed [³H]-choline-labeled cells are required. (This amount equates to 50 mL of confluent HL60 cells [labeled] as a start material.)

2. A cocktail containing SLO (0.4 IU/mL final), MgATP (1 m*M* final), and Ca²⁺ (100 n*M* buffered with 100 µ*M* EGTA final) in 1 mL is added to the labeled cells.

3. At timed permeabilization intervals (0, 2, 4, 8, 12, 16, 20, 25, 30, 40, 45, 60 min), four aliquots of cells are withdrawn (50 µL) and transferred to duplicate assay tubes containing 50 µL of Ca²⁺ (pCa 5 [10 µ*M*] buffered with 3 m*M* EGTA final), MgATP (1 m*M* final), MgCl₂ (2 m*M*) ± GTPγS (10 µ*M* final).

4. Assay tubes are incubated at 37°C for a further 30 min to monitor the extent of GTPγS-stimulated PLD activity.

5. At the end of the incubation, the reactions are quenched as described previously in **Subheading 3.1., step 5** for acutely permeabilized cells.

6. The data are plotted as the extent of the GTPγS-stimulated PLD activity as a function of the permeabilization interval. The run-down of activity is seen as the

permeabilization interval increases and the optimum time for run-down determined (*see* **Notes 1–4**).

3.3. Reconstitution of GTPγS-Stimulated PLD by Cytosolic Factors in Cytosol-Depleted Cells

Having established the optimum period for observing run-down, the restoration of GTPγS-stimulated PLD activity can be performed using exogenously added cytosol or purified proteins.

1. Four milliliters of washed [^{3}H]choline-labeled HL60 cells in permeabilization buffer are incubated with a cocktail containing SLO (0.4 IU/mL final), MgATP (1 mM final), and Ca^{2+} (100 nM buffered with 100 μM EGTA final) in 1 mL for the appropriate time that achieves run-down (10–40 min).
2. After permeabilization, the cells are diluted with 40 mL of ice-cold permeabilization buffer and centrifuged at 2000g for 5 min at 4°C to pellet the cells.
3. The cells are resuspended in ice-cold permeabilization buffer and 50-μL aliquots are transferred to assay tubes on ice. Assay tubes contain 50 μL of Ca^{2+} (pCa 5 [10 μM] buffered with 3 mM EGTA final), MgATP (1 mM final), MgCl$_2$ (2 mM) ± GTPγS (10 μM final), and rat brain cytosol (1–3 mg/mL) or purified proteins (such as ARF, Rho, or PKC).
4. Assay tubes are transferred to a water bath and further incubated at 37°C for 30 min to monitor the extent of GTPγS-stimulated PLD activity.
5. At the end of the incubation, the reactions are transferred to ice and reactions quenched as described previously in **Subheading 3.1.**, **step 5** for acutely permeabilized cells.

4. Notes

1. Depletion of proteins from the cytosol is dependent on their interactions with membranes or cytoskeleton. Truly cytosolic proteins exit with a faster time-course compared with some proteins, which are loosely anchored in cells. For example, release of ARF1 occurs within 5 min, whereas release of ARF6 can vary between 5 min and incomplete at 30 min depending on cell type *(11,13)*. PKC release is also complete within 5 to 10 min, but Rac and Rho proteins remain cell-associated despite prolonged permeabilization *(18)*. PLD enzymes are tightly membrane-associated because of palmitoylation and therefore remain cell-associated. Therefore, the length of time used for the depletion of cytosolic proteins is important because it will dictate which proteins leak out. For any known protein of interest, it is worthwhile to track the protein by Western blotting using appropriate antibodies in the supernatants obtained after pelleting the permeabilized cells. It should never be assumed that a protein that is recovered in the cytosol when cells are homogenized will leak out of the cells. Rho-GDI is one such protein that does not leak out of extensively permeabilized cells *(18)*.
2. During the step for cytosol depletion, we routinely have MgATP (1 mM) present. It is possible that in its presence some proteins may be retarded if their phosphorylation state is important in attachment to intracellular structures. In addi-

tion, in the presence of MgATP, the pool of PI(4,5)P$_2$ is maintained *(20)* and many proteins are tethered to membranes by their association with this lipid. Depletion of proteins can also be carried out in the absence of MgATP which does influence the time-course of run-down.

3. "Run-down is variable; therefore, it is important to work under well-defined cell densities and SLO concentrations routinely. The concentration of SLO can be increased to 0.6 IU/mL if run-down is insufficient. Normally, run-down of PLD activity is partial and routinely ranges from a loss of 70 to 90% of the response seen in acutely-permeabilized cells.

4. This protocol can be applied to any cellular response and not just PLD. We have applied this protocol for purifying proteins required for phospholipase C regulated by either G protein-coupled receptors or by receptor tyrosine kinases *(21)* and proteins required for exocytosis *(15)*. In this case, run-down of the secretory response is dependent on the absence of MgATP during the cytosol-depletion step. In the presence of MgATP, run-down is slower compared with in the absence of MgATP.

Acknowledgments

The work in the author's laboratory is supported by The Wellcome Trust.

References

1. Cockcroft, S. (1997) Phospholipase D: regulation by GTPases and protein kinase C and physiological relevance. *Prog. Lipid Res.* **35**, 345–370.
2. Cockcroft, S. (2001) Signalling roles of mammalian phospholipase D1 and D2. *Cell Mol. Life Sci.* **58**, 1674–1687.
3. Exton, J. H. (2002) Phospholipase D-structure, regulation and function. *Rev. Physiol. Biochem. Pharmacol.* **144**, 1–94.
4. Nakamura, S., Akisue, T., Jinnai, H., et al. (1998) Requirement of GM2 ganglioside activator for phospholipase D activation. *Proc. Natl. Acad. Sci. USA* **95**, 12,249–12,253.
5. Sarkar, S., Miwa, N., Kominami, H., et al. (2001) Regulation of mammalian phospholipase D2: interaction with and stimulation by G(M2) activator. *Biochem. J.* **359**, 599–604.
6. Lee, S., Park, J. B., Kim, J. H., et al. (2001) Actin directly interacts with phospholipase D, inhibiting its activity. *J. Biol. Chem.* **276**, 28,252–28,260.
7. Park, J. B., Kim, J. H., Kim, Y., et al. (2000) Cardiac phospholipase D2 localizes to sarcolemmal membranes and is inhibited by a-actinin in an ADP-ribosylation factor-reversible manner. *J. Biol. Chem.* **275**, 21,295–21,301.
8. Jenco, J. M., Rawlingson, A., Daniels, B., and Morris, A. J. (1998) Regulation of phospholipase D2-selective inhibition of mammalian phospholipase D isozymes by alpha- and beta-synucleins. *Biochemistry* **37**, 4901–4909.
9. Cullen, P. J., Cozier, G. E., Banting, G., and Mellor, H. (2001) Modular phosphoinositide-binding domains—their role in signalling and membrane trafficking. *Curr. Biol.* **11**, R882–R893.

10. Sciorra, V. A., Rudge, S. A., Prestwich, G. D., Frohman, M. A., Engebrecht, J., and Morris, A. J. (1999) Identifcation of a phosphoinositide binding motif that mediates activation of mammalian and yeast phospholipase D isoenzymes. *EMBO J.* **20,** 5911–5921.

11. Way, G., O'Luanaigh, N., and Cockcroft, S. (2000) Activation of exocytosis by cross-linking of the IgE receptor is dependent on ARF-regulated phospholipase D in RBL-2H3 mast cells: Evidence that the mechanism of activation is via regulation of PIP₂ synthesis. *Biochem. J.* **346,** 63–70.

12. O'Luanaigh, N., Pardo, R., Fensome, A., et al. (2002) Continual production of phosphatidic acid by phospholipase D is essential for antigen-stimulated membrane ruffling in cultured mast cells. *Mol. Biol. Cell* **13,** 3730–3746.

13. Skippen, A., Jones, D. H., Morgan, C. P., Li, M., and Cockcroft, S. (2002) Mechanism of ADP-ribosylation factor-stimulated phosphatidylinositol 4,5-bisphosphate synthesis in HL60 cells. *J. Biol. Chem.* **277,** 5823–5831.

14. Cockcroft, S., Thomas, G. M. H., Fensome, A., et al. (1994) Phospholipase D: A downstream effector of ARF in granulocytes. *Science* **263,** 523–526.

15. Fensome, A., Cunningham, E., Prosser, S., et al. (1996) ARF and PITP restore GTPγS-stimulated protein secretion from cytosol-depleted HL60 cells by promoting PIP₂ synthesis. *Curr. Biol.* **6,** 730–738.

16. Sarri, E., Pardo, R., Fensome-Green, A., and Cockcroft, S. (2003) Endogenous phospholipase D2 localizes to the plasma membrane of RBL 2H3 mast cells and can be distinguished from ADP ribosylation factor-stimulated phospholipase D1 activity by its specific sensitivity to oleic acid. *Biochem. J* . **369,** 319–329.

17. Fensome, A., Whatmore, J., Morgan, C. P., Jones, D., and Cockcroft, S. (1998) ADP-ribosylation factor and Rho proteins mediate fMLP-dependent activation of phospholipase D in human neutrophils. *J. Biol. Chem.* **273,** 13,157–13,164.

18. Leino, L., Forbes, L., Segal, A., and Cockcroft, S. (1999) Reconstitution of GTPgammaS-induced NADPH oxidase activity in streptolysin-O permeabilised neutrophils by specific cytosol fractions. *Biochem. Biophys. Res. Commun.* **265,** 29–37.

19. Tatham, P. E. R. and Gomperts, B. D. (1990) Cell permeabilisation, in *Peptide Hormones—A Practical Approach* (Siddle, K. and Hutton, J. C., eds.). IRL Press, Oxford, pp. 257–269.

20. Geny, B. and Cockcroft, S. (1992) Synergistic activation of phospholipase D by a protein kinase C- and a G-protein-mediated pathway in streptolysin O-permeabilized HL60 cells. *Biochem. J.* **284,** 531–538.

21. Thomas, G. M. H., Cunningham, E., Fensome, A., et al. (1993) An essential role for phosphatidylinositol transfer protein in phospholipase C-mediated inositol lipid signalling. *Cell* **74,** 919–928.

II

SPECIFIC TOPICS

E. Genomics and Proteomics

19

Analysis of Global Gene Expression Profiles Activated by Chemoattractant Receptors

Fernando O. Martinez and Massimo Locati

Summary

Microarrays are made by immobilizing to a solid support thousands of DNA probes that detect soluble complementary target sequences using the hybridization pairing rules of nucleic acids. Receptor triggering induces a cascade of signaling events that often involves the modulation of gene expression. In the last decade, the development of microarrays has provided scientists with an innovative tool to interrogate the cell transcriptional profile at a global level and to characterize genes according to their behavior in different conditions. This chapter outlines the use of microarrays as an innovative approach to study the global effect of transmembrane-receptor triggering. The effect of formyl peptides receptors activation on the gene transcriptional program of human monocytes is described as a model.

Key Words: GeneChip; Affymetrix; transcriptome; chemotactic factor; G protein-coupled receptor; microarray; signaling; receptor; gene expression.

1. Introduction

Cells detect a vast group of environmental changes using highly specialized membrane receptors, which initiate intracellular signaling events aimed at achieving appropriate cellular responses. In several cases, a main component of this adaptive response is the reconfiguration of the pool of expressed genes *(1–3)*.

Among the numerous membrane receptor families, the seven-transmembrane domain heterotrimeric G protein-coupled receptors constitute the largest of all, including more than 1000 distinct members in humans. These receptors are classified in several families according to their sequence homology and function. Chemoattractant receptors represent a distinct family that mediates leukocyte chemotaxis in response to agonist gradients by selectively coupling

From: *Methods in Molecular Biology, vol. 332: Transmembrane Signaling Protocols, Second Edition*
Edited by: H. Ali and B. Haribabu © Humana Press Inc., Totowa, NJ

to the Gα_i subunit containing G proteins. The human formyl-peptide receptor (FPR) was the first receptor of this family to be discovered (4) and can be considered the prototype (5). FPR and the related FPR-like receptor (FPRL)-1 are activated by peptides containing a formylated methionine residue (i.e., the tripeptide formyl-Leu-Met-Phe [fMLP]), a feature exclusively present in bacterial or mitochondrial proteins, that suggests an additional role for these receptors as "pattern recognition" receptors to detect the presence of bacterial infection or tissue damage (6). FPR and FPRL-1 are involved in phagocyte recruitment and activation in inflammatory foci, and also support the induction of transcripts for inflammatory cytokines and acute phase proteins (7).

Microarrays are an organized arrangement of probes immobilized on a solid support that, using the hybridization base-pairing rules, provide a tool for matching thousand of known nucleic acid simultaneously (8). For gene expression studies, transcripts levels are interrogated using a high number of immobilized complementary DNA (cDNA) probes. The use of different surfaces and methods for spotting or synthesizing the probes influences the experimental design principles and samples preparation procedures. This chapter will describe the use of the Affymetrix GeneChip® Human Genome U133A 2.0 array, which interrogates 14,500 well-characterized human genes (see **Note 1**). Although the experimental design may vary considerably among different studies, the data generated by microarray analysis can always be viewed as a matrix of expression levels, organized by samples vs genes, and reported in tabular format (9). Each sample represents separate microarray hybridization and generates a set of expression levels, one for each gene. The expression profile of a gene is formed by the vector of expression levels across the different samples.

This chapter describes the use of microarrays as an innovative tool to investigate the effects of transmembrane receptors triggering at the level of global gene expression. As an example, the simplest experimental protocol, a comparison of two biological conditions, each represented by a set of replicate samples, will be used to analyze the transcriptional program induced by fMLP in human monocytes.

2. Materials

This section lists all the reagents needed for this procedure, organized according to their order of use. In parentheses will be indicated the part of the procedure where each reagent is used. Please note that enzymes should not be stored in a frost-free freezer.

2.1. Total RNA Isolation, Nucleic Acid Synthesis, and Clean-Up (see Subheadings 3.2.–3.4.)

1. RNeasy Mini Kit (QIAGEN, cat. no. 74104). Store at room temperature (RT).
2. T7-oligo (dT) Promoter Primer Kit (Affymetrix, cat. no. 900375). Store at –20°C.

3. SuperScript™ Double-Stranded cDNA Synthesis Kit (Invitrogen, cat. no. 11917-010). Store at –20°C.
4. Enzo Bioarray™ HighYield™ RNA Transcript Labeling Kit (Affymetrix, cat. no. 900182). Store at –20°C.
5. GeneChip Sample Cleanup Module (Affymetrix, cat. no. 900371). Store at RT.

2.2. Eucaryotic Target Hybridization, Washing, and Staining (see Subheading 3.5.)

1. 50 mg/mL Bovine serum albumine (BSA) solution (Invitrogen Life Technologies, cat. no. 15561-020). Store at –20°C.
2. 10 mg/mL herring sperm DNA (Promega Corporation, cat. no. D1811). Store at –20°C.
3. GeneChip® Eukariotic Hybridization Control Kit (Affymetrix, cat. no. 900454). Store at –20°C.
4. 12X N-morpholinoethane sulfonic acid (MES) stock buffer (*see* **Note 2**): 64.61 g of MES hydrate SigmaUltra (Sigma-Aldrich, cat. no. M-5287); 193.3 g of MES sodium salt (Sigma-Aldrich, cat. no. M-5057); and molecular biology-grade water up to 800 mL. Mix, set pH at 6.5, and adjust volume to 1 L with molecular biology-grade water. Store at 4°C.
5. 2X Hybridization buffer (*see* **Note 2**): 8.3 mL of 12X MES stock buffer; 17.7 mL of 5 M NaCl (Ambion, cat. no. 9760G); 4.0 mL of 0.5 M ethylenediaminetetraacetic acid (EDTA; Sigma-Aldrich, cat. no. E-7889); 0.1 mL of Surfact-Amps 20 (Pierce Chemicals, cat. no. 28320); 19.9 mL of molecular biology-grade water. Store at 4°C.
6. Wash buffer A (nonstringent wash buffer; *see* **Note 2**): 300 mL of 20X sodium chloride/sodium phosphate/EDTA (Cambrex, cat. no. 51214); 1.0 mL of 10% Surfact-Amps 20 (Pierce Chemicals, cat. no. 28320); molecular biology-grade water to 1 L. Store at 4°C.
7. Wash buffer B (stringent wash buffer; *see* **Note 2**): 83.3 mL of 12X MES stock buffer (*see* **Subheading 3.5.**); 5.2 mL of 5 M NaCl (Ambion, cat. no. 9760G); 1.0 mL of 10% Surfact-Amps 20 (Pierce Chemicals, cat. no. 28320); and molecular biology-grade water to 1 L. Store at 4°C.
8. 2X Stain buffer: 41.7 mL of 12X MES stock buffer; 92.5 mL of 5 M NaCl (Ambion, cat. no. 9760G); 2.5 mL of 10% Surfact-Amps 20 (Pierce Chemicals, cat. no. 28320); and molecular biology-grade water to 250 mL. Store at 4°C.
9. Goat anti-immunoglobulin (Ig)G (Sigma-Aldrich, cat. no. I-5256) at 10 mg/mL in 150 mM NaCl. Store at 4°C.
10. R-phycoerythrin streptavidin (Molecular Probes, cat. no. S-866). Store at 4°C.
11. Anti-streptavidin antibody, biotinylated (Vector Laboratories, cat. no. BA-0500). Store at 4°C.

2.3. Miscellaneous Reagents and Supplies

1. Phosphate-buffered saline, pH 7.2 (Invitrogen Life Technologies, cat. no. 20012-019). Store at RT.
2. 10X TBE buffer (Cambrex, cat. no. 50843). Store at RT.

3. Molecular biology-grade water (Ambion, cat. no. 9915G). Store at RT.
4. 3 *M* Sodium acetate (Sigma-Aldrich, cat. no. S-7899). Store at 4°C.
5. Ethidium bromide (Sigma-Aldrich, cat. no. E-8751). Store at RT.
6. Dimethyl sulfoxide (Sigma-Aldrich, cat. no. D-5879). Store at RT and protect from moisture.
7. Absolute ethanol (molecular biology-grade). Store at RT.
8. 80% Ethanol aqueous solution. Store at RT.
9. 70% Ethanol aqueous solution. Store at RT.
10. 1 *N* NaOH (molecular biology-grade). Store at RT.
11. 1 *N* HCl (molecular biology-grade). Store at RT.
12. 50 m*M* MgCl$_2$ (molecular biology-grade). Store at RT.
13. 0.5 *M* EDTA (molecular biology-grade). Store at RT.
14. Sterile, RNase-free, microcentrifuge vials, 1.5 mL (Eppendorf, cat. no. 0030.121.589).
15. Aerosol-barrier RNase-free pipet tips (Ambion, cat. no. 12640).
16. Tygon tubing, 0.04 inner diameter (Cole-Parmer, cat. no. H-06418-04).
17. Tough-spots label dots (USA Scientific, cat. no. 9185-0000).

3. Methods

3.1. Overview

Transcripts detection in Affymetrix GeneChip array rely on oligoprobes (probe feature) directly synthesized on a glass support using a technology that merges combinatorial chemistry and photolithography. To be prepared for hybridization, the sample undergoes a multistep process that begins with the isolation of high quality total RNA from cells, used in a sond step as template for the synthesis of double-stranded cDNA. Subsequently an in vitro transcription reaction is used to produce biotin-labeled cRNA that must be fragmented before hybridization. After fragmentation the cRNA is end-labeled with biotin by terminal transferase and included in a hybridization cocktail. Each target biotin-labeled cRNA, derived from the corresponding mRNA present in the sample, is recognized by a distinct probe set represented by 11 perfect-match oligoprobes whose specificity is controlled by a corresponding set of 11 single-mismatch oligoprobes (**Fig. 1**). After hybridization, the array goes through computerized washing and staining protocols in a dedicated fluidics station. After hybridization, the array is scanned using the GeneArray® Scanner at 570 nm wavelength, and light emission of every probe cell, which is directly proportional to the abundance of the gene interrogated, is recorded (**Fig. 1**). The software defines the position of the different probe cells and computes their intensity level. Each image is stored in a file accessible from the Microarray Suite Expression Analysis platform. The following sections describe in detail the operative protocol, from RNA isolation to data managing (*see* **Note 3**).

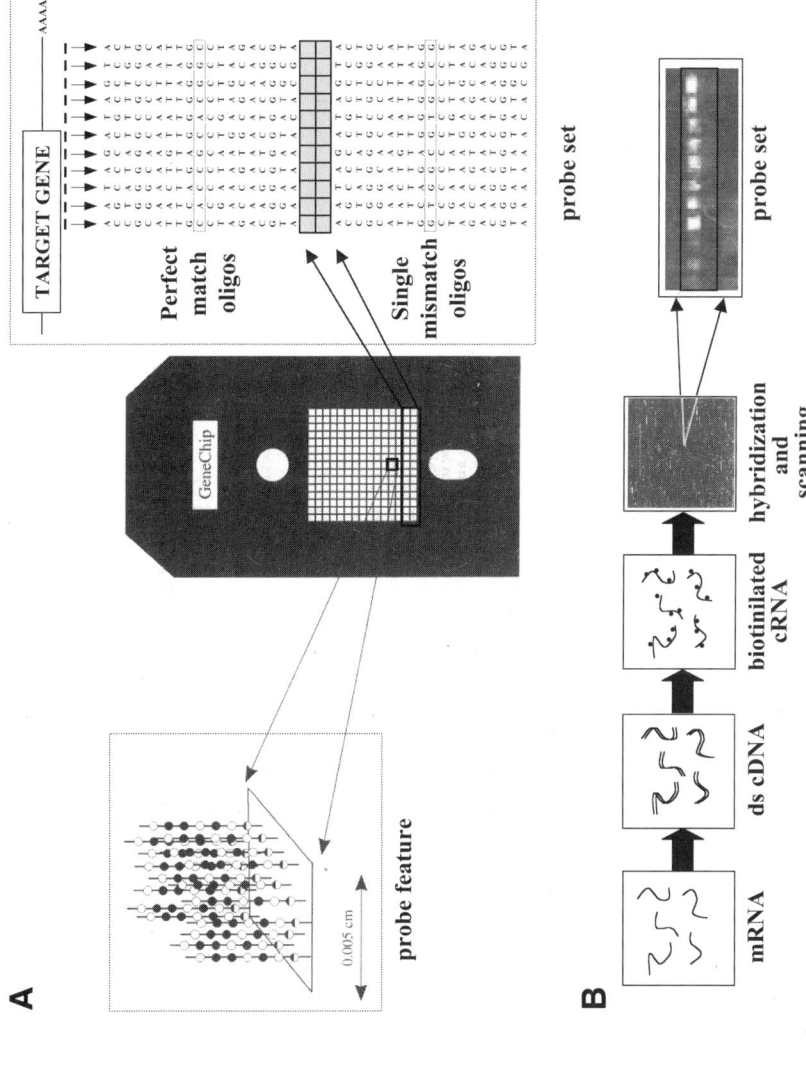

Fig. 1. (A) General organization of GeneChip arrays. The position of probe feature and probe set on a chip are schematically depicted. (B) Sequential steps in sample preparation and analysis.

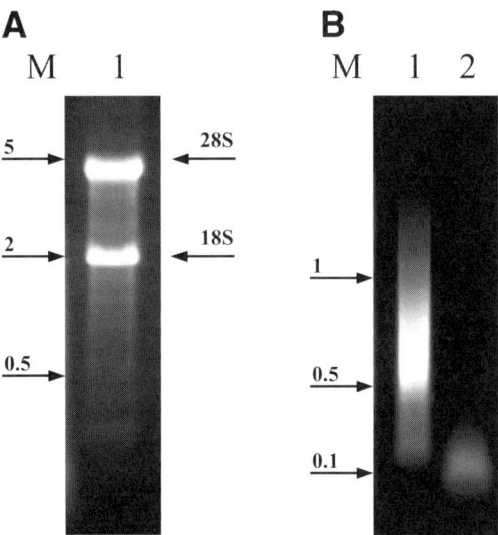

Fig. 2. (**A**) A 1-μg dose of total RNA on a 1.5% denaturing agarose gel. Arrows indicate the position of 18S and 28S ribosomal RNA bands (**right**) and RNA size markers (M) in kb (**left**). (**B**) cRNA (500 ng) before (**1**) and after (**2**) fragmentation on a 1% agarose gel. Size markers (M) in kb (left).

3.2. Total RNA Isolation and Quantification

The GeneChip sample synthesis protocol requires a total RNA quantity ranging from 5 to 20 μg. RNA should be at a minimum concentration of 0.5 μg/μL, as determined by absorbance at 260 nm on a spectrophotometer (1 absorbance unit = 40 μg/mL for RNA). The A260/A280 ratio should be between 1.9 and 2.1 (*see* **Note 4**). The RNA quality should be checked before proceeding by running it on an agarose gel. The ribosomal RNA bands should be clear, without any obvious smearing patterns from degradation (**Fig. 2A**). For high-quality total RNA from mammalian cells (up to 100 μg total RNA from 1×10^7 cells), use the RNeasy Mini Kit according to the following protocol:

1. Pellet cells by centrifugation.
2. Discard supernatant removing completely all media.
3. Loosen cell pellet by flicking and add buffer RLT as follows:

Buffer RLT (μL)	Number of cells
350	Up to 5×10^6
600	5×10^6 to 1×10^7

4. Resuspend pellet by pipetting to lyse cells (*see* **Note 5**).
5. Add one volume of 70% ethanol and mix well by pipetting.

6. Load up to 700 µL of sample, including any precipitate, on an RNeasy column sitting in a 2-mL collection tube, centrifuge at 8000g for 15 s at RT, and discard flow-through.
7. Pipet 700 µL of buffer RW1 and centrifuge at 8000g for 15 s at RT.
8. Transfer the column to a new tube, add 500 µL of buffer RPE, and centrifuge at 8000g for 15 s at RT.
9. Pipet 500 µL of buffer RPE onto the column and centrifuge at 8000g for 1 min at RT.
10. Open the cap of the spin column and centrifuge at 25,000g for 2 min in order to dry the membrane (*see* **Note 6**).
11. Transfer RNeasy column into a new 1.5-mL collection tube and pipet 11 to 50 µL of molecular biology-grade water directly onto the membrane.
12. Heat tube for 1 min at 65°C.
13. To elute centrifuge at 8000g for 1 min at RT.
14. Repeat if the expected yield is more than 30 µg.
15. Use immediately for synthesis of double-stranded cRNA or store at –80°C (*see* **Note 7**).

3.3. Synthesis of Double-Stranded cDNA

This step requires 5 to 20 µg of total RNA. Use the GeneChip T7-oligo (dT) Promoter Primer Kit and a T7-$(dT)_{24}$ primer (high-quality high-performance liquid chromatography purified) for priming first-strand cDNA synthesis, following this protocol.

3.3.1. First-Strand Synthesis

1. Before starting cDNA synthesis, determine the volumes of SuperScript II RT (SSII RT) required as follows:

Starting amount of total RNA (µg)	SSII RT (µL)
5–8	1
8.1–16	2
16.1–24	3

2. Calculate the volume of molecular biology-grade water to bring the final first strand synthesis volume to 20 µL, taking into account the volume of total RNA and SSII enzyme to be added.
3. Add the following to an RNase-free tube:

T7-$(dT)_{24}$ primer	1 µL
Total RNA	X µL
Molecular biology-grade water	12 - X - µL SSII RT

4. Incubate for 10 min at 70°C.
5. Spin briefly then place on ice for 1 min.
6. Add the following kit components in order (*see* **Note 8**):

5X first-strand cDNA Buffer	4 µL
0.1 *M* dithiothreitol (DTT)	2 µL
10 m*M* dinucleotide triphosphate mix	1 µL

7. Incubate for 2 min at 42°C.
8. Add SSII RT as calculated at point 1.
9. Incubate for 1 h at 42°C.
10. Quickly chill on ice and centrifuge to collect sample.
11. Use immediately for second-strand synthesis reaction or store at –20°C.

3.3.2. Second-Strand Synthesis

1. Place first-strand reactions on ice and centrifuge briefly to bring sample down.
2. Add to the first-strand reaction (20 µL) the following kit components (*see* **Note 8**):

5X sond-strand reaction buffer	30 µL
10 m*M* dinucleotide triphosphate mix	3 µL
10 U/µL *Escherichia coli* DNA ligase	1 µL
10 U/µL *E. coli* DNA polymerase I	4 µL
2 U/µL *E. coli* RNase H	1 µL
Molecular biology-grade water to a final volume of 150 µL	

3. Tap tube to mix and briefly spin to remove condensation.
4. Incubate for 2 h at 16°C.
5. Add 2 µL of (10 U) T4 DNA polymerase.
6. Incubate for 5 min at 16°C.
7. Stop reaction with 10 µL of 0.5 *M* EDTA.
8. Use immediately for clean-up procedure for cDNA or store at –20°C.

3.3.3. Double-Stranded cDNA Clean-Up (see **Note 9**)

1. Add 600 µL of cDNA binding buffer to the 162 µL of final double-stranded cDNA product.
2. Vortex for 3 s.
3. Apply 500 µL of the sample to the cDNA cleanup spin column sitting on a 2-mL collection tube.
4. Centrifuge at 8000*g* for 1 min at RT and discard flow-through.
5. Reload the spin column with the rest of the mixture (262 µL).
6. Centrifuge at 8000*g* for 1 min at RT and discard flow-through and collection tube.
7. Transfer column into a 2-mL collection tube.
8. Pipet 750 µL of cDNA wash buffer.
9. Centrifuge at 8000*g* for 1 min at RT and discard flow-through.
10. To dry the membrane open the cap of the spin column and centrifuge at 25,000*g* for 5 min at RT (*see* **Note 6**).
11. Discard flow-through and collection tube and transfer column to a 1.5-mL collection tube.
12. Pipet 14 µL of cDNA elution buffer directly onto the column membrane.
13. Incubate at RT for 1 min.
14. To elute, centrifuge at 25,000*g* for 1 min at RT.

15. Dilute 1 µL of eluate 1:100 and measure absorbance at 260 and 280 nm on a spectrophotometer. The A260/A280 ratio should be greater than 1.5 to proceed with the protocol.
16. Use immediately for the synthesis of biotin-labeled cRNA or store at –20°C.

3.4. Synthesis of Biotin-Labeled cRNA

3.4.1. In Vitro Transcription

1. Use the following table to determine the amount of cDNA to be used for each in vitro transcription (IVT) reaction, assuming that 12 µL was recovered in the previous reaction:

Starting amount of total RNA (µg)	Volume of cDNA (µL)
5–8	10
8.1–16	5
16.1–24	3.3

2. To avoid precipitation of DTT keep tubes at RT while adding the different components following this order (*see* **Note 8**):

cDNA	from previous table
10X HY reaction buffer	4 µL
10X biotin-labeled ribonucleotides	4 µL
10X DTT	4 µL
10X RNase inhibitor mix	4 µL
20X T7 RNA polymerase	2 µL
Molecular biology-grade water up to 40 µL	

3. Carefully mix reagents and collect mixture by brief centrifugation.
4. Incubate for 4 h at 37°C, mixing every 30 min.
5. Use immediately for cRNA purification procedure or store at –20°C (–80°C for long-term storage).

3.4.2. Biotin-Labeled cRNA Clean-Up and Quantification (see **Note 9**)

Save an aliquot of the unpurified IVT product for analysis by gel electrophoresis. All steps of the protocol should be performed at RT. During the procedure, work without interruption.

1. To the IVT reaction add 60 µL of molecular biology-grade water and vortex for 3 s.
2. Add 350 µL of IVT cRNA binding buffer and vortex for 3 s.
3. Add 250 µL of absolute ethanol and mix well by pipetting. Do not centrifuge.
4. Apply sample (700 µL) to an IVT cRNA cleanup spin column sitting on a 2-mL collection tube.
5. Centrifuge at 8000g for 15 s.
6. Discard flow-through and collection tube and transfer the column into a new 2-mL collection tube.
7. Pipet 500 µL of IVT cRNA wash buffer onto the column.

8. Centrifuge at 8000*g* for 15 s.
9. Discard flow-through and pipet 500 μL of 80% ethanol onto the column.
10. Centrifuge at 8000*g* for 15 s and discard flow-through.
11. Open the cap of the column and centrifuge at 25,000*g* for 5 min to dry the membrane.
12. Discard flow-through and transfer column into a new 1.5-mL collection tube.
13. Pipet 11 μL of molecular biology-grade water directly onto the column membrane.
14. Centrifuge at 25,000*g* for 1 min.
15. Pipet 10 μL of molecular biology-grade water directly onto the column membrane.
16. Centrifuge at 25,000*g* for 1 min.
17. Dilute 1 μL of eluate 1:100 in water and measure absorbance at 260 nm and 280 nm on a spectrophotometer (1 absorbance unit = 40 μg/mL cRNA). The A260/A280 ratio should range between 1.9 and 2.1 to proceed with the protocol.
18. Calculate cRNA concentration adjusting for remnant of unlabeled starting material (adjusted cRNA) by applying the following formula, assuming a 100% carryover:

$$\text{adjusted cRNA} = \text{RNAm} - (\text{total RNAi}) \times (y)$$

where RNAm = amount of cRNA measured after IVT (μg); total RNAi = starting amount of total RNA (μg); and y = fraction of cDNA reaction used in IVT (i.e., 1, 1/2, …)
20. Adjust cRNA concentration to 0.6 μg/μL before proceeding with the fragmentation.
21. Control yield and size distribution of labeled transcripts by running 1% of each sample on a 1% agarose gel (**Fig. 2B**). Prepare samples for electrophoresis by mixing with loading dye and ethidium bromide (final concentration 0.5 μg/mL) and heating for 5 min at 65°C before loading.

3.4.3. Sample Fragmentation

The fragmentation buffer, present in the GeneChip sample clean-up module, has been optimized to break full-length cRNA to 35 to 200 base fragments (**Fig. 2B**). The final concentration of cRNA in the fragmentation mix can range from 0.5 to 2 μg/μL. A minimum of 20 μg of fragmented cRNA is required for quality control checking and a GeneChip hybridization.

1. Add the following reagents (scale up volumes if required):

cRNA (20 μg)	X μL
5X Fragmentation buffer	8 μL
Molecular biology-grade water	32-X μL

2. Incubate for 35 min at 94°C.
3. Chill on ice.
4. Store undiluted, fragmented sample cRNA at –20°C until ready to perform the hybridization (*see* **Note 10**).

3.5. GeneChip Hybridization and Scanning

3.5.1. Hybridization

1. Equilibrate array to RT 15 min before use (*see* **Note 11**).
2. For each target prepare the hybridization cocktail by mixing the following (scale up volumes for multiple targets):

Fragmented cRNA	15 µL
3 n*M* Control oligonucleotide B2	5 µL
20X Eukaryotic hybridization controls	15 µL
10 mg/mL Herring sperm	3 µL
50 mg/mL BSA	3 µL
2X Hybridization buffer	150 µL
Dimethyl sulfoxide	30 µL

   Molecular biology-grade water to final volume of 300 µL

3. Heat the hybridization cocktail for 5 min to 99°C in a heat block.
4. Fill the probe array through one of the septa with 250 µL of 1X hybridization buffer.
5. Incubate it for 10 min at 45°C with rotation (*see* **Note 12**).
6. Transfer the hybridization cocktail for 5 min at 45°C in a heat block.
7. Spin hybridization cocktail in a centrifuge at maximum speed for 5 min to remove insoluble material.
8. Remove the hybridization buffer from the probe array cartridge and fill with appropriate volume of the clarified hybridization cocktail.
9. Place probe array into the hybridization oven set at 45°C.
10. Hybridize at 60*g* for 16 h at 45°C (*see* **Note 13**).

3.5.2. Washing and Staining

1. Prime the fluidic station and load buffers in the respective lanes (*see* **Note 14**).
2. Start the MicroArray Suite 5 (MAS5) program.
3. Click fluidic icon on the tool bar and select *Run* and *Fluidics*.
4. Select *Experiment* and *N°* probe array in the experiment drop down menu.
5. *Select Protocol* and *Prime* in the drop down menu.
6. Select *Run*.
7. Go over **steps 3** to **6** for each module that will be used.

3.5.3. Scanning

1. Switch on the laser 10 to 20 min before the fluidic station protocol ends.
2. Select *Scanner* and then *Laser ON* in the drop down menu.
3. Eject cartridge when wash is complete.
4. Keep array dark before scanning (*see* **Note 15**).
5. Start the automatic clean-out procedure.
6. Clean the array surface with a delicate tissue before scanning.
7. Select *Run scanner*.

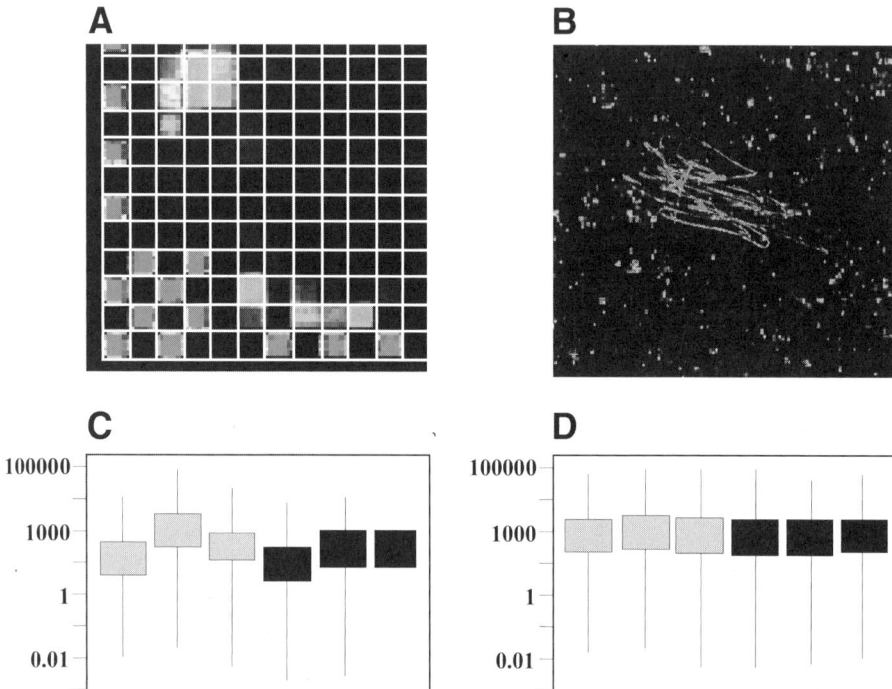

Fig. 3. (**A**) Correct grid alignment. (**B**) Detail of a scratched area. (**C** and **D**) Box-plot of average expression levels before and after scaling, respectively.

8. Once scanning is complete, select *Scanner* and then *Laser OFF* in the drop-down menu.

3.6. Data Analysis

Numerous software packages, both free and commercial *(10,11)*, presently are available for microarray data management. The purpose of this section is to provide some rudiments for microarray data analysis in the Affymetrix platform, that uses MAS5 software for image detection, expression values generation and basic statistics, and Data Mining Tool 2 (DMT2) software for second-level statistics and data mining. Description of different algorithms and software, indicated in particular for complex experimental designs, is beyond the scope of this chapter and can be found in specialized textbooks (*see* **Note 1**).

3.6.1. First-Level Data Analysis: Detection and Normalization

Data analysis starts by visual inspection of the array images to confirm the correct alignment of the detection grid and to exclude the presence of artifacts, such as scratches or dust particles (**Fig. 3A,B**). In the presence of a small damaged area, the involved probe cells can be masked; if instead it is extended, the arrays should be excluded for further analysis. Next, the expression values for all samples are computed using the *Analysis* toolbar of MAS5. The software generates a report file that includes parameters that describe the quality of the samples (background, scaling factor, noise, 3'/5' glyseraldehyde-3-phosphate dehydrogenase and actin ratios: *see* **Notes 16–18**). MAS5 accompanies the expression levels of each gene with statistical quality metrics and a qualitative detection call (present/absent). Before proceeding with the analysis, examine the behavior of the different metrics as well as the frequency distributions of the expression levels in the different samples to detect poor quality data.

The expression levels of the different arrays are not homogeneous, and before being compared, they must be normalized. MAS5 includes two types of normalization procedures relying on the assumption that, being the quantity of initial mRNA identical for all samples, the overall expression (intended as brightness) should be the same in all arrays. In the first procedure, called normalization, the software compares an experimental array with a baseline array and normalizes the average intensity of the former to the average intensity of the selected baseline; in the second type, called scaling, the operator designates an arbitrary target signal and the software scales the average intensity of all genes on each array, to the target signal specified, allowing comparison of multiple arrays within a data set. Because the average of expression levels is around 500 for all arrays data set, scaling to that target value will be applied (**Fig. 3C,D**).

3.6.2. Second-Level Data Analysis: Gene Filtering and Clustering

Microarray comparative data analysis can be separated into two extensive categories: grouping of genes to discover broad patterns of biological behavior, and filtering of genes to identify specific genes of interest *(9)*. Although gene grouping mainly is addressed by cluster analysis *(12)*, the gene-filtering task mainly relies on hypothesis testing. Although cluster analysis techniques are defined and reproducible powerful procedures, different algorithms, normalizations or distance metrics will place the objects (samples or genes) into different positions. The reliability of clustering techniques greatly benefits from complex experimental design, such as time–course or dose–response experiments to different or combined stimuli, that allow the formation of more defined expression vectors. As stated previously, our experimental design corresponds to a simple analysis of two biological conditions tested in replicate, which generates a data set more appropriate for differential expression analysis rather than expression patterns assessment.

Statistical tests have been performed in the DMT2 software platform (*see* **Note 1**). The comparison relies on the *t*-test that measures the difference

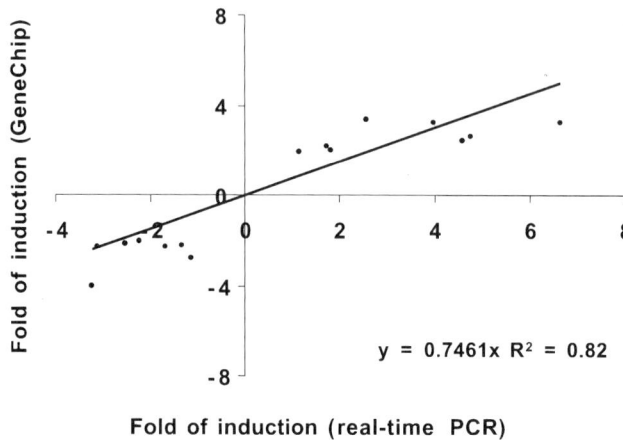

Fig. 4. Correlation of fold of induction detected by GeneChip microarray and quantitative real-time polymerase chain reaction.

between the two sample means, based on the amount of variability in the sample means. The assumption of unequal variance would seem to be more appropriate for gene expression analysis, especially if the active genes have greater variability in gene expression than inactive ones have. Statistical analysis allows one to filter the data set, selecting genes whose values are significantly different in treated and control conditions. Statistically generated lists of genes frequently include a large number of molecules with minor expression changes. To select genes with a satisfactory statistical test and fold change, usually regarded as genes with higher probability of biological relevance, one frequently applies a second filter based on cutoff values. There are not universal cutoff values for filtering genes in microarray analysis, and the operator should validate different values and select the most informative. However, in most experiments based on GeneChip microarrays a cutoff of $p > 0.05$ and a fold change greater than 2 has been applied, leading to results that usually were in good agreement with results obtained by independent methods for transcript analysis. Filtering the experimental database with indicated cutoff values identified a total of 123 fMLP-responsive transcripts (0.85% of investigated transcripts). To confirm validity of microarray analysis, 15 fMLP-responsive transcript identified by GeneChip analysis were investigated by means of quantitative real-time polymerase chain reaction. As shown in **Fig. 4**, the two methods displayed a good agreement ($r^2 = 0.82$), indicating that filtering parameters were adequate. The slope value ($m = 0.72$) also demonstrates the tendency of GeneChip analysis to underestimate fold of gene induction, most likely as a consequence of the normalization procedures.

To better understand the results, the information on individual genes retrieved in public databases is of invaluable help. DMT2 adds to the gene list some known identifiers, but additional information can be obtained through the NetAffx web interface (*see* **Note 1** and **ref. *13***), that allows to import a .txt file containing the list of interest and produces a detailed annotation table containing for each probe set gene and protein detailed characterization, functional information, metabolic pathway, and disease association.

4. Notes

1. Additional information about the Affymetrix technology and softwares can be found at www.affymetrix.com. For a more comprehensive review on microarrays technology and analysis tools, visit http://ihome.cuhk.edu.hk/%7Eb400559/array.html.
2. Filter through a 0.2-μm filter; do not autoclave. Store at 4°C, and shield from light. Discard solution if yellow.
3. During this laboratory procedure, only powder-free gloves should be worn. To avoid contamination with exogenous nucleases, all reagents and supplies used must be molecular biology grade. It is recommended to read all information and instructions accompanying reagents and kits because they are updated incessantly. Before using the different reagents, centrifuge them briefly to guarantee that the components remain at the bottom of the tube. When working with spin columns, dispense directly onto the membrane to avoid low eluate recovery.
4. RNA amounts between 5 and 20 μg guarantees sufficient quantity of labeled cRNA for target assessment and hybridization to the expression probe arrays. It also avoids the risk of overloading the clean-up columns caused by excessive starting material.
5. If cells grow in monolayer, the lysis for RNA isolation can de done directly by adding Buffer RLT to the wells (maximum diameter, 10 cm).
6. If open cap centrifugation is required place columns into the centrifuge alternating buckets. Point caps over the neighboring bucket, oriented in the opposite direction to the rotation.
7. There are several stopping points in the assay, coinciding with the quality checkpoints (RNA isolation, cRNA synthesis, cRNA fragmentation). Purity, quantity, and electrophoretic size distribution must be controlled because they constitute sensitive indicators of problematic labeling procedures and/or starting materials. **Fig. 2** provides exemplary gel pictures.
8. Preparing a master mix is often used. Although not ideal, this avoids sample-to-sample variation.
9. In the clean-up procedures, all steps must be carried out at RT and without pause. IVT cRNA and cDNA washing buffers are supplied concentrated. Before using them for the first time, add 20 and 24 mL of absolute ethanol, respectively, as indicated on each bottle. The IVT cRNA binding buffer may form a precipitate that can be dissolved by warming to 30°C, and then placing it at RT.
10. cRNA targets can be stored safely for at least 1 yr at –80°C.

11. It is important to allow the arrays to equilibrate to RT completely. Specifically, if the rubber septa are not equilibrated, they may be prone to cracking, which can lead to leaks. It is necessary to use two pipet tips when filling the probe array cartridge: one for filling and the second to allow venting of air from the hybridization chamber.

12. To avoid stress to the motor load probe arrays balanced around the axis. Rotate at 60g.

13. Some reagents are required immediately after completion of hybridization. Please check **Subheading 3.5.2.** and prepare them during the last part of the incubation.

14. The scanner must be switched on before the MicroArray Suite is launched. The fluidics station must be primed when it is first started, when wash solutions are changed, before washing, when a shutdown has been performed or when the liquid crystal display instructs you to prime.

15. The array can be stored for up to 8 h at 4°C if needed.

16. The 3'/5' signal intensity ratio gives a global indication of starting RNA integrity, first strand cDNA synthesis efficiency, and cRNA IVT efficiency. This parameter should be less than three for most of the tissues. It is appropriate to document the 3'/5' ratios of all samples and select the results that diverge.

17. The maximum background accepted for a sample should be less than 600. A high background implies the presence of impurities, such as cell debris and salts that are fluorescing at 570 nm. High background creates an overall loss of sensitivity in the experiment, and in particular, transcripts present at very low levels in the sample may be incorrectly called as absent. Sample with a noise greater than 21 should be discharged.

18. Spiked controls (BioB, bioC, bioD, and cre) are added during hybridization at staggered concentrations. BioB is at the detection limit for most expression arrays and should be present at least 70% of the time; other controls should be called present all of the time, with increasing signal expression values (BioC, BioD, and cre, respectively). Absent calls, or relatively low signal values, indicate a potential problem with the hybridization reaction or subsequent washing and staining steps.

References

1. Lemon, B. and Tjian, R. (2000) Orchestrated response: a symphony of transcription factors for gene control. *Genes Dev.* **14**, 2551–2569.
2. Emerson, B. M. (2002) Specificity of gene regulation. *Cell* **109**, 267–270.
3. Brivanlou, A. H. and Darnell, J. E., Jr. (2002) Signal transduction and the control of gene expression. *Science* **295**, 813–818.
4. Boulay, F., Tardif, M., Brouchon, L., and Vignais, P. (1990) The human N-formylpeptide receptor. Characterization of two cDNA isolates and evidence for a new subfamily of G-protein-coupled receptors. *Biochemistry* **29**, 11,123–11,233.
5. Murphy, P. M. (1994) The molecular biology of leukocyte chemoattractant receptors. *Annu. Rev. Immunol.* **12**, 593–633.

6. Le, Y., Murphy, P. M., and Wang, J. M. (2002) Formyl-peptide receptors revisited. *Trends Immunol.* **23,** 541–548.
7. Arbour, N., Tremblay, P., and Oth, D. (1996) N-formyl-methionyl-leucyl-phenylalanine induces and modulates IL-1 and IL-6 in human PBMC. *Cytokine* **8,** 468–475.
8. Southern, E., Mir, K., and Shchepinov, M. (1999) Molecular interactions on microarrays. *Nat. Genet.* **21,** 5–9.
9. Nadon, R. and Shoemaker, J. (2002) Statistical issues with microarrays: processing and analysis. *Trends Genet.* **18,** 265–271.
10. Gautier, L., Cope, L., Bolstad, B. M., and Irizarry, R. A. (2004) affy—analysis of Affymetrix GeneChip data at the probe level. *Bioinformatics* **20,** 307–315.
11. Barash, Y., Dehan, E., Krupsky, M., et al. (2004) Comparative analysis of algorithms for signal quantitation from oligonucleotide microarrays. *Bioinformatics* **20,** 839–846.
12. Lyons-Weiler, J., Patel, S., and Bhattacharya, S. (2003) A classification-based machine learning approach for the analysis of genome-wide expression data. *Genome Res.* **13,** 503–512.
13. Liu, G., Loraine, A. E., Shigeta, R., et al. (2003) NetAffx: Affymetrix probesets and annotations. *Nucleic Acids Res.* **31,** 82–86.

20

Genetic Reconstitution of Bone Marrow for the Study of Signal Transduction Ex Vivo

Martha S. Jordan

Summary

Introducing genes into cells by retroviral transduction has greatly increased the ability to study signal transduction pathways in primary cells. Retroviral transduction has proven to be an efficient method to express genes of interest in cells that are difficult to manipulate using standard transfection techniques. This technology also can be coupled with classic protocols for generating bone marrow chimeras. Murine bone marrow cells can be infected with a retrovirus expressing wild-type or mutant forms of a gene of interest and subsequently transplanted into irradiated recipient hosts. The requirement for a gene of interest in hematopoietic cell development, as well as its role in specific signal transduction pathways, can then be studied. This chapter provides protocols for the production of high-titer replication-incompetent retrovirus, retroviral infection of murine bone marrow, the generation of bone marrow chimeras, and analysis of chimeras by flow cytometry.

Key Words: Chimeras; retroviral transduction; 5-fluorouracil; MIGR1; MCSV; spin infection; GFP.

1. Introduction

Studying signal transduction in primary cells typically is hampered by the relative inefficiency of primary cells to be transfected and cultured and often is limited to the study of mature cell populations. These issues have been circumvented to some degree through the use of transgenic and knockout technologies; however, these techniques can be costly and time consuming. Genetic reconstitution of bone marrow cells by retroviral transduction followed by the generation of bone marrow chimeras is becoming an increasingly popular way to study the role of signaling molecules in the development of hematopoietic cells, as well as their function in specific molecular pathways *(1–3)*.

From: *Methods in Molecular Biology, vol. 332: Transmembrane Signaling Protocols, Second Edition*
Edited by: H. Ali and B. Haribabu © Humana Press Inc., Totowa, NJ

Retroviral transduction is an efficient way of stably introducing nonviral genes into a variety of cell types. Retroviral vectors are generated easily by using standard recombinant DNA technologies. In general, these vectors contain a promoter to drive transcription of an inserted gene of interest, a RNA packaging signal to direct packaging of this RNA, and all of the viral sequences required for proper integration of proviral sequences. Importantly, these vectors lack the full compliment of viral packaging genes that are required for the production of replication-competent retrovirus. Instead, these structural genes are provided in *trans* in cells used for the production of retrovirus. This method allows for the production of retrovirus containing a gene of interest that is capable of entering a target cell and stably integrating into the host genome, but incapable of replicating because it lacks the packaging machinery.

One challenge for achieving efficient infection into the cell type of interest has been generating high-titer retrovirus. Initially, producer cell lines were generated by the stable transfection of retroviral structural genes, gag, pol, and env followed by a second stable transfection with a retroviral vector containing the gene of interest *(4)*. In efforts to increase viral titers and decrease the time associated with retrovirus production, two transient transfection methods for the production of high-titer retrovirus have been described. One involves co-transfection of the retroviral vector containing a complementary DNA (cDNA) of interest with a helper packaging plasmid, which encodes for the viral packaging genes, into 293 or 293T cells *(5–7)*. The second utilizes a 293T cell derived packaging line that stably expresses the genes required for viral packaging such that only transfection with the retroviral vector is required *(8)*. Both methods yield high-titer retrovirus that is replication-incompetent.

This chapter first describes the generation of high-titer retrovirus by co-transfection of helper plasmid and the replication-incompetent retroviral vector MIGR1 into 293 or 293T cells *(9)*. A protocol for transducing murine bone marrow with retrovirus and the generation of bone marrow chimeras follows. Finally, this chapter includes a basic protocol for flow cytometry because retrovirally transduced cells using the MIGR1 plasmid can be detected by their green fluorescent protein (GFP) fluorescence.

It is important to note that although the production of replication-incompetent virus is described, recombination events may lead to the production of replication-competent virus. Retrovirus produced according to this protocol specifically targets rodent cells; however, one should take precautions when handling and storing the virus and transduced cells especially if amphotropic plasmids are used.

2. Materials

2.1. Construction of Retroviral Vector

1. MIGR1 plasmid (Warren Pear, University of Pennsylvania).
2. Restriction enzymes, T4 DNA ligase.
3. *Escherichia coli.*
4. Agarose and DNA gel electrophoresis equipment.

2.2. Generation of High-Titer Retrovirus

1. 293 Cells.
2. Trypsin and phosphate-buffered saline (PBS) without Ca^{2+} or Mg^{2+}.
3. MIGR1 plasmid with cDNA of interest.
4. 25 mM Chloroquine (1000X stock).
5. Sterile water.
6. 0.45-µm Syringe filters (Millipore).
7. Helper plasmid (Imgenex, pCL-ECO).
8. 2.5 M Calcium chloride.
9. 6-cm Tissue culture dishes (Corning).
10. 1-mL or 1.5-mL cryovials.
11. Transfection cocktail: combine 10 µg of retroviral vector DNA, 5 µg of pCL-ECO helper virus, 50 µL of 2.5 M CaCl, and sterile H_2O to a final volume of 500 µL.
12. 2X HEPES-buffered saline: 50 mM HEPES, 10 mM KCl, 12 mM dextrose, 280 mM NaCl, 1.5 mM Na_2HPO_4, pH 7.05.
13. 293/3T3 cell media: Dulbecco's modified Eagle's medium, 10% fetal bovine serum (FBS), 2 mM L-glutamine, 50 U/mL penicillin, 50 µg/mL streptomycin.
14. 3T3 cells.
15. Polybrene (American Bioanalytical).

2.3. Retroviral Transduction of Murine Bone Marrow

1. Donor mice, approx 6- to 12-wk old.
2. Fluorouracil (5-FU; Roche, NDC 0004–1977–01, Item no. 72660) 5-FU is a toxic chemical and should be handled with care. Protect from light.
3. PBS.
4. Culture medium: Iscove's modified Dulbecco's medium (IMDM) + 15% FBS, 2 mM L-glutamine, 50 U/mL penicillin, 50 µg/mL streptomycin, 50 µM β-mercaptoethanol.
5. Stimulation media: culture media containing 10 ng/mL interleukin-3, 10 ng/mL interleukin-6, 50 ng/mL stem cell factor, and 5% WEHI media (optional). Cytokines purchased from R&D Systems or Peprotech.
6. Polybrene (American Bioanalytical).
7. 70-µm Nylon cell strainer (BD Falcon, cat. no. 352350).

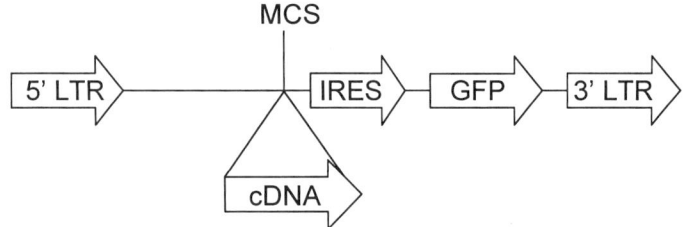

Fig. 1. Schematic of the MIGR1 vector.

8. Six-well tissue culture plate.
9. 10-mL Syringe and 30-gage needle.
10. Red blood cell lysis buffer (Sigma).

2.4. Reconstitution of Recipient Mice

1. Recipient mice, approx 6- to 10-wk old.
2. Cesium source.
3. Neomycin sulfate (Sigma) and polymixin B (Schein, Florham Park, NJ) or trimethoprim/sulfamethoxazole (Animal facility veternarian).

2.5. Ex Vivo Analysis by Flow Cytometry

1. 96-Well V-bottom plates.
2. Fluorescence-activated cell sorting (FACS) buffer: PBS + 2% FBS + 0.02% azide.
3. Fluorescently labeled antibodies.
4. Flow cytometer.

3. Methods

3.1. Construction of the Retroviral Vector

The MIGR1 (MSCV, IRES, GFP, retrovirus-1) vector *(9)* is a replication-incompetent retroviral vector derived from the murine stem cell virus (MSCV) vector MSCV 2.2 *(10)*. It contains a 5' long-term repeat followed by multiple cloning sites into which a cDNA of interest can be inserted. Following the multiple cloning sites is an internal ribosomal entry site (IRES) that drives transcription of the enhanced GFP cDNA, allowing for coexpression of a gene of interest with GFP as a marker for transduced cells. This vector also contains a 3' long-term repeat and the selectable marker ampicillin (**Fig. 1**).

Cloning of a gene of interest into the MIGR1 vector is performed by standard recombinant technology and is not described here. Other vectors have been used for generating retrovirus capable of infecting bone marrow and can be modified to accommodate different experimental protocols. For example, instead of GFP, cell surface proteins can be coexpressed to identify transduced cells *(11)*. It is important to ensure that expression of the "marker" does not alter function of the transduced cells. This can be achieved by using a marker of a different species or a truncated form of the protein *(11)*. Another consideration is the level of expression of your protein of interest. cDNAs downstream of an internal ribosomal entry site typically are expressed at lower levels as compared with those expressed off of the retroviral promoter *(11,12)*. However, in many instances, although the actual expression levels may differ, the relative expression of the marker and inserted cDNA can be correlated.

3.2. Generation of High-Titer Retrovirus

Herein is a protocol for co-transfection of 293 cells with a retroviral vector and helper packaging plasmid (293T-cells also can be used following this same protocol). It is based on calcium phosphate transfection for gene delivery as described previously *(8)*. Although transfection by calcium phosphate is described here, high-titer retrovirus can be achieved using other methods of transfection including commercially available kits. For all sections of this protocol, cells should be cultured at 37°C with 5% CO_2 and centrifuge spins are based on spinning in a tabletop Sorval RT7 or Beckman GS-6R unless otherwise noted.

3.2.1. Growing 293 and 3T3 Cells

1. Remove media, rinse plate once with PBS (no Ca^{2+} or Mg^{2+}) add 2 mL of trypsin. Allow cells to come off the plate. Pipet up and down to generate a single cell suspension. Add 5 mL of 293/3T3 complete media to quench the trypsin. Pellet cells at 300*g* for 7 min and resuspend in complete 293/3T3 media.
2. Split cells 1:4 or 1:5 for passaging when they reach 90% confluence.

3.2.2. Making Retrovirus

1. Day 1: Harvest 293 cells when they are roughly 60 to 80% confluent. Plate 2.5×10^6 cells in a 6-cm tissue culture plate in 4 mL of complete media. Culture overnight (*see* **Note 1**).
2. Day 2: Cells should be roughly 60 to 80% confluent. Replace media with 3 mL of media containing 25 µ*M* chloroquine (*see* **Note 2**).
3. Using a pipet to generate bubbles, slowly add 500 µL of 2X HEPES-buffered saline to the transfection cocktail. Pipet to mix and immediately add drop-wise evenly over the plate and gently move plate from side-to-side and forward-backward to mix. Return plate to incubator.
4. Seven to 10 h later, replace media with fresh 293 media without chloroquine.
5. Day 3: Replace media with 4 mL of fresh 293 media. To increase the viral titer, 3 mL instead of 4 mL can be added.
6. Day 4: Harvest viral supernatant. Between 16 and 24 h after the last media change, collect supernatant and filter through a 0.45-µm filter or spin at 300*g* for 7 min to remove any remaining cells.
7. Aliquot retroviral supernatant into 1-mL aliquots, saving 120 µL for use in determining the relative retroviral titer.

8. If using the retrovirus for immediate infection, place vials on ice, otherwise, snap-freeze retroviral supernatants in an ethanol/dry ice bath and store at –80°C. Do not freeze and thaw more than once. Virus can be stored at –80°C for at least 1 yr.

3.2.3. Determining Relative Titer of Retroviral Supernatants

1. Day 1: Plate 2×10^5 NIH 3T3 cells in a 6-cm tissue culture plate in 4 mL of complete media. Place in incubator for 24 h. As with 293 cells, the 3T3 cells should be evenly distributed.
2. Day 2: Cells should be approx 50% confluent.
3. Replace media with 2 mL of media containing 4 µg/mL polybrene.
4. Thaw the 120 µL test vial of retroviral supernatant in a 37°C water bath.
5. Once thawed, immediately add 120 µL of retroviral supernatant and gently move plate from side-to-side and forward–backward to mix.
6. Return plate to incubator. Three to 5 h later, add 0.5 mL of media to dilute the polybrene.
7. Day 3: Replace media with 3 mL of fresh 3T3 media without polybrene.
8. Day 4: Harvest cells and determine the percent of GFP-positive cells by flow cytometry. Retroviral supernatants that give 30% or greater GFP-positive NIH 3T3 cells in this assay are sufficient for transduction of murine bone marrow (*see* **Note 3**).

3.3. Retroviral Transduction of Murine Bone Marrow

There are several issues that must be considered when generating bone marrow chimeras. The first is the histocompatibility of the donor and recipient mice. Donor and recipient mice should be of the same genetic strain to avoid graft-vs-host disease. Graft-vs-host disease is mediated by mature T-cells from the donor marrow, which can mount an immune response against allogeneic host cells. Although the bone marrow only contains 1 to 5% mature T-cells, investigators should deplete the marrow of mature T-cells if sygeneic marrow is not used *(13)*. A second factor that may confound results is the potential for autoreconstitution by recipient bone marrow. In theory, after irradiation and reconstitution, the resulting hematopoietic system should be donor derived. However, it is possible for host cells to contribute to repopulation of the bone marrow. When using a retroviral vector that coexpresses a marker, such as GFP, comparisons can be made between GFP+ cells from mice transduced with the gene of interest or vector alone. However, it is often helpful to compare nontransduced (GFP–) donor cells with autoreconstituting recipient cells. This comparison often is achieved by using congenic differences between the donor and recipient mice in markers such as CD45 (expressed on all nucleated blood cells) or Thy-1 (expressed on all T-cells) when looking at the T-cell compartment.

The protocol below details harvesting bone marrow cells from 5-FU treated mice (*see* **Note 4**), culturing them in vitro *(14)* and transduction by spin infection *(15,16)*.

3.3.1. 5-FU Treatment of Donor Mice

1. Select mice that are 6 to 12 wk of age. Estimate roughly one donor mouse per every two recipient mice.
2. Day 1: Inject each donor mouse intraperitoneally with 200 µL of a 25 mg/mL solution of 5-FU diluted in PBS (5 mg/mouse).

3.3.2. Preparation of Donor Bone Marrow

1. Day 5: Harvest bone marrow from donor mice 4 d after 5-FU treatment.
2. Sacrifice mouse and place it on its back. Dissect through the skin, making a long cut down the length of the leg. Once the majority of the muscle has been cut away from the bone, cut the tibia just above the ankle joint and the femur as close to the hip socket as possible.
3. Place bones in a sterile tissue culture dish containing complete IMDM media. Place the dish on ice if harvesting several bones.
4. Under sterile conditions, load a 10-mL syringe with complete IMDM and attach a 30-gage needle.
5. Using the needle and syringe, flush out the bone marrow by forcing media through the bone cavity. Collect marrow in a sterile conical. Flush the bones from both ends. Bones will appear white when all of the marrow has been removed.
6. Once all of the bone marrow has been harvested, filter it through a nylon strainer, breaking up clumps with the end of the pipet.
7. Pellet cells at 300g for 7 min at 4°C. Resuspend pellet in approx 1 mL per mouse harvested. Remove 10 µL of cells and mix with 90 µL of red blood cell lysis buffer. Count cells on a hemacytometer. Expect between 3 and 10 million cells per mouse.
8. Pellet cells and resuspend in stimulation media at a concentration of 2–5 × 10^6 cells/mL. Culture of bone marrow cells in these cytokines will cause them to cycle, making them susceptible to retroviral infection.
9. Culture overnight.

3.3.3. Spin Infection of Bone Marrow

1. Day 6: Harvest cells from the tissue culture plate into a conical tube. Wash the plate with culture media until all non-adherent cells have been removed. Pellet cells at 300g for 7 min. To conserve cytokines, keep the initial harvest separate from the washes and use it for the infection step.
2. Spin infect cells in one well of a 6-well plate in a final volume of 4 mL and a final concentration of cytokines and FBS as listed previously in **step 5** of **Subheading 2.3.** Initially plate 3 mL of bone marrow cells into one well of a six-well plate. Thaw 1 mL of retroviral supernatant and quickly add it to the well. The final cell concentration should be approx 2–5 × 10^6 cells/mL (*see* **Note 5**).
3. Add 1.6 µL of a 10 mg/mL polybrene stock to each well for a final concentration of 4 µg/mL.
4. Spin in a tabletop centrifuge at 1300g for 90 min at 24°C.
5. Return cells to the incubator following the spin infection. Two to three hours later, gently resuspend any pelleted cells. Return to the incubator for overnight culture.

6. Day 7: Pull off 1 mL of media from each well. Spin at 300*g* for 7 min.
7. Aspirate supernatant and resuspend any pelleted cells in 1 mL of freshly thawed retroviral supernatant. Add to the appropriate well. Add 1.6 µL of a 10 mg/mL polybrene stock for a final concentration of 4 µg/mL.
8. Spin in a tabletop centrifuge at 1300*g* for 90 min at 24°C.
9. Return cells to the incubator following the spin infection. Let cells incubate for at least 4 h before injecting into recipient mice. Cells can be cultured overnight after gentle resuspension for injection the following morning.

3.4. Reconstitution of Donor Mice

1. Lethally irradiate recipient mice the night before or the morning of reconstitution. Irradiate BALB/c mice with 900 rads and C57BL/6 with 1100 rads of γ-irradiation with a Cesium source (*see* **Note 6**). Maintain irradiated mice on antibiotic water (2 mg/mL of neomycin sulfate and 100 U/mL polymixin B or trimethoprim [40 mg]/sulfamethoxazole [200 mg] per water bottle) for 2 wk after irradiation. Antibiotics should be changed twice a week.
2. Harvest transduced bone marrow cells and wash at least two times with PBS.
3. Resuspend cells in PBS for injection. Inject 200 µL/recipient mouse intravenously either in the tail vein or retro-orbitally. As few as $0.2–0.5 \times 10^6$ cells are necessary to reconstitute lethally irradiated mice; however, injection of more cells is not harmful and may ensure engraftment.
4. The amount of time required for full reconstitution depends on the hematopoietic lineage of interest. For T-cells, 8 wk is recommended; neutrophils, macrophages, and platelets can be studied by 6 wk *(1,17)*.

3.5. Ex Vivo Analysis

The type of experimentation that can be performed with cells from retroviral chimeras is limited to those that are based on single cell analysis. An exception is if the donor marrow does not support development of a particular lineage and cells of that lineage only arise as a result of expression of the introduced cDNA. Because flow cytometry is appropriate for the evaluation of any hematopoietic lineage, a basic flow cytometry protocol follows. Other assays that may be useful will depend on the cell type of interest and the molecular pathway of interest. Such assays may include, but are certainly not limited to, the measurement of Ca^{2+} flux by flow cytometry, analysis of cell spreading in response to various stimuli as visualized by microscopy, and examination of the upregulation of activation markers by flow cytometry. It also is possible to flow sort transduced from nontransduced cells, which relieves the requirement for single cells assays but is time consuming and provides only limited amounts of material.

Figure 2 is an example of flow cytometric analysis of cells harvested from retroviral bone marrow chimeras. Mice that are deficient in SH2 domain-

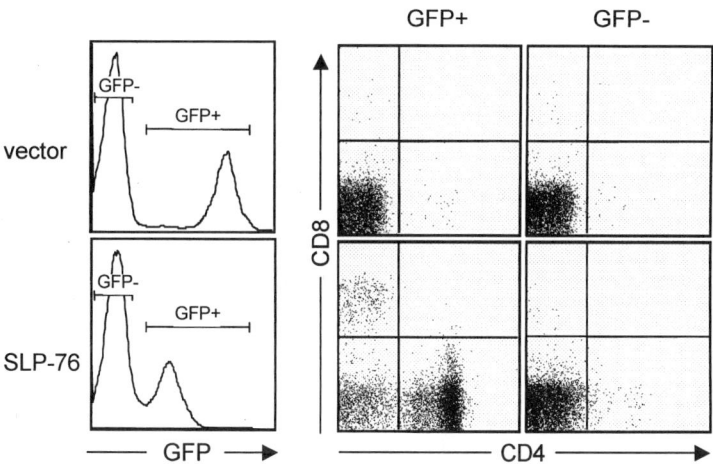

Fig. 2. Analysis of splenocytes from mice reconstituted with retrovirally transduced SLP-76-deficient bone marrow. Rag-1-deficient mice were injected with SLP-76-deficient bone marrow that was retrovirally transduced with either wild-type SLP-76 or vector alone. Splenocytes were stained with anti-CD8-phycoerythin and anti-CD4-allophycocyanin. Dot plots shown the expression of CD4 and CD8 on green fluorescent protein (GFP)+ and GFP– populations.

containing leukocyte protein of 76kDa (SLP-76) fail to generate T-cells *(18)*. In this experiment, bone marrow from SLP-76-deficient mice was transduced with wild-type SLP-76 in MIGR1 or with the empty MIGR1 vector and injected into Rag-1-deficient mice. Rag-1-deficient mice do not contain T- or B-cells because of an inability to express antigen receptors. The ability of wild-type SLP-76 reconstituted bone marrow to restore the development of T-cells was analyzed by flow cytometry. Splenocytes from vector only and SLP-76 reconstituted chimeras were stained with antibodies to the T-cell markers CD4 and CD8; GFP+ and GFP– populations were analyzed. No T-cells were present in the vector only reconstituted mouse; however, wild-type SLP-76 restores T-cell development as the GFP+ population contained both CD4+ and CD8+ T-cells but the GFP-population did not.

1. Harvest tissue(s) containing cell type of interest and generate a single cell suspension.
2. Wash cells twice in FACS buffer.
3. Count cells and resuspend in FACS buffer at 1×10^7 cells/mL.
4. Plate 100 µL of cells in one well of a 96-well V-bottom plate. Spin for 3 min at 350*g* in a tabletop centrifuge (*see* **Note 7**).
5. After spinning, rid of the FACS buffer by quickly turning the V-bottom plate upside down with deliberate force. Cells should remain in a pellet in the plate.

6. Resuspend the cell pellet with 50 µL of FACS buffer containing the appropriate amounts of fluorescently labeled antibodies. Avoid using fluorescein isothiocyanate-conjugated antibodies if the transduced cells express GFP. Antibody dilutions must be determined empirically. Generate compensation samples, that is, samples stained with only one of each of the fluorochomes used, as well as an unstained sample. Again, if the transduced cells express a fluorescent tag, a nontransduced mouse may be required to generate appropriate compensations.

7. Place cells on ice or at 4°C for 30 min; protected from light.

8. Wash cells three times with FACS buffer.

9. Resuspend cells in 200 to 400 µL of FACS buffer. Read samples on a flow cytometer.

4. Notes

1. Cells should be evenly distributed when plating 293 cells. Although these cells are adherent, they may lift off the plate after being transfected so care should be taken when changing the media.

2. Chloroquine increases the pH of the lysosomal and endosomal compartments and can inhibit the degradation of transfected DNA in the endosome as it transits to the nucleus *(19)*. In BOSC cells, a 293T-cell derivative containing viral packaging genes, chloroquine has been shown to double the infectious titer *(8)*. Others have reported no effect of chloroquine treatment on retroviral production in 293T-cells; instead, addition of 10 m*M* sodium butyrate to 293T-cells 17 h after transfection increased titers significantly. In this study, cells were treated with sodium butyrate for 12 h, washed, and re-fed with fresh media before harvesting viral supernatant 48 h after transfection *(6)*.

3. A lower infectious titer may be sufficient depending on the experimental design. For example, when culturing bone marrow in vitro for differentiation into various cell types which can then be sorted or selected for successful transduction, lower titer virus may suffice. However, for the generation of bone marrow chimeras, 30% or greater is desirable. Also, the level of GFP expression may be lower in 3T3 cells transduced with retrovirus coexpressing a gene of interest, as compared with cells only expressing GFP.

4. 5-FU is an anticancer drug that targets dividing cells. The rationale for its use in this context is that 5-FU depletes the bone marrow of rapidly dividing cells. Hematopoietic stem cells are resistant to 5-FU treatment *(20)*; therefore, their relative frequency in the bone marrow increases making it more likely that they will be infected by the retrovirus.

5. Efficient transduction can be achieved using a higher cell concentration. However, increasing the virus to cell ratio can increase infection efficiencies. This increase can be accomplished by plating fewer cells per well or by concentrating the viral supernatant. To concentrate virus, centrifuge the retroviral supernatant at 14,500*g* (Sorvall, RC5B, SA600 rotor) overnight at 4°C *(21)*.

6. Mice can be irradiated with a single dose of irradiation; however, if doses of 1000 rads or greater are used, split doses (e.g., two doses of 550 rads for C57BL/6) given 3 to 6 h apart increases survival by diminishing damage to the gut and lung.

Death of mice between 10 and 14 d after transplant typically is attributable to a failure to reconstitute, whereas death within the first week after irradiation is likely the result of a bacterial infection *(17)*.

7. The number of cells stained will depend on the efficiency of transduction and the number of events required for analysis. It is often beneficial to run some cells through the flow cytometer before staining to determine the transduction frequency by assessing the percent GFP⁺ cells. Bleeding the mice before the experiment and staining the blood for the cell type of interest also can be helpful in determining those mice that have a high level of retroviral transduction.

Acknowledgments

The author thanks Drs. Gary Koretzky, Jennifer N. Wu, and J. Todd Lawrence for helpful discussions and Dr. Andrew L. Singer for assistance with the experimental data. M. S. J. is supported by a grant from the Cancer Research Institute.

References

1. Judd, B. A., Myung, P. S., Obergfell, A., et al. (2002) Differential requirement for LAT and SLP-76 in GPVI versus T cell receptor signaling. *J. Exp. Med.* **195,** 705–717.

2. Gugasyan, R., Quilici, C., I Stacey, T. T., et al. (2002) Dok-related protein negatively regulates T cell development via its RasGTPase-activating protein and Nck docking sites. *J. Cell Biol.* **158,** 115–125.

3. Izon, D. J., Punt, J. A., Xu, L., et al. (2001) Notch1 regulates maturation of CD4+ and CD8+ thymocytes by modulating TCR signal strength. *Immunity* **14,** 253–264.

4. Mann, R., Mulligan, R. C., and Baltimore, D. (1983) Construction of a retrovirus packaging mutant and its use to produce helper-free defective retrovirus. *Cell* **33,** 153–159.

5. Finer, M. H., Dull, T. J., Qin, L., Farson, D., and Roberts, M. R. (1994) kat: a high-efficiency retroviral transduction system for primary human T lymphocytes. *Blood* **83,** 43–50.

6. Soneoka, Y., Cannon, P. M., Ramsdale, E. E., et al. (1995) A transient three-plasmid expression system for the production of high titer retroviral vectors. *Nucleic Acids Res.* **23,** 628–633.

7. Naviaux, R. K., Costanzi, E., Haas, M., and Verma, I. M. (1996) The pCL vector system: rapid production of helper-free, high-titer, recombinant retroviruses. *J. Virol.* **70,** 5701–5705.

8. Pear, W. S., Nolan, G. P., Scott, M. L., and Baltimore, D. (1993) Production of high-titer helper-free retroviruses by transient transfection. *Proc. Natl. Acad. Sci. USA* **90,** 8392–8396.

9. Pear, W. S., Miller, J. P., Xu, L., et al. (1998) Efficient and rapid induction of a chronic myelogenous leukemia-like myeloproliferative disease in mice receiving P210 bcr/abl-transduced bone marrow. *Blood* **92,** 3780–3792.

10. Hawley, R. G., Lieu, F. H., Fong, A. Z., and Hawley, T. S. (1994) Versatile retroviral vectors for potential use in gene therapy. *Gene Ther.* **1,** 136–138.
11. Saitoh, S., Odom, S., Gomez, G., et al. (2003) The four distal tyrosines are required for LAT-dependent signaling in FcepsilonRI-mediated mast cell activation. *J. Exp. Med.* **198,** 831–843.
12. Mizuguchi, H., Xu, Z., Ishii-Watabe, A., Uchida, E., and Hayakawa, T. (2000) IRES-dependent second gene expression is significantly lower than cap-dependent first gene expression in a bicistronic vector. *Mol. Ther.* **1,** 376–382.
13. Korngold, B. and Sprent, J. (1978) Lethal graft-versus-host disease after bone marrow transplantation across minor histocompatibility barriers in mice. Prevention by removing mature T cells from marrow. *J. Exp. Med.* **148,** 1687–1698.
14. Bodine, D. M., Karlsson, S., and Nienhuis, A. W. (1989) Combination of interleukins 3 and 6 preserves stem cell function in culture and enhances retrovirus-mediated gene transfer into hematopoietic stem cells. *Proc. Natl. Acad. Sci. USA* **86,** 8897–8901.
15. Kotani, H., Newton, P. B., 3rd, Zhang, S., et al. (1994) Improved methods of retroviral vector transduction and production for gene therapy. *Hum. Gene Ther.* **5,** 19–28.
16. Pear, W. S. (1996) Transient transfection methods for preparation of high-titer retroviral supernatants, in *Current Protocols in Molecular Biology* (Chanda, V. B., ed.), Vol. 2, John Wiley & Sons, New York, pp. 9.11.10–19.11.11.
17. Spangrude, G. J. (1994) Assesment of lymphocyte development in radiation bone marrow chimeras, in *Current Protocols in Immunology* (Coico, R., ed.), Vol. 1, pp. 4.6.1–4.6.7. John Wiley & Sons, New York.
18. Clements, J. L., Yang, B., Ross-Barta, S. E., et al. (1998) Requirement for the leukocyte-specific adapter protein SLP-76 for normal T cell development. *Science* **281,** 416–419.
19. Luthman, H. and Magnusson, G. (1983) High efficiency polyoma DNA transfection of chloroquine treated cells. *Nucleic Acids Res.* **11,** 1295–1308.
20. Van Zant, G. (1984) Studies of hematopoietic stem cells spared by 5-fluorouracil. *J. Exp. Med.* **159,** 679–690.
21. Huppa, J. B., Gleimer, M., Sumen, C., and Davis, M. M. (2003) Continuous T cell receptor signaling required for synapse maintenance and full effector potential. *Nat. Immunol.* **4,** 749–755.

21

Proteomic Analysis of Human Neutrophils

George Lominadze, Richard A. Ward, Jon B. Klein, and Kenneth R. McLeish

Summary

Proteomics is the study of the set of proteins, or proteome, expressed by a cell under specific conditions. Proteomics methodology consists of protein extraction, protein separation, and protein identification. Currently, two-dimensional gel electrophoresis (2DE) and matrix-assisted laser-desorption ionization time of flight mass spectrometry are the most widespread methods for proteomic studies. The recent introduction of precast immobilized pH gradient gel strips, precast gradient sodium dodecyl sulfate-polyacrylamide gel electrophoresis gels, and well-designed electrophoresis equipment has made 2DE a highly reproducible and relatively simple method for protein separation. Inherent limitations of the procedure, however, require approaches in sample preparation that may be cell- or tissue-dependent. This chapter describes a methodology for proteomic analysis of human neutrophils and discusses its applications.

Key Words: Proteomics; two-dimensional gel electrophoresis; mass spectrometry; neutrophils.

1. Introduction

Proteomics aims at filling the gap of knowledge between the genome of a cell and the set of proteins, or proteome, expressed by the cell under specific conditions (1). Often, two proteomes are compared by a subtractive analysis (2) in which differences arising from a drug treatment (3), genetic variation (4), or culture conditions (5) are observed. The differences in protein expression under different conditions may lead to the discovery of drug targets or to elucidation of protein functions or interactions. Proteomics can be used in protein expression studies by analyzing whole-cell proteomes and organelle subproteomes, or it can be applied to signaling studies by analyzing protein complexes and protein posttranslational modifications.

From: *Methods in Molecular Biology, vol. 332: Transmembrane Signaling Protocols, Second Edition*
Edited by: H. Ali and B. Haribabu © Humana Press Inc., Totowa, NJ

Proteomics methodology can be divided into three sequential steps: protein extraction, deconvolution of the protein mixture, and protein identification. Protein extraction methods typically entail solubilization of cells in lysis buffers containing detergents and/or chaotropes. The next step is to reduce sample complexity to simplify subsequent protein identification by mass spectrometry. Currently, there are two major approaches for accomplishing this task: gel-based separation of proteins and liquid chromatography-based separation of proteolytic peptides derived from proteins. Liquid chromatography methods are newer, less widespread, and more expensive than gel-based methods and will be discussed briefly at the end of the chapter. The most common method of separation and analysis of proteins for proteomics studies is high-resolution two-dimensional gel electrophoresis (2DE), which has gained widespread use since O'Farrell described the method in 1975 *(6)*. 2DE separates complex protein mixtures based on two independent variables. The first dimension (isoelectric focusing [IEF]) separates proteins by charge, and the second dimension (sodium dodecyl sulfate-polyacrylamide gel electrophoresis [SDS-PAGE]) separates proteins by mass. The proteins subsequently are visualized by staining the gel, and individual protein spots are identified by mass spectrometry (MS) methods. A commonly used MS method is matrix-assisted laser-desorption ionization time of flight mass spectrometry (MALDI-TOF MS). This method provides the investigator with mass spectra of peptides resulting from a protease digestion of the protein, which can be compared against theoretical spectra obtained from primary-sequence databases *(7)*. Recent introduction of pre-cast immobilized pH gradient (IPG) gel strips *(8)*, pre-cast gradient SDS-PAGE gels, and well-designed electrophoresis equipment has made 2DE a highly reproducible and relatively simple method for proteomic studies. However, inherent limitations of the procedure exist, including the intolerance of IEF to the presence of ions in the sample, inability to separate hydrophobic proteins, such as membrane proteins, as well as very large and very small proteins (those with sizes of >180 kDa or <12 kDa). Thus, understanding these limitations is necessary for this technology to answer specific scientific questions. This chapter describes a methodology for proteomic analysis of human neutrophils and discusses some of its applications to the study of neutrophil biology.

2. Materials

1. 9 *M* Chaotrope lysis buffer: 7 *M* urea, 2 *M* thiourea, 2% CHAPS, 0.5% Triton X-100 (v/v), 50 m*M* dithiothreitol (DTT), 0.005% bromophenol blue, 5.0% pH 3.0–10.0 ampholytes (Genomic Solutions, Ann Arbor, MI).
2. Rehydration buffer: 7 *M* urea, 2 *M* thiourea, 2% CHAPS, 0.5% Triton X-100 (v/v), 50 m*M* DTT, 0.005% bromophenol blue, 1.2% pH 3.0–10.0 ampholytes (Genomic Solutions).
3. Nonionic detergent lysis buffer: 20 m*M* Tris-HCl, pH 7.5, 150 m*M* NaCl, 1% Triton X-100, 0.5% NP-40, and protease inhibitor cocktail (Sigma, St. Louis, MO).

4. Chloroform.
5. Methanol.
6. Equilibration buffer I: 0.1 *M* Tris-HCl, pH 8.9, 6 *M* urea, 20% glycerol, 2% SDS, 50 m*M* DTT, 0.005% bromophenol blue.
7. Equilibration buffer II: 0.1 *M* Tris-HCl, pH 8.9, 6 *M* urea, 20% glycerol, 2% SDS, 100 m*M* iodoacetamide, 0.005% bromophenol blue.
8. Fixative solution: 10% methanol in 7% acetic acid.
9. SealPAK pouches (Kapak Corporation, St. Louis Park, MN).
10. Coomasie brilliant blue colloidal stain (Molecular Probes, Eugene, OR).
11. SYPRO Ruby (Genomic Solutions).
12. Parafilm (American National Can; Chicago, IL).
13. Lyophilized sequencing grade modified trypsin (Promega, Madison, WI): 20 µg in 20 µL of enclosed acetate buffer (store at –20°C).
14. 50 m*M* ammonium bicarbonate: 0.040 g in 10 mL of ultrapure water.
15. α-Hydroxycinnamic acid (α-CN) solution I: 10 mg of α-CN in 1 mL of acetone.
16. α-CN acid solution II: 10 mg of α-CN in 1 mL of 0.1% trifluoroacetic acid in 1:1 solution of acetonitrile and water.
17. Nitrocellulose solution: dissolve 10 mg of nitrocellulose in 1 mL of 1:1 solution of acetone and isopropanol.
18. MALDI-MS steel plate (Micromass, UK).
19. 2% Formic acid: 20 µL of formic acid in 980 µL of double-deionized water.

3. Methods

The described methods outline the extraction of protein from neutrophils, preparation of the extracts for IEF, separation of proteins by 2DE, excision and in-gel digestion of protein spots from gels, preparation of digests for MALDI-TOF MS, and identification of proteins from MALDI-TOF MS data by online search engines. Mass spectrometry techniques are beyond the scope of this chapter, and they are typically performed by a core or reference laboratory.

3.1. Neutrophil Protein Extraction

Regardless of the method of protein extraction used, proteolysis is a serious problem when working with neutrophil lysates. Because of the presence of large amounts of potent proteases, pretreatment with 10 µ*M* diisopropylfluorophosphate (DFP) before cell lysis is required (*see* **Note 1**). Extraction of proteins for 2DE analysis can be accomplished by direct solubilization of cells in 9 *M* chaotrope lysis buffer or by cell lysis with nonionic detergents (Triton X-100 and NP-40). The proteins extracted by each method vary, and the choice of the method depends on the final aim of the analysis. Extraction with 9 *M* chaotrope lysis buffer results in whole-cell lysate, whereas cell lysis in nonionic detergent lysis buffer allows the investigator to obtain a lysate largely devoid of nuclear and Triton X-insoluble cytoskeletal proteins.

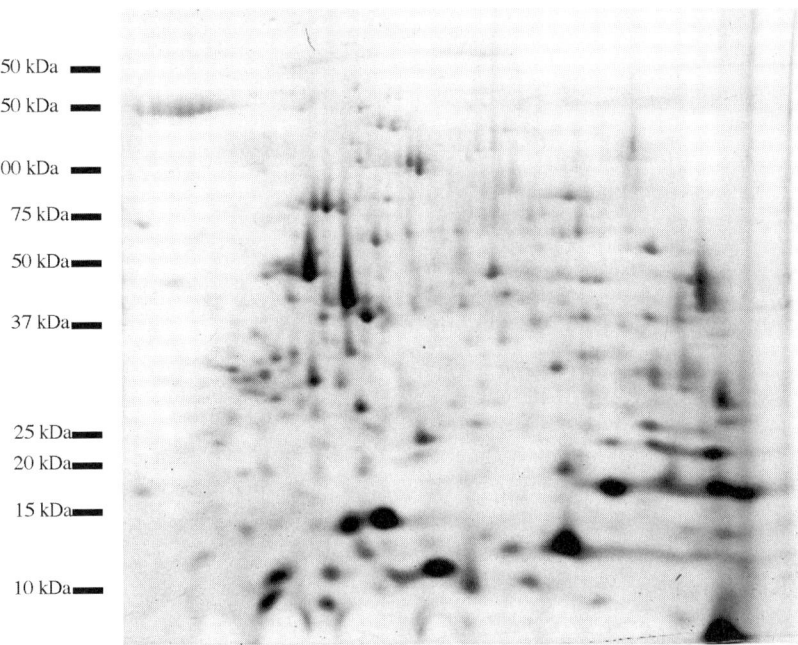

Fig. 1. An example of a gel showing the separation of proteins from a neutrophil lysate using two-dimensional electrophoresis (2DE). 5×10^7 neutrophils were lysed in 500 µL of nonionic detergent-based lysis buffer. A total of 130 µL of rehydration buffer was added to 25 µL of the lysate and proteins were separated by 2DE using 7 cm 3–10 ioselectric point range immobilized pH gradient strips for first dimension and 4–12% *bis*-Tris polyacrylamide gel (7 × 8 cm) in the second dimension. The gel was stained with colloidal Coomassie stain. Some 200 spots representing more than 80 proteins were observed.

3.1.1. Protein Extraction With 9 M Chaotrope Lysis Buffer

The extraction of proteins using the 9 *M* chaotrope lysis buffer is the simplest method for preparing the cell lysate for 2DE. This method allows preparation of the lysate with a high protein concentration and minimal ion content, thus avoiding additional sample preparation steps prior to the separation of proteins by IEF. An example of a 2D gel showing the separation of neutrophil lysate proteins is presented in **Fig. 1**. The method consists of simple mixing of cells in 9 *M* chaotrope lysis buffer, and centrifugation of the particulate matter to obtain a clarified supernatant.

1. Pellet the isolated neutrophils (1×10^7 cells will yield about 400 µg of protein) in a 1.5-mL centrifuge tube.
2. Add 300 µL of the lysis buffer to the cell pellet, mix thoroughly, and rotate at room temperature for 1 h (*see* **Note 2**).

3. Centrifuge the solution at 20,000*g* for 20 min at room temperature in a tabletop centrifuge to obtain a clarified supernatant (*see* **Note 3**).
4. Remove the supernatant and freeze at –70°C until further use (lysates can be kept in this condition for more than 1 yr).

3.1.2. Preparation of Protein Extracts Using Nonionic Detergents

Disruption of neutrophils with nonionic detergent-based lysis buffer leaves nuclei and the Triton X-insoluble cytoskeleton intact, whereas cytosolic proteins, plasma membrane and subcellular organelles are solubilized. The use of small volumes of buffer in the method results in a highly concentrated protein sample; however, the sample will contain large amounts of ions that require removal before IEF. Although many membrane proteins are solubilized in this buffer, they are not denatured and will be lost by precipitation upon their denaturation in the rehydration buffer and during IEF. The protocol for the initial extraction of proteins is very similar to the one described for 9 *M* chaotrope lysis buffer.

1. Pellet the isolated neutrophils (5×10^7 cells) in a 1.5-mL centrifuge tube.
2. Add 500 µL of ice-cold non-ionic detergent lysis buffer to the cell pellet, mix thoroughly, and rotate in the cold room for 1 h.
3. Centrifuge the solution at 20,000*g* for 20 min at 4°C in a tabletop centrifuge to pellet the cytoskeleton and nuclei.
4. Remove the supernatant and freeze at –80°C until further use. The sample can be kept for several months.

3.2. Preparation of the Sample for IEF

The protein extract prepared using the 9 *M* chaotrope lysis buffer (*see* **Subheading 3.1.1.**) does not require desalting and can be used directly for loading on the IEF precast gel strips. The samples prepared by extraction with nonionic detergent lysis buffer, however, require removal of ions before IEF.

3.2.1. Preparation of the Sample Extracted by 9 M Chaotrope Lysis Buffer for IEF

The urea–thiourea lysis buffer is compatible with IEF; therefore, it can be used with the dilution of up to 2:1 with the rehydration buffer. Direct loading of the sample in the 9 *M* chaotrope lysis buffer onto the IPG strip is not recommended because the ampholyte concentration (5%) is higher than the optimal concentration for IEF (0.5–2.5%).

1. Thaw out the sample in 9 *M* chaotrope lysis buffer. Invert the sample intermittently until all solid material dissolves (*see* **Note 4**).
2. Dilute the sample in rehydration buffer to a protein concentration of 0.5 µg/µL.
3. Proceed to the rehydration of the sample into the IPG strip (*see* **Note 5**).

3.2.2. Preparation of the Sample Extracted
by Nonionic Detergent Lysis Buffer for IEF

This lysis buffer contains large amounts of ions that require removal before IEF. The maximal concentration of ions in the sample should not exceed 20 mM. Removal of ions can be accomplished by desalting precipitation or by buffer exchange using ultrafiltration.

3.2.2.1. Preparation of the Sample Using Desalting Precipitation

Desalting precipitation is the fastest, cheapest, and the least laborious method for desalting and concentrating proteins, and it results in complete removal of ions and formation of easily solubilized protein pellet (*see* **Note 6**). The more concentrated the original protein solution, the more effective the precipitation procedure. Some loss of protein will occur because of solvation of the protein in organic phases.

1. Thaw the sample in the nonionic detergent lysis buffer at room temperature.
2. To 1 volume of sample, add 3 volumes of 100% methanol and mix by vortexing.
3. Add 1 volume chloroform and mix by vortexing.
4. Add 4 volumes water, vortex, and incubate on ice for 30 min with intermittent vortexing.
5. Centrifuge the resultant emulsion at 10,000g for 2 min at room temperature. Precipitated protein forms at the top of the chloroform layer, whereas salt is in the upper aqueous layer.
6. Remove aqueous phase carefully, without disturbing the protein layer, and replace with 4 volumes of methanol and vortex.
7. Pellet the protein at 10,000g for 2 min, remove the supernatant, and dry the pellet in air for 5 h or until it forms a thin crust on the bottom of the tube.
8. Dissolve the pellet in rehydration buffer and proceed to rehydration of the sample into IPG strip.

3.2.2.2. Preparation of the Sample Using Buffer Exchange by Ultrafiltration

Removal of ions by ultrafiltration is a more costly, laborious, and time-consuming procedure than desalting precipitation of proteins. The protein loss, however, is less with this procedure. Ultrafiltration is superior to dialysis, as it allows buffer exchange directly with rehydration buffer, and minimizes the possibility of protein dilution or loss caused by precipitation that can occur during desalting by dialysis of concentrated protein samples.

1. Thaw the sample in non-ionic detergent lysis buffer at room temperature.
2. Dilute the sample in rehydration buffer at the ratio of 1:5 or to 500 µL.
3. Place 500 µL of the solution in 3 to 10 kDa cutoff ultrafiltration device (such as Nanosep Omega from Millipore) and centrifuge at 14,000g at room temperature for 15 min in a tabletop centrifuge (*see* **Note 7**).

4. Mix the retentate solution by pipetting it up and down and centrifuge it again. Continue the procedure until retentate volume reaches approx 100 µL.
5. Add 400 µL of rehydration buffer to the 100 µL of retentate and repeat **steps 3** and **4** from the aforementioned procedure.
6. Bring the volume of each 100 µL sample to desired volume for loading on the IPG strips, as described in manufacturer's instructions (*see* **Note 5**).

3.3. IEF and SDS-PAGE

Both the first dimension separation of proteins by IEF and the second dimension separation by SDS-PAGE can be conducted using a number of commercially available systems. Large format systems (18-cm IPG strips and 20-cm second-dimension gels) are more tolerant to the presence of ions in the rehydration buffer than the 7-cm IPG strips and 8-cm minigels. Small-format systems require less protein, provide results more rapidly, and are easier to handle. For both types of systems, gradient second-dimension gels provide optimal results. IEF conditions for the two formats are different. For 7-cm IPG strips, the following parameters are used during the run: 200 V for 20 min, 450 V for 15 min, 750 V for 15 min, and 2000 V for 40 min. For the 18-cm IPG strips, the voltage is increased gradually and the focusing is stopped when specific number of volt-hours have accumulated. The running parameters are as follows: maximum voltage, 5000 V; maximum current, 80 mA per strip; accumulated volt-hours, 80,000. The details of the operation of various 2DE systems differ substantially and depend on the manufacturer. The basic steps of operation, however, are similar and consist of rehydration, IEF, equilibration with SDS-containing sulfhydryl-reducing and alkylating buffers (*see* **Note 8**), and second-dimension SDS-PAGE.

1. Rehydrate the IPG strip overnight at room temperature by applying the sample in rehydration buffer to the strip in a container (10-mL pipet can be used as a container and both ends closed with parafilm).
2. Perform IEF according to the instructions of the manufacturer of the electrophoresis equipment, observe the migration of the dye (*see* **Note 9**).
3. Equilibrate the strip in equilibration buffer I (contains DTT) at room temperature by placing it in an equilibration tray included with the electrophoresis equipment and by gently shaking for 5 to 10 min.
4. Remove the buffer and equilibrate the strip in equilibration buffer II (contains iodoacetamide) for 10 min at room temperature.
5. Place the strip on top of the second-dimension polyacrylamide gel and perform SDS-PAGE.
6. Fix the gel in 10% methanol, 7% acetic acid fixative for 30 min to prepare it for staining (*see* **Note 10**).
7. Stain the gel in colloidal Coomassie brilliant blue or SYPRO Ruby stain by incubating the gel with the stain overnight on shaker (*see* **Note 11**). To avoid tearing, there should be enough stain to ensure that the gel is not dragging on the bottom of the box.

8. Destain the Coomassie-stained gel by incubating it in fixative until the background is clear (SYPRO-stained gel does not require destaining). Repeat the incubation if necessary.

9. Incubate the gel in double-deionized water for 15 min. At this stage gels can be documented by scanning on a gel scanner. Gels can be left in water for days, or bagged in small plastic bags and kept at 4°C for up to 2 mo before spot excision and MALDI-TOF MS analysis.

3.4. In-Gel Digestion and Sample Preparation for MALDI-TOF MS

After scanning the gel, protein spots can be excised and prepared for MALDI-TOF MS analysis. During spot excision and in-gel digestion, it is imperative to minimize the chances of contaminating gels with keratin. Therefore, all manipulations must be done wearing gloves and face masks.

3.4.1. Excision and In-Gel Digestion of Spots

During the excision of spots, the ratio of acrylamide to the protein must be minimized. This is accomplished by using a cutter that can be made in a laboratory. A 200-µL pipet tip is cut at the pointed end using a razor blade to widen its opening to the diameter of 0.5 mm (approx one-fifth of the length of the tip is removed). This shortened tip is firmly attached to a Pasteur pipet, which acts as its handle. Spot excision is performed by simply pressing down the cutter on the spot.

1. Place the gel on a transilluminator cleaned with 70% ethanol (to remove keratin).
2. Excise spots and transfer the excised pieces to labeled 0.6-mL tubes by squeezing the pipet bulb of the cutter.
3. Add 40 µL of 50 m*M* NH$_4$HCO$_3$ and incubate at room temperature for 10 min, vortex intermittently. Leave the solution in the tube for the next step.
4. Add 50 µL of acetonitrile and incubate for 10 min at room temperature, vortex intermittently.
5. Remove the solution and repeat **steps 3** and **4**.
6. Remove the solvent and dry the gel plugs either in a speedvac, or by placing the open tubes at room temperature for 6 h to overnight (*see* **Note 12**).
7. Prepare trypsin working solution (20 ng/µL sequencing grade modified trypsin in 50 m*M* NH$_4$HCO$_3$) by adding 10 µL of the stock solution (1 µg/µL) to 500 µL of 50 m*M* ammonium bicarbonate solution.
8. Add 3.5 to 5 µL of working solution of sequencing grade-modified trypsin to the gel plug. Let the gel plug hydrate with the solution for 10 min. If needed, add more trypsin solution to completely cover the gel plugs.
9. Incubate the tubes at 37°C overnight (14–18 h). The next day the liquid around the gel plug will contain peptides that will be analyzed by MALDI-TOF MS.

3.4.2. Preparation of Peptides for MALDI-TOF MS

After protein digestion the obtained peptide solutions are processed to prepare for MALDI-MS analysis. Freezing the solutions is not recommended.

1. Mix nitrocellulose solution and α-CN solution I in 1:4 ratio (20/80 μL).
2. Deposit 1.5 μL of mixture onto the target circle on a MALDI-MS plate. A thin, tan film must form on it in approx 15 s (*see* **Note 13**).
3. Mix 1.5 μL of the peptide sample with 1.5 μL of the α-CN solution II by depositing them very close to each other (solutions should be in contact) on the surface of parafilm and mixing them by pipetting up and down.
4. Deposit 2 μL of this mixture onto the thin film on the MALDI plate by first letting the liquid droplet hang on the pipet tip and then touching the droplet onto the spotted support (*see* **Note 14**).
5. Let the deposited peptide samples dry at room temperature.
6. Add 1 μL of 2% formic acid to each spot and remove this solution after 1 min by touching with a Kimwipe.
7. Let the samples dry at room temperature. The samples are now ready for MALDI-TOF MS analysis.

3.5. Analysis of Obtained Peptide Masses

MALDI-TOF MS analysis typically is conducted in a core laboratory or a reference laboratory. MALDI-TOF MS provides the investigator with a list of masses of singly charged peptides for each excised protein spot that can be searched against an online database of human proteins to allow the identification of the protein from which the peptides were derived. A web-based search engine Mascot (www.matrixscience.com) is a valuable tool for identifying proteins. Other search engines, such as Profound can also be used (*see* **Note 15**). In Mascot, if the score exceeds the significance level, there is less than a 1 in 20 chance that the match is incorrect.

The following parameters should be used for the peptide fingerprint analysis using Mascot when the digestion has been performed as outlined above:

> Fixed modification—carbamidomethylation (Cys).
> Variable modification—oxidation (Met; *see* **Note 16**).
> Error—100 to 150 ppm.
> Mass values—MH+.
> Missed cleavages—1.

The database of choice for the identification of the protein is usually the National Center for Biotechnology Information.

3.6. Applications

3.6.1. Expression Studies

Proteomics has been used successfully for analysis of differential protein expression in control and activated neutrophils. Proteins from lipopolysaccharide-treated and control neutrophils have been separated and protein expression profiles compared using 2DE and MS *(9)*. Furthermore, by using a pharmacological inhibitor of p38 mitogen-activated protein kinase, a putative regulatory role was assigned to this kinase in the expression of 18% of lipopolysaccharide-regulated proteins.

Neutrophil subproteomes also can be investigated. Our laboratory examined neutrophil granules obtained by separation of postnuclear lysates on Percoll gradients. The isolated granule fractions were subjected to 2DE. Enrichment of granules greatly increases the likelihood of identifying low abundance proteins that are unlikely to be identified with whole-cell lysate analysis.

Investigation of subproteomes can be further extended to the study of protein complexes and protein–protein interactions. Thus, large protein complexes can be isolated by size-exclusion chromatography and analyzed by 2DE. Similarly, a protein can be immunoprecipitated and coprecipitated proteins can be visualized by 2DE.

3.6.2. Signaling Studies

The described methodology has been applied to the study of signaling pathways in human neutrophils. For example, phorbol myristate acetate-induced protein phosphorylation in [^{32}P]-orthophosphate-labeled neutrophils has been assessed using 2DE *(10)*. In this study, cells were labeled with ^{32}P-orthophosphate and the patterns of phosphorylated proteins on 2D gels from phorbol 12-myristate 13-acetate- and vehicle-treated cells were compared by autoradiography.

Differentially phosphorylated proteins from two conditions also can be assessed by Western blotting of gels and probing of membranes with phosphoprotein-specific, or antiphospho-Ser/Thr and antiphospho-Tyr antibodies. In this method, proteins are separated by 2DE using two parallel gels for each condition, and one set of gels is stained, whereas the other set is subjected to Western blotting. Probing the blot with antiphosphoamino acid antibodies provides the investigator with a pattern of phosphorylated proteins that can be compared with the pattern of proteins from the stained gels, and may allow identification of differentially phosphorylated proteins *(11)*.

As an alternative approach to the use of phosphoprotein-specific or antiphosphoamino acid antibodies, blots can be probed with an antibody directed against a particular protein, and the appearance of a new spot with a more acidic isoelectric point (pI) (an acidic shift) can be observed on phosphorylation of the protein. The identity of the kinase responsible for the protein phosphorylation can be elucidated using an appropriate kinase inhibitor and observing the reduction of the shift in response to the pretreatment of cells with the inhibitor prior to their activation by an agonist *(12,13)*.

In addition to Western blotting, radioactive labeling of lysates can successfully lead to identification of protein kinase substrates. Neutrophil lysates can be used for a kinase reaction using [^{32}P]-adeonsine triphosphate and a recombinant kinase before separation of proteins by 2DE. Comparison of radiograms of gels of lysate alone, kinase alone, and lysate plus kinase conditions reveals radioactive spots that are unique to kinase plus lysate condition. Matching of these spots from radiograms to the stained spots from the same gel allows for the identification of the putative substrate *(12,13)*.

3.7. Limitations and Alternative Approaches

Theoretically, 2DE can separate more than 10,000 proteins from the source material *(14)*. In practice, however, analysis of hydrophobic proteins and small molecular weight proteins from complex samples by 2DE has been found to be challenging. Small proteins of less than 20 kDa are difficult to resolve by 2DE and only a few small proteins have been successfully analyzed by this method *(15)*. Hydrophobic proteins, such as α-helical transmembrane proteins and large globular proteins, are also grossly underrepresented on 2D gels *(16,17)*. Transmembrane sections of the membrane proteins and the interior portions of large globular proteins are hydrophobic and as a result these proteins are sparingly soluble in solubilization buffers used for 2DE. During IEF, hydrophobic proteins precipitate upon approaching their respective pIs. Use of more powerful zwitterionic detergents and fractionation of samples prior to subjecting them to 2DE, improve results of the analysis of membrane proteins by 2DE, but do not completely solve the problem *(18)*.

To circumvent the protein solubility problem, peptide chromatography-based methods have been introduced into proteomics. These approaches are based on iterative liquid chromatographic separation of peptides obtained from global proteolytic digestion of a complex sample, and analysis of peptides by tandem MS *(19)*. This method turns insoluble proteins into soluble peptides and allows for identification of a large number of proteins, however, it is difficult to conduct quantitative analysis of the samples. Recent introduction of differential isotope labeling of peptides has addressed this problem. The method uses labeling of peptides with an isotope-coded affinity tag reagent *(20)*. Subsequent MS analysis reveals mass differences imparted on peptides by the isotope-coded tags, and allows for the comparison of the relative abundance of isotope-labeled peptides in the sample *(21)*.

For identification of a phosphoproteome, another method that supplements chromatography-based proteomic approaches has been developed. This method, termed immobilized metal-affinity chromatography, requires the use of immobilized metal columns for the enrichment of phosphorylated peptides and is useful for the analysis of phosphoproteins from complex mixtures *(22,23)*. It should be noted, however, that chromatographic methods are relatively new and unrefined, require more expensive equipment, and may not be as reproducible as gel-based methods. Thus, gel-based proteomics remains an important tool because of its simplicity and high degree of reproducibility.

4. Notes

1. DFP is a highly toxic chemical. Cells are to be treated with DFP using all required precautions. As many as 5×10^7 cells per milliliter can be treated with 10 μM DFP *(24)*.

2. Ampholytes form complexes with released chromatin, allowing for effective pelleting of nuclear material. Do not use sonication to disrupt neutrophils in 9 *M* chaotrope lysis buffer. This will release large amounts of chromatin into the solution and will result in irreversible gelling of the sample.

3. It is important to maintain the ratio of 9 *M* chaotrope lysis buffer to neutrophils at 200–500 µL per 5×10^7 cells, as the use of more cells will result in gelling of the sample.

4. Avoid warming the samples in urea-containing buffers to temperatures greater than 30°C for more than 5 min because protein carbamylation can occur, which will alter the protein pI *(25)*.

5. For IPG strip rehydration, prepare 22 to 25 µL of solution per 1 cm of the strip.

6. To concentrate and desalt dilute protein solutions prior to IEF, dialysis of the sample followed by the reduction of volume by ultrafiltration prior to desalting precipitation, can be used. However, it must be noted that dialysis against buffers with low ion and detergent content can result in protein precipitation.

7. Other ultrafiltration devices with various molecular-weight cutoffs also can be used. However, the investigator must ensure that the membrane is compatible with the high concentration of urea.

8. It is important to equilibrate the strips in DTT-containing buffer to reduce disulfides, and then to alkylate the cysteines in an iodoacetamide-containing buffer to avoid the formation of intermolecular disulfide bonds during the second dimension run. Both of these buffers contain SDS to aid in the migration of the proteins from the strip into the second dimension gel.

9. During the normal course of IEF the bromophenol blue dye will migrate to the acidic end of the strip, and will change color to green and then to yellow. Uneven or incomplete migration of the dye is indicative of the presence of ions in the sample.

10. Using higher concentrations of acetic acid or methanol is not recommended because they affect the MALDI analysis of the protein.

11. SYPRO Ruby is a fluorescent stain and requires appropriate scanners for visualization of the gel. It has an advantage over Coomassie blue because of its superior dynamic range of sensitivity *(26)*. Therefore, it is excellent for comparison of protein-spot intensities on different gels. However, it is necessary to use ultraviolet transilluminator to visualize protein spots on SYPRO-stained gels for excision, which makes the procedure cumbersome compared to Coomassie-stained gels.

12. Dried gel plug will be opaque white, very small, and brittle. One should be aware that it is easy to crush it by accident with a pipet tip, or lose it from the tube.

13. If the film is white, increase the ratio of α-CN solution I to nitrocellulose solution to obtain a film of tan color and smooth surface. If the film is almost transparent and crystalline, add more nitrocellulose solution to the mixture.

14. This solution should not turn yellow and should not dissolve matrix. If these occur, the NH_4HCO_3 powder is old and must be replaced. In addition, solutions used for preparation of peptides for MALDI should not be kept for longer than 1 wk at 4°C and should not be frozen.

15. Profound is available at http://prowl.rockefeller.edu/. This search engine allows for the input of observed molecular-weight and pI for the protein in question. Input of the following parameters is suggested when using Profound: Cysteine modified by iodoacetamide, maximum missed cleavage sites, 1; charge state, MH+; tolerance unit, ppm; mass tolerance, ±150 ppm.

16. Fixed modification means that peptides with all applicable residues will be searched with that modification, whereas variable modifications mean that the peptide will be analyzed by looking at a possibility of either one or all applicable residues being considered as modified.

References

1. Wilkins, M. R., Pasquali, C., Appel, R. D., et al. (1996) From proteins to proteomes: large scale protein identification by two-dimensional electrophoresis and amino acid analysis. *BioTechnology* **14**, 61–65.
2. Aebersold, R. and Leavitt, J. (1990) Sequence analysis of proteins separated by polyacrylamide gel electrophoresis: Towards an integrated protein database. *Electrophoresis* **11**, 517–527.
3. Anderson, N. L. and Anderson, N. G. (1998) Proteome and proteomics: new technologies, new concepts, and new words. *Electrophoresis* **19**, 1853–1861.
4. Klose, J. (1999) Genotypes and phenotypes. *Electrophoresis* **20**, 643–652.
5. Quadroni, M., Staudenmann, W., Kertesz, M., and James, P. (1996) Analysis of global responses by protein and peptide fingerprinting of proteins isolated by two-dimensional gel electrophoresis. Application to the sulfate-starvation response of Escherichia coli. *Eur. J. Biochem.* **1**, 773–881.
6. O'Farrell, P. H. (1975) High resolution two-dimensional electrophoresis of proteins. *J. Biol. Chem.* **50**, 4007–4021.
7. Choudhary, J. S., Blackstock W. P., Creasy D.M, and Cottrell J. S. (2001) Matching peptide mass spectra to EST and genomic DNA databases. *TRENDS Biotech.* **19**(suppl), S17–S22.
8. Görg, A., Obermaier, C., Boguth, G., Harder, A., Scheibe, B., Wildgruber, R., and Weiss, W. (2000) The current state of two-dimensional electrophoresis with immobilized pH gradients. *Electrophoresis* **21**, 1037–1053.
9. Fessler, M. B., Malcolm, K. C., Duncan, M. W., and Worthen S. G. (2002) A genomic and proteomic analysis of activation of the human neutrophil by lipopolysaccharide and its mediation by p38 mitogen-activated protein kinase. *J. Biol. Chem.* **277**, 31,291–31,302.
10. Hayakawa, T., Suzuki, K., Suzuki, S., Andrews, P., and Babior, B. (1986) Possible role of protein phosphorylation in the activation of the respiratory burst in human neutrophils. *J. Biol. Chem.* **261**, 9109–9115.

11. Kaufmann, H., Bailey, J. E., and Fussenegger, M. (2001) Use of antibodies for detection of phosphorylated proteins separated by two-dimensional gel electrophoresis. *Proteomics* **1**, 194–199.

12. Powell, D. W., Rane, M. J., Joughin, B. A., et al. (2003) Proteomic identification of 14-3-3ζ as a mitogen-activated protein kinase-activated protein kinase 2 substrate: role in dimer formation and ligand binding. *Mol. Cell. Biol.* **23**, 5376–5387.

13. Singh, S., Powell, D. W., Rane, M. J., et al. (2003) Identification of the p16 Arc subunit of the Arp2/3 complex as a substrate of MAPK-activated protein kinase 2 by proteomic analysis. *J. Biol. Chem.* **278**, 36,410–36,417.

14. Klose, J. and Kobalz, U. (1995) Two-dimensional electrophoresis of proteins: an updated protocol and implications for a functional analysis of the genome. *Electrophoresis* **16**, 1034–1059.

15. Schrader, M. and Schulz-Knappe, P. (2001) Peptidomics technologies for human body fluids. *Trends Biotech.* **19**(Suppl), S55–S60.

16. Santoni, V., Molloy, M., and Rabilloud, T. (2000) Membrane proteins and proteomics: un amour impossible? *Electrophoresis* **21**, 1054–1070.

17. Gygi, S. P., Rist, B., and Aebersold, R. (2000) Measuring gene expression by quantitative proteome analysis *Curr. Opin. Biotech.* **11**, 396–401.

18. Molloy, M. (2000) Two-dimensional electrophoresis of membrane proteins using immobilized pH gradients. *Analyt. Biochem.* **280**, 1–10.

19. Yates, J. R., III, Carmack, E., Hays, L., Link, A. J., and Eng, J. K. (1999) Automated protein identification using microcolumn liquid chromatography-tandem mass spectrometry (review). *Methods Mol. Biol.* **112**, 553–569.

20. Gygi, S. P., Rist, B., Gerber, S. A., Turecek, F., Gelb, M. H., and Aebersold, R. (1999) Quantitative analysis of complex protein mixtures using isotope-coded affinity tags. *Nat. Biotechnol.* **17**, 994–999.

21. Moseley, M. A. (2001) Current trends in differential expression proteomics: isotopically coded tags. *Trends Biotech.* **19**(suppl), S10–S16.

22. Ficarro, S. B., McCleland M. L., Stukenberg P. T., et al. (2002) Phosphoproteome analysis by mass spectrometry and its application to *Saccharomyces cerevisiae*. *Nat. Biotechnol.* **20**, 301–305.

23. Nuhse, T. S., Stensballe A., Jensen O. N., and Peck S. C. (2003) Large scale analysis of in vivo phosphorylated membrane proteins by immobilized metal ion affinity chromatography and mass spectrometry. *Mol. Cell. Proteomics* **2**, 1234–1243.

24. Amrein, P. C. and Stossel, T. P. (1980) Prevention of degradation of human polymorphonuclear leukocyte proteins by diisopropylfluorophosphate. *Blood* **56**, 442–447.

25. Gianazza, E. (1995) Isoelectric focusing as a tool for the investigation of posttranslational processing and chemical modifications of proteins. *J. Chromatogr. A.* **705**, 67–87.

26. Patton, W. F. (2000) A thousand points of light: the application of fluorescence detection technologies to two-dimensional electrophoresis and proteomics. *Electrophoresis* **21**, 1123–1144.

Index